Molecular Biology: Structure and Dynamics

Molecular Biology: Structure and Dynamics

Edward Walker

MURPHY & MOORE
www.murphy-moorepublishing.com

Molecular Biology: Structure and Dynamics
Edward Walker
ISBN: 978-1-63987-374-6 (Hardback)

MURPHY & MOORE
Published by Murphy & Moore Publishing,
1 Rockefeller Plaza,
New York City, NY 10020, USA

Cataloging-in-Publication Data

Molecular biology : structure and dynamics / Edward Walker.
p. cm.
Includes bibliographical references and index.
ISBN 978-1-63987-374-6
1. Molecular biology. 2. Biomolecules. 3. Biochemistry. I. Walker, Edward.
QH506 .M65 2022
572.8--dc23

For more information regarding Murphy & Moore Publishing and its products, please visit the publisher's website www.murphy-moorepublishing.com

Table of Contents

Preface

The branch of biology which is concerned with the study of biological activity in and between cells at a molecular level is known as molecular biology. Areas of focus under this discipline are synthesis and functions of DNA, RNA and other essential proteins. The guiding principles of molecular biology are the processes which govern how DNA is transcribed into RNA and finally translated into proteins. This field is closely related to the subjects of biochemistry and genetics. Some prominent methods used in molecular biology to study the structures of proteins are molecular cloning, gel electrophoresis, polymerase chain reaction, microarrays, allele specific oligonucleotide, macromolecule blotting and probing. This book is compiled in such a manner, that it will provide in-depth knowledge about the theory and practice of molecular biology. While understanding the long-term perspectives of the topics, it makes an effort in highlighting their impact as a modern tool for the growth of the discipline. This book aims to serve as a resource guide for students and experts alike and contribute to the growth of the discipline.

A detailed account of the significant topics covered in this book is provided below:

Chapter 1- The field of biology which deals with the molecular basis of biological activities such as composition, structure and interactions of cellular molecules is known as molecular biology. It includes the study of various molecules like nucleic acids and proteins that carry out biological processes. This is an introductory chapter which will introduce briefly all the significant aspects of molecular biology.

Chapter 2- DNA stands for deoxyribonucleic acid. It is made up of two polynucleotide chains that wind around one another to form double helix. DNA carries the genetic instructions for the growth, development and reproduction of living organisms. This chapter has been carefully written to provide an easy understanding of the varied facets of DNA such as its structure, replication, damage, repair, recombination and sequencing.

Chapter 3- RNA stands for ribonucleic acid. It is an important macromolecule which is present in all cells and is genetic material of some viruses. RNA is responsible for protein synthesis and carries instructions from DNA. It is of three types; rRNA, mRNA and tRNA. The aim of this chapter is to explore these types of RNAs. These topics are crucial for a complete understanding of the subject.

Chapter 4- Some of the diverse techniques which are used within the domain of molecular biology are blotting, northern blotting, cationic electrophoresis and eastern blotting, and DNA sequencing. This chapter closely examines these key techniques of molecular biology to provide an extensive understanding of the subject.

Chapter 5- The process of transferring macromolecules like proteins, DNA and RNA from gel onto a carrier such as nitrocellulose or nylon membrane is known as blotting. In order to completely understand macromolecule blotting and probing, it is necessary to understand the processes related to it. The following chapter elucidates the varied processes and mechanisms associated with this area of study.

Chapter 6- The process by which the information encoded in the DNA is translated into a functional product is known as gene expression. This chapter elucidates the concepts of GeneCalling, genetic code, antitermination, artificial gene synthesis, gene silencing, RNA splicing, alternative splicing and epitranscriptome. The topics elaborated in this chapter will help in gaining a better perspective about gene expression.

Chapter 7- Molecular Biology is an interdisciplinary field of biology which makes it essential to understand its related fields. Some of the topics related to the study of molecular biology such as molecular modeling, molecular engineering and molecular genetics are explained in a detailed manner. The significant sub-disciplines of molecular biology have been thoroughly discussed in this chapter.

Chapter 8- Molecular biology plays a significant role in understanding the actions and regulations of various cells of the body which can be used to develop target drugs. The diverse applications of molecular biology such as molecular medicine, bioinformatics and central dogma in the current scenario have been thoroughly discussed in this chapter.

It gives me an immense pleasure to thank our entire team for their efforts. Finally in the end, I would like to thank my family and colleagues who have been a great source of inspiration and support.

Edward Walker

1

An Introduction to Molecular Biology

The field of biology which deals with the molecular basis of biological activities such as composition, structure and interactions of cellular molecules is known as molecular biology. It includes the study of various molecules like nucleic acids and proteins that carry out biological processes. This is an introductory chapter which will introduce briefly all the significant aspects of molecular biology.

Molecular biology is a field of science concerned with studying the chemical structures and processes of biological phenomena that involve the basic units of life, molecules. The field of molecular biology is focused especially on nucleic acids (e.g., DNA and RNA) and proteins—macromolecules that are essential to life processes—and how these molecules interact and behave within cells. Molecular biology emerged in the 1930s, having developed out of the related fields of biochemistry, genetics, and biophysics; today it remains closely associated with those fields.

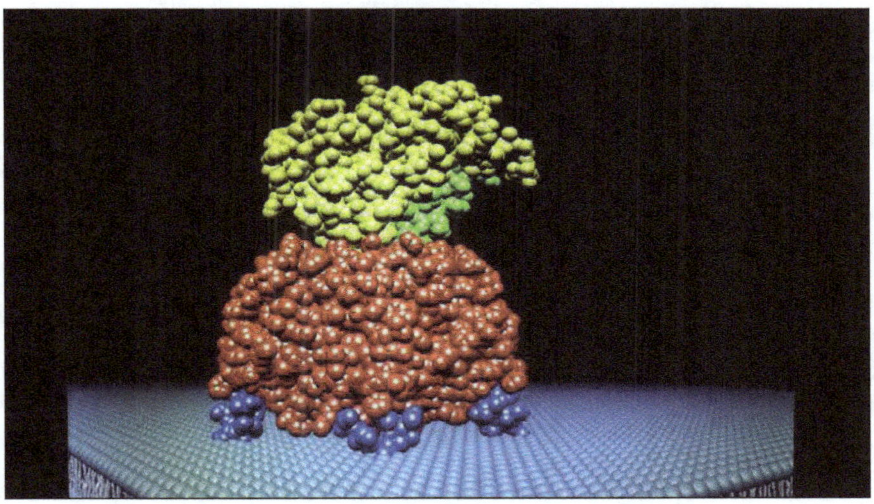

X-ray protein crystallography in molecular biology.

Molecular Biology Techniques

The discovery and characterisation of a number of key enzymes have permitted the development of various techniques for the analysis and manipulation of DNA. In particular, the enzymes termed type II restriction endonucleases have come to play a key role in all aspects of molecular biology.1 These enzymes recognise specific DNA sequences, usually 4–6 base pairs (bp) in length, and cleave them in a defined manner. The sequences recognised are palindromic or of an inverted repeat nature, that is, they read the same in both directions on each strand. When cleaved they leave a flush-ended or staggered (also termed a cohesive-ended) fragment depending on the particular enzyme used.

An important property of staggered ends is that those produced from different molecules by the same enzyme are complementary (or 'sticky') and so will anneal to each other. The annealed strands are held together only by hydrogen bonding between complementary bases on opposite strands. Covalent joining of ends on each of the two strands may be brought about by the enzyme DNA ligase. This is widely exploited in molecular biology to allow the construction of recombinant DNA, i.e. the joining of DNA fragments from different sources.

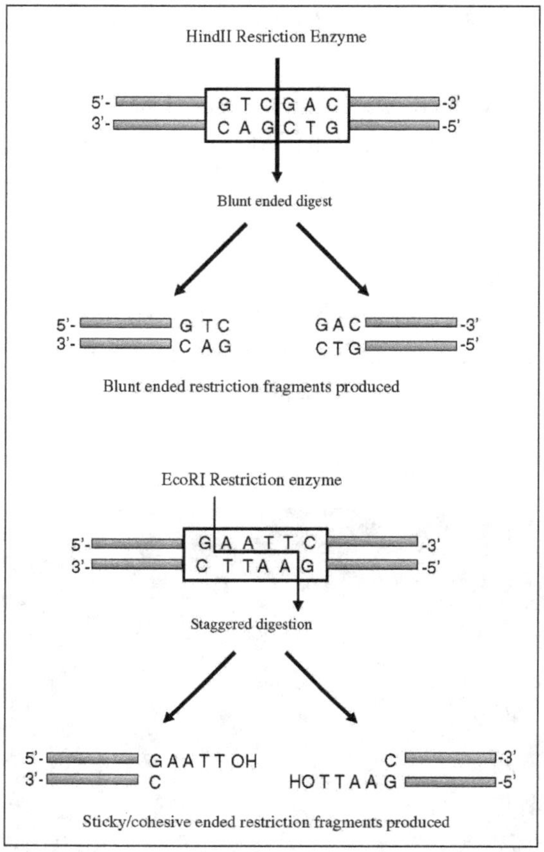

Examples of digestion of DNA by restriction endonucleases. The upper panel indicates the result of a restriction digestion forming blunt fragments with the enzyme HindIII. The bottom panel indicates the cohesive fragments produced by digestion with the enzyme EcoR1.

Approximately 500 restriction enzymes have been characterised that recognise over 100 different target sequences. A number of these, termed isoschizomers, recognise different target sequences but produce the same staggered ends or overhangs. A number of other enzymes have proved to be of value in the manipulation of DNA, as summarised in Table, and are indicated at appropriate points within the text.

Table: Examples of restriction endonucleases that recognise different target sequences and the resulting fragments following digestion.

Name	Recognition Sequence	Digestion Products	
	↓ Four Nucleotide Recognition Sequence		
HaeIII	5'-GGCC-3'	5'-GG CC-3'	Blunt End
	3'-CCGG-5'	3'-CC GG-5'	Digestion
HpaII	5'-CCGG-3'	5'-C CGG-3'	Cohesive End
	3'-GGCC-5'	3'-GGC C-5'	Digestion

	Six Nucleotide Recognition Sequence		
	↓		
BamHI	5'-GGATTC-3'	5'-G GATCC-3'	
	3'-GGCC-5'	3'-CCTAG G-5'	
	↓		
EcoRI	5'-GAATTC-3'	5'-G AATCC-3'	
	3'-CTTAAG-5'	3'-CTTAA G-5'	
	↓		
HindIII	5'-AAGCTT-3'	5'-A AGCTT-3'	
	3'-TTCGAA-5'	3'-TTCGA A-5'	
	Eight Nucleotide Recognition Sequence		
	↓		
NotI	5'-GCGGCCGC-3'	5'-GC GGCCGC-3'	
	3'-CGCCGGCG-5'	3'-CGCCGG CG-5'	

Table: Comparison of the various labelling methods for DNA.

Labelling method	Enzyme	Probe Type	Specific Activity
5' end labelling	Alkaline Phosphatase Polynucleotide Kinase	DNA	Low
3' end labelling	Terminal Transferase	DNA	Low
Nick Translation	D Nase I	DNA	High
	DNA Polymerase I	DNA	
Random Hexamer	DNA Polymerase I	DNA	High
PCR	Taq DNA Polymerase	DNA	High
Riboprobes (cRNA)	RNA Polymerase	RNA	High

Isolation and Separation of Nucleic Acids

Isolation of DNA

The use of DNA for analysis or manipulation usually requires that it is isolated and purified to a certain extent. DNA is recovered from cells by the gentlest possible method of cell rupture to prevent the DNA from fragmenting by mechanical shearing. This is usually in the presence of EDTA which chelate the Mg^{2+} ions needed for enzymes that degrade DNA termed DNase. Ideally, cell walls, if present, should be digested enzymatically (e.g. lysozyme treatment of bacteria) and the cell membrane should be solubilised using detergent. If physical disruption is necessary, it should be kept to a minimum and should involve cutting or squashing of cells, rather than the use of shear forces. Cell disruption (and most subsequent steps) should be performed at 4°C, using glassware and solutions which have been autoclaved to destroy DNase activity.

After release of nucleic acids from the cells, RNA can be removed by treatment with ribonuclease (RNase) which has been heat treated to inactivate any DNase contaminants; RNase is relatively stable to heat as a result of its disulfide bonds, which ensure rapid renaturation of the molecule on cooling. The other major contaminant, protein, is removed by shaking the solution gently with water-saturated phenol or with a phenol–chloroform mixture, either of which will denature proteins but not nucleic acids. Centrifugation of the emulsion formed by this mixing produces a lower, organic phase, separated from the upper, aqueous phase by an interface of denatured protein. The aqueous solution is recovered and deproteinised repeatedly, until no more material is seen at the interface. Finally, the deproteinised DNA preparation is mixed with two volumes of absolute ethanol and the DNA allowed precipitating out of solution in a freezer. After centrifugation, the DNA pellet is redissolved in a buffer containing EDTA to inactivate any DNases present. This solution can be stored at 4 1C for at least 1 month. DNA solutions can be stored frozen, although repeated freezing and thawing tend to damage long DNA molecules by shearing.

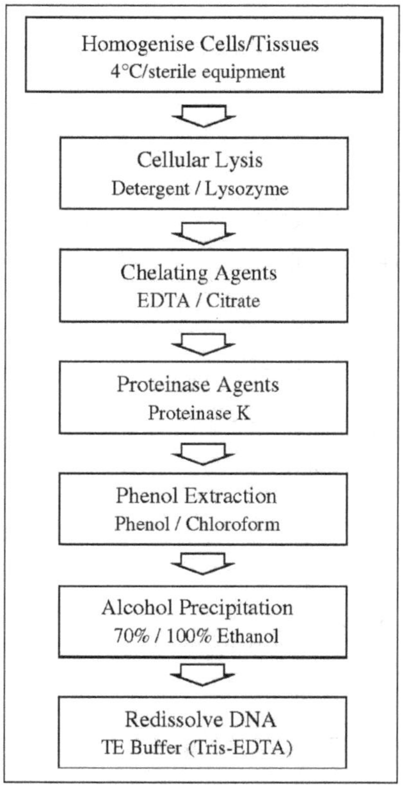

Flow diagram of the main steps involved in the extraction of DNA.

The procedure described above is suitable for total cellular DNA. If the DNA from a specific organelle or viral particle is needed, it is best to isolate the organelle or virus before extracting its DNA, since the recovery of a particular type of DNA from a mixture is usually rather difficult. Where a high degree of purity is required, DNA may be subjected to density gradient ultracentrifugation through caesium chloride, which is particularly useful for the preparation of plasmid DNA.[2] It is possible to check the integrity of the DNA by agarose gel electrophoresis and determine the concentration of the DNA by using the fact that 1 absorbance unit equates to 50 $\mu g\ ml^{-1}$ of DNA and so $50 \times A_{260}$ = concentration of DNA sample ($\mu g\ ml^{-1}$).

Contaminants may also be identified by scanning UV spectrophotometry from 200 to 300 nm. A ratio of 260 nm:280 nm of approximately 1.8 indicates that the sample is free of protein contamination, which absorbs strongly at 280 nm.

Isolation of RNA

The methods used for RNA isolation are very similar to those described above for DNA; however,

RNA molecules are relatively short and therefore less easily damaged by shearing, so cell disruption can be more vigorous. RNA is, however, very vulnerable to digestion by RNases. which are present endogenously in various concentrations in certain cell types and exogenously on fingers. Gloves should therefore be worn and a strong detergent should be included in the isolation medium to denature immediately any RNases. Subsequent deproteinisation should be particularly rigorous, since RNA is often tightly associated with proteins.[3] DNase treatment can be used to remove DNA and RNA can be precipitated by ethanol. One reagent in which is commonly used in RNA extraction is guanadinium thiocyanate, which is both a strong inhibitor of RNase and a protein denaturant. It is possible to check the integrity of an RNA extract by analysing it by agarose gel electrophoresis. The most abundant RNA species are rRNA molecules, 23S and 16S for prokaryotes and 18S and 28S for eukaryotes. These appear as discrete bands on the agarose gel and indicate that the other RNA components are likely to be intact. This is usually carried out under denaturing conditions to prevent secondary structure formation in the RNA. The concentration of the RNA may be estimated by using UV spectrophotometry. At 260 nm, 1 absorbance unit equates to 40μg ml^{-1} of RNA and therefore 40 $\times A_{260}$ = concentration of RNA sample (μg ml^{-1}).

Contaminants may also be identified in the same way as for DNA by scanning UV spectrophotometry; however, in the case of RNA a 260 nm:280 nm ratio of approximately 2 would be expected for a sample containing no contamination.

In many cases, it is desirable to isolate eukaryotic mRNA, which constitutes only 2–5% of cellular RNA, from a mixture of total RNA molecules. This may be carried out by affinity chromatography on oligo (dT)-cellulose columns. At high salt concentrations, the mRNA containing poly(A) tails binds to the complementary oligo(dT) molecules of the affinity column and so mRNA will be retained; all other RNA molecules can be washed through the column with further high-salt solution. Finally, the bound mRNA can be eluted using a low concentration of salt. Nucleic acid species may also be subfractionated by more physical means such as electrophoretic or chromatographic separations based on differences in nucleic acid fragment sizes or physicochemical characteristics.

Electrophoresis of Nucleic Acids

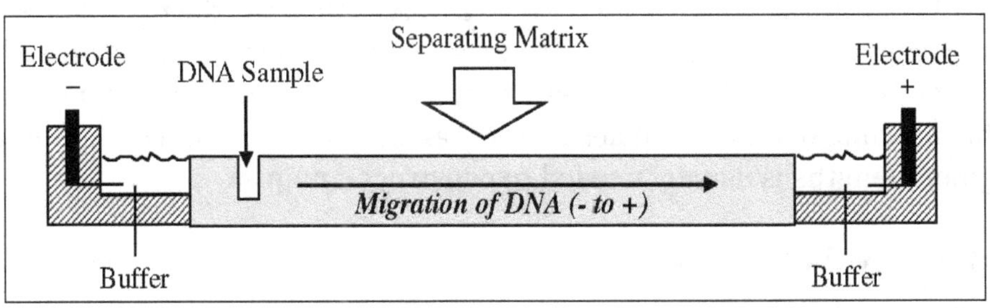

A typical setup required for agarose gel electrophoresis of DNA. The upper panel indicates a cross-section of the unit used for gel electrophoresis.

Electrophoresis in agarose or polyacrylamide gels is the most usual way to separate DNA molecules according to size. The technique can be used analytically or preparatively and can be qualitative or quantitative. Large fragments of DNA such as chromosomes may also be separated by a modification of electrophoresis termed pulsed field gel electrophoresis (PFGE). The easiest and most widely applicable method is electrophoresis in horizontal agarose gels, followed by staining with ethidium bromide. This dye binds to DNA by insertion between stacked base pairs (intercalation) and it exhibits a strong orange/red fluorescence when illuminated with ultraviolet light. Very often electrophoresis is used to check the purity and intactness of a DNA preparation or to assess the extent of a enzymatic reaction during, for example, the steps involved in the cloning of DNA. For such checks 'mini-gels' are particularly convenient, since they need little preparation, use small samples and give results quickly. Agarose gels can be used to separate molecules larger than about 100 bp. For higher resolution or for the effective separation of shorter DNA molecules, polyacrylamide gels are the preferred method.

When electrophoresis is used preparatively, the piece of gel containing the desired DNA fragment is physically removed with a scalpel. The DNA may be recovered from the gel fragment in various ways. This may include crushing with a glass rod in a small volume of buffer, using agarase to digest the agarose thus leaving the DNA, or by the process of electroelution. In the latter method, the piece of gel is sealed in a length of dialysis tubing containing buffer and is then placed between two electrodes in a tank containing more buffer. Passage of an electric current between the electrodes causes DNA to migrate out of the gel piece, but it remains trapped within the dialysis tubing and can therefore be recovered easily. More commonly, commercial spin columns can be used which contain an isolating matrix used in conjunction with a bench-top microcentrifuge. The use of such standardised 'kits' in molecular biology is now commonplace. An alternative to conventional analysis of nucleic acids by electrophoresis is through the use of microfluidic systems. These are carefully manufactured chip-based units where microlitre volumes may be used and with the aid of computer analysis provide much of the data necessary for analysis. Their advantage lies in the fact that the sample volume is very small, allowing much of an extract to be used for further analysis.

Restriction Mapping of DNA Fragments

Restriction mapping involves the size analysis of restriction fragments produced by several restriction enzymes individually and in combination.6 The principle of this mapping is illustrated, in which the restriction sites of two enzymes, A and B, are being mapped. Cleavage with A gives fragments 2 and 7 kilobases (kb) from a 9 kb molecule, hence we can position the single A site 2 kb from one end. Similarly, B gives fragments 3 and 6 kb, so it has a single site 3 kb from one end; but it is not possible at this stage to say if it is near to A's site or at the opposite end of the DNA. This can be resolved by a double digestion. If the resultant fragments are 2, 3 and 4 kb, then A and B cut at opposite ends of the molecule; if they are 1, 2 and 6 kb, the sites are near each other. Not surprisingly, the mapping of real molecules is rarely as simple as this and computer analysis of the restriction fragment lengths is usually needed to construct a map.

Nucleic Acid Analysis Methods

There are numerous methods for analysing DNA and RNA; however, many of them are solution based or more recently include the use of chip-based array systems. Indeed, the lab-on-a-chip approach is developing rapidly and it is possible to envisage many detection and analysis methods being developed in this format in the future. However, traditional methods are still employed in many laboratories and much is still made of producing a hard copy of digested and separated singlestranded DNA fragments attached to a matrix such as nylon for analysis with an appropriate labelled probe.

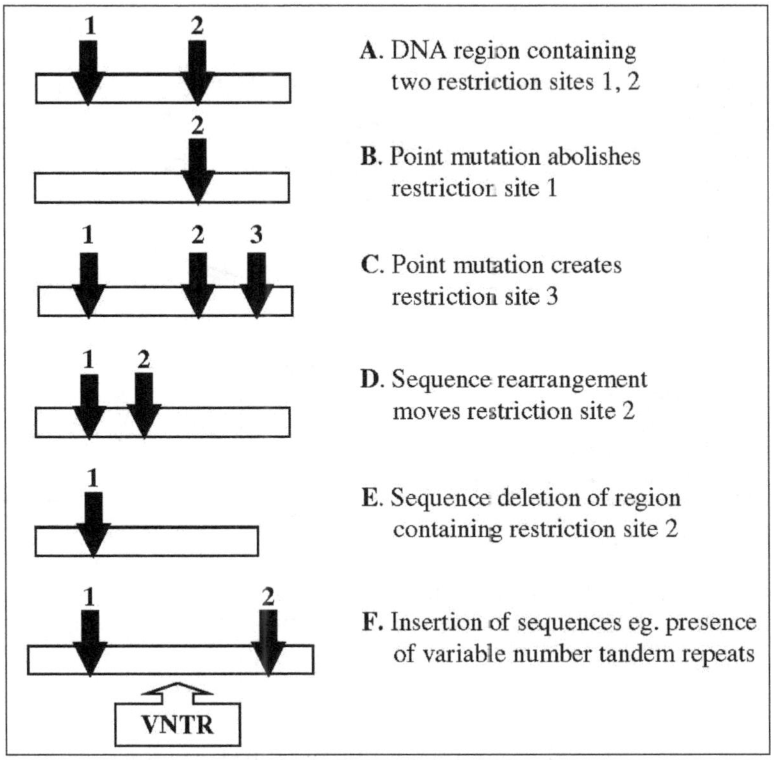

Restriction fragment length polymorphisms (RFLP). The schematic panels A–F indicate the various fragments obtained following digestion as a result of differences in the position of restriction endonuclease target sequences.

DNA Blotting

Electrophoresis of DNA restriction fragments allows separation based on size to be carried out; however, it provides no indication as to the presence of a specific, desired fragment among the complex sample. This can be achieved by transferring the DNA from the intact gel on to a piece of nitrocellulose or nylon membrane placed in contact with it.8 This provides a more permanent record of the sample since DNA begins to diffuse out of a gel that is left for a few hours. First the gel is soaked in alkali to render the DNA single stranded. It is then transferred to the membrane so that the DNA becomes bound to the it in exactly the same pattern as that originally on the gel. This transfer, named a Southern blot after its inventor Ed Southern, can be performed electrophoretically or by drawing large volumes of buffer through both gel and membrane, thus transferring DNA from one to the other by capillary action. The point of this operation is that the membrane can now be treated with a labelled DNA molecule, for example a gene probe. This single-stranded DNA probe will hybridise under the right conditions to complementary fragments immobilised on the membrane. The conditions of hybridisation, including the temperature and salt concentration, are critical for this process to take place effectively. This is usually referred to as the stringency of the hybridisation and it is particular for each individual gene probe and for each sample of DNA. A series of washing steps with buffer are then carried out to remove any unbound probe and the membrane is developed, after which the precise location of the probe and its target may be visualised. It is also possible to analyse DNA from different species or organisms by blotting the DNA and then using a gene probe representing a protein or enzyme from one of the organisms. In this way, it is possible to search for related genes in different species. This technique is generally termed Zoo blotting.

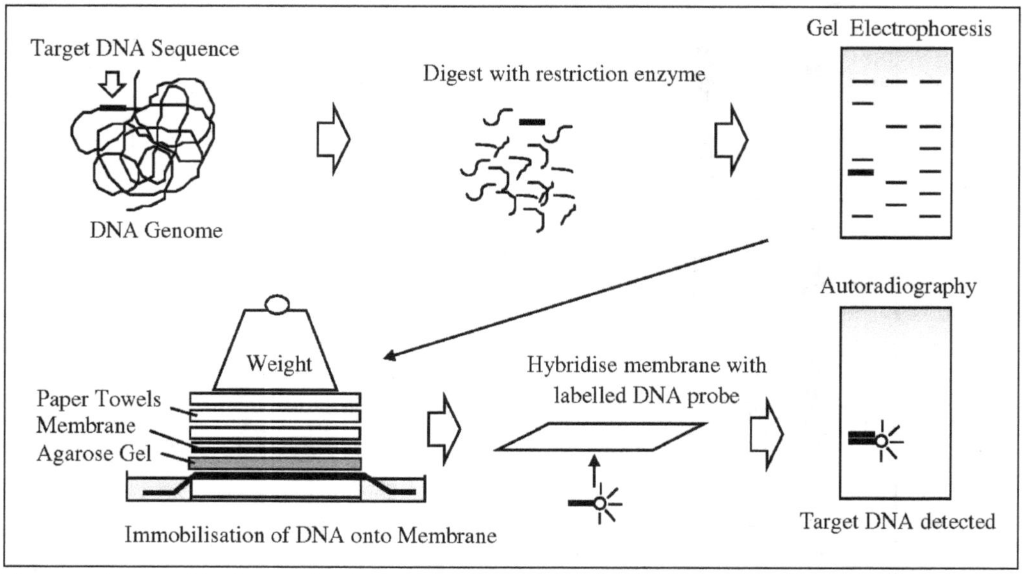

The steps involved in the production of a Southern blot and the subsequent detection of a specific DNA sequence following hybridisation with a complementary labelled gene probe.

RNA Blotting

The same basic process of nucleic acid blotting can be used to transfer RNA from gels on to similar membranes. This allows the identification of specific mRNA sequences of a defined length by hybridisation to a labelled gene probe and is known as Northern blotting.[9] With this technique it is not only possible to detect specific mRNA molecules but it may also be used to quantify the relative amounts of the specific mRNA. It is usual to separate the mRNA transcripts by gel electrophoresis under denaturing conditions since this improves resolution and allows a more accurate estimation of the sizes of the transcripts. The format of the blotting may be altered from transfer from a gel to direct application to slots on a specific blotting apparatus containing the nylon membrane. This is termed slot or dot blotting and provides a convenient means of measuring the abundance of specific mRNA transcripts without the need for gel electrophoresis; it does not, however, provide information regarding the size of the fragments.

A further method of RNA analysis that overcomes the problems of RNA blotting is termed the ribonuclease protection assay. Here the RNA from a sample is extracted and then mixed with a probe representing the sequence of interest in solution. The probe and the appropriate RNA fragment hybridise to form a double-stranded sequence. RNase is then added, which cleaves any single-stranded RNA present but leaves the double-stranded RNA intact. The intact RNA can then be separated by electrophoresis and an indication of the size of the fragment generated. The efficient removal of the background of RNA and the improved sensitivity make the ribonuclease protection assay a popular choice for the analysis of specific RNA molecules.

An important step in the field of RNA analysis was the development of RNAi (RNA interference), which inhibits gene expression. Here double-stranded DNA promotes the degradation of mRNA. Doublestranded RNA in the cell is cleaved by a dicer enzyme, resulting in the formation of small 21–25 bp interfering RNAs (siRNA). The siRNA are complementary to a target RNA strand. Small RNAi proteins are guided by the siRNA to the appropriate mRNA, where the target is then cleaved and is unable to be translated. Many areas are now benefiting from the adoption of this technique in the molecular biology and biotechnology fields.

Gene Probe Derivation

The availability of a gene probe is essential in many molecular biology techniques, yet in many cases is one of the most difficult steps. The information needed to produce a gene probe may come from many sources, but with the development and sophistication of genetic databases this is usually one of the first stages. There are a number of genetic databases throughout the world and it is possible to search these over the internet and identify particular sequences relating to a specific gene or protein. In some cases it is possible to use related proteins from the same gene family to gain information on the most useful DNA sequence. Similar proteins or DNA sequences but from different species may also provide a starting point with which to produce a so-called heterologous gene probe. Although in some cases probes have already been produced and cloned, it is possible, armed with a DNA sequence from a DNA database, to synthesise chemically a single-stranded oligonucleotide probe. This is usually undertaken by computer-controlled gene synthesisers which link dNTPs together based on a desired sequence. It is essential to carry out certain checks before probe production to determine that the probe is unique, is not able to self-anneal or is self-complementary, all of which may compromise its use.

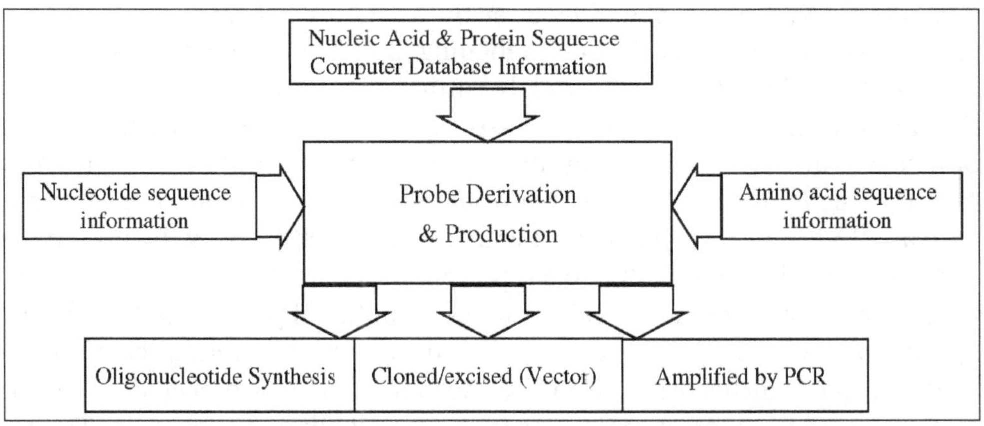

Alternative strategies for designing and producing a gene probe.

Where little DNA information is available to prepare a gene probe, it is possible in some cases to use the knowledge gained from analysis of the corresponding protein. Thus it is possible to isolate and purify proteins and sequence part of the N-terminal end of the protein. From our knowledge of the genetic code, it is possible to predict the various DNA sequences that could code for the protein and then synthesise appropriate oligonucleotide sequences chemically. Due to the degeneracy of the genetic code, most amino acids are coded for by more than one codon, hence there will be more than one possible nucleotide sequence which could code for a given polypeptide. The longer the polypeptide, the greater is the number of possible oligonucleotides which must be synthesised. Fortunately, there is no need to synthesise a sequence longer than about 20 bases, since this should hybridise efficiently with any complementary sequences and should be specific for one gene. Ideally, a section of the protein should be chosen which contains as many tryptophan and methionine residues as possible, since these have unique codons and there will therefore be fewer possible base sequences which could code for that part of the protein. The synthetic oligonucleotides can then be used as probes in a number of molecular biology methods.

Labelling DNA Gene Probe Molecules

An essential feature of a gene probe is that it can be visualised by some means. In this way, a gene probe that hybridises to a complementary sequence may be detected and identify that desired sequence from a complex mixture. There are two main ways of labelling gene probes; traditionally it has been carried out using radioactive labels, but gaining in popularity are non-radioactive labels. Perhaps the most often used radioactive label is phosphorus-32 (^{32}P), although for certain techniques sulfur-35 (^{35}S) and tritium (^{3}H) are used. These may be detected by the process of autoradiography, where the labelled probe molecule, bound to sample DNA, located for example on a nylon membrane, is placed in contact with an X-ray-sensitive film. Following exposure, the film is developed and fixed just as a black and white negative and reveals the precise location of the labelled probe and therefore the DNA to which it has hybridised.

Non-radioactive labels are increasingly being used to label DNA gene probes. Until recently, radioactive labels were more sensitive than their non-radioactive counterparts. However, recent developments have led to similar sensitivities, which, when combined with their improved safety, have led to their greater acceptance.

The labelling systems are termed either direct or indirect. Direct labelling allows an enzyme reporter such as alkaline phosphatase to be coupled directly to the DNA. Although this may alter the characteristics of the DNA gene probe, they offer the advantage of rapid analysis since no intermediate steps are needed. However, indirect labelling is at present more popular. This relies on the incorporation of a nucleotide which has a label attached. At present, three of the main labels in use are biotin, fluorescein and digoxygenin. These molecules are covalently linked to nucleotides using a carbon spacer arm of 7, 14 or 21 atoms. Specific binding proteins may then be used as a bridge between the nucleotide and a reporter protein such as an enzyme. For example, biotin incorporated into a DNA fragment is recognised with a very high affinity by the protein streptavidin. This may be either coupled or conjugated to a reporter enzyme molecule such as alkaline phosphatase. This is able to convert a colourless substrate, p-nitrophenol phosphate (PNPP), into a yellow compound, p-nitrophenol (PNP), and also offers a means of signal amplification. Alternatively labels such as digoxygenin incorporated into DNA sequences may be detected by monoclonal antibodies, again conjugated to reporter molecules including alkaline phosphatase. Thus, rather than the detection system relying on autoradiography, which is necessary for radiolabels, a series of reactions resulting in either a colour or a light or chemiluminescent reaction takes place. This has important practical implications since autoradiography may take 1–3 days, whereas colour and chemiluminescent reactions take minutes.

End Labelling of DNA Molecules

The simplest form of labelling DNA is by 5' or 3' end labelling; 5' end labelling involves a phosphate transfer or exchange reaction where the 5' phosphate of the DNA to be used as the probe is removed and in its place a labelled phosphate, usually ^{32}P, is added. This is usually carried out by using two enzymes; the first, alkaline phosphatase, is used to remove the existing phosphate group from the DNA. Following removal of the released phosphate from the DNA, a second enzyme, polynucleotide kinase, is added, which catalyses the transfer of a phosphate group (^{32}P-labelled) to the 5' end of the DNA. The newly labelled probe is then purified, usually by chromatography through a Sephadex column, and may be used directly.

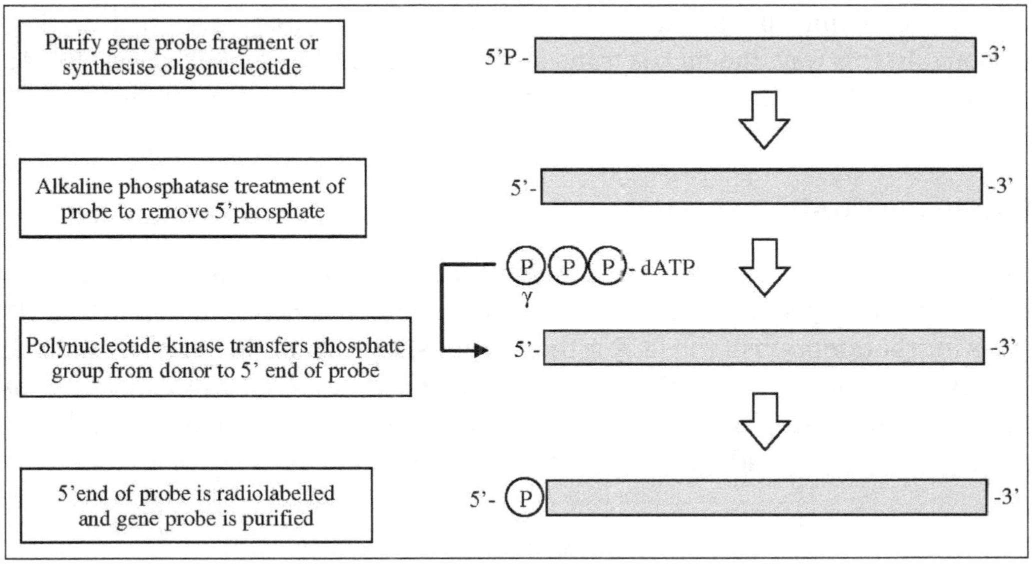

The steps involved in the production of a 50 -labelled gene probe.

Using the other end of the DNA molecule, the 3' end, is slightly less complex. Here a new dNTP which is labelled (e.g. [^{32}P]adATP or biotinlabelled dNTP) is added to the 3' end of the DNA by the enzyme terminal transferase. Although this is a simpler reaction, a potential problem exists because a new nucleotide is added to the existing sequence and so the complete sequence of the DNA is altered, which may affect its hybridisation to its target sequence. End labelling methods also suffer from the fact that only one label is added to the DNA so they are of a lower activity in comparison with methods that incorporate label along the length of the DNA.

Random Primer Labelling

In random primer labelling the DNA to be labelled is first denatured and then placed under renaturing conditions in the presence of a mixture of many different random sequences of hexamers or hexanucleotides. These hexamers will, by chance, bind to the DNA sample wherever they encounter a complementary sequence and so the DNA will rapidly acquire an approximately random sprinkling of hexanucleotides annealed to it. Each of the hexamers can act as a primer for the synthesis of a fresh strand of DNA catalysed by DNA polymerase since it has an exposed 30 -hydroxyl group. The Klenow fragment of DNA polymerase is used for random primer labelling because it lacks a 5' –3' exonuclease activity. This is prepared by cleavage of DNA polymerase with subtilisin, giving a large enzyme fragment which has no 5' to 3' exonuclease activity, but which still acts as a 5' to 3' polymerase. Thus, when the Klenow enzyme is mixed with the annealed DNA sample in the presence of dNTPs, including at least one which is labelled, many short stretches of labelled DNA will be generated. In a similar way to random primer labelling, the polymerase chain reaction may also be used to incorporate radioactive or non-radioactive labels.

Nick Translation

A traditional method of labelling DNA is by the process of nick translation. Low concentrations of DNase I are used to make occasional single-strand nicks in the double-stranded DNA that is to be used as the gene probe. DNA polymerase then fills in the nicks, using an appropriate deoxyribonucleoside triphosphate (dNTP), at the same time making a new nick to the 3' side of the previous one. In this way, the nick is translated along the DNA. If labelled dNTPs are added to the reaction mixture, they will be used to fill in the nicks and so the DNA can be labelled to a very high specific activity.

The Polymerase Chain Reaction

There have been a number of key developments in molecular biology techniques. However, one that has had the most impact in recent years has been the polymerase chain reaction (PCR). One of the reasons for the adoption of the PCR is the elegant simplicity of the reaction and relative ease of the practical manipulation steps. Frequently this is one of the first techniques to be used when analysing DNA and RNA and in its quantitation it has opened up the analysis of cellular and molecular processes to those outside the field of molecular biology.

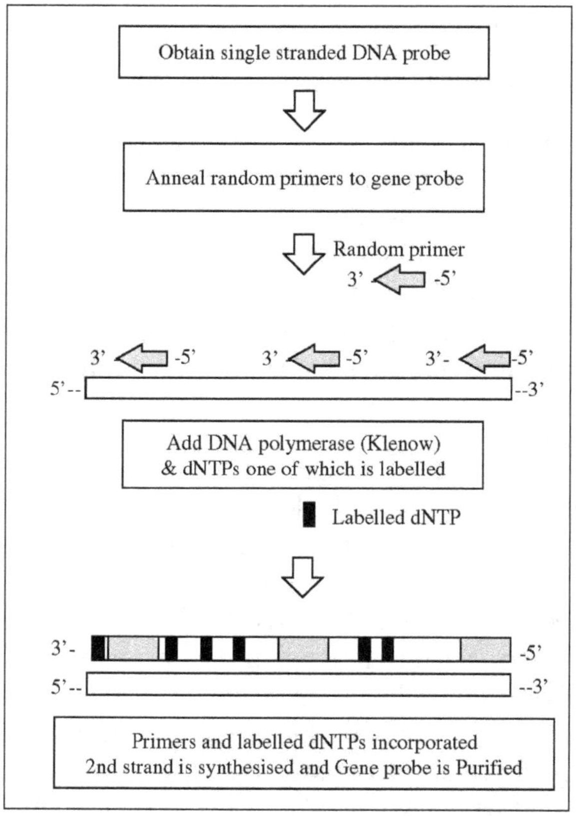

The steps involved in the production of a gene probe produced by the random hexamer method.

The PCR is used to amplify a precise fragment of DNA from a complex mixture of starting material, usually termed the template DNA, and in many cases requires little DNA purification. It does require the knowledge of some DNA sequence information that flanks the fragment of DNA to be amplified (target DNA). From this information, two oligonucleotide primers may be chemically synthesised, each complementary to a stretch of DNA to the 3' side of the target DNA, one oligonucleotide for each of the two DNA strands. The result is an amplification of a specific DNA fragment which obviates the need for more time-consuming cloning procedures. Further developments in molecular biology and biotechnology have allowed numerous genomes to be analysed and genes identified. It is not surprising that this has been aided by the developments in the area of bioinformatics and whole genome analysis. DNA databases and other nucleic acid sequence and protein analysis software may all be accessed over the internet given the relevant software and authority. This is now relatively straightforward with web browsers that provide a user-friendly graphical interface for sequence manipulation. Consequently, the new expanding and exciting areas of bioscience research are those that analyse genome and DNA sequence databases, (genomics) and also their protein counterparts (proteomics). This is sometimes referred to as in silico research, and there is no doubt that for basic and biotechnological research it is as important to have Internet and database access as it is to have equipment and reagents for laboratory molecular biology.

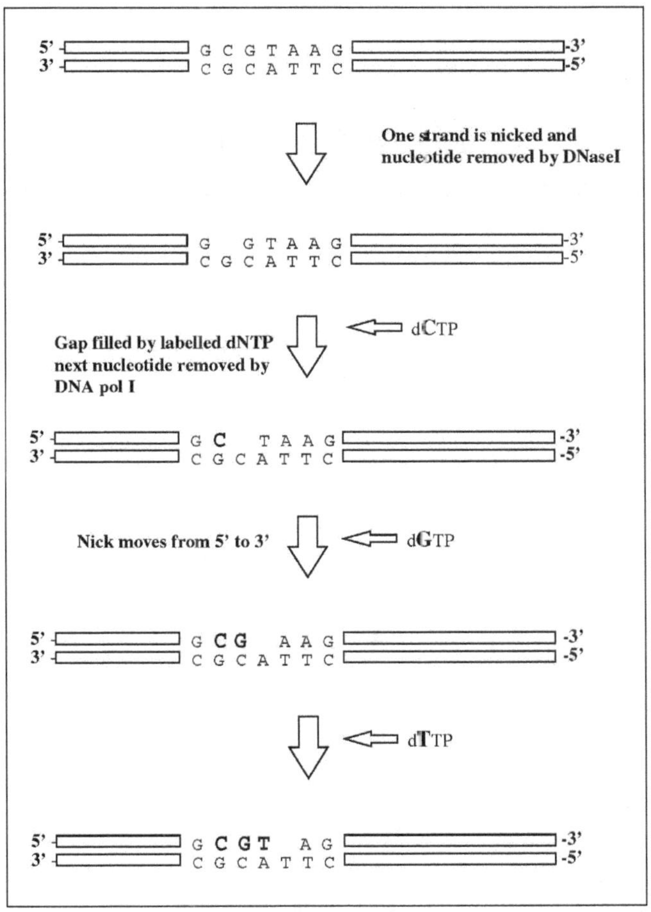

The steps involved in the production of a gene probe by the nick translation method.

Table: Nucleic acid and protein database resources.

Database or Resource	
General DNA Sequence Databases	
EMBL	European genetic database
GenBank DDBJ	US genetic database Japanese genetic database
Protein Sequence Databases	
Swiss-Prot	European protein sequence database
TREMBL	European protein sequence database

PIR	US protein information resource
Protein Structure Databases	
PDB	Brookhaven protein database
NRL-3D	Protein structure database
Genome Project Databases	
Human Mapping Database Johns Hopkins, USA	
dbEST (cDNA and partial sequences)	
Genethon Genetic maps based on repeat markers	
Whitehead Institute (YAC and physical maps)	

References

- Molecular-biology, science: britannica.com, Retrieved 03, April 2020
- Basic-Molecular-Biology-Techniques-50364897: researchgate.net, Retrieved 30, January 2020

The Basics of DNA

DNA stands for deoxyribonucleic acid. It is made up of two polynucleotide chains that wind around one another to form double helix. DNA carries the genetic instructions for the growth, development and reproduction of living organisms. This chapter has been carefully written to provide an easy understanding of the varied facets of DNA such as its structure, replication, damage, repair, recombination and sequencing.

DNA is abbreviation of deoxyribonucleic acid, organic chemical of complex molecular structure that is found in all prokaryotic and eukaryotic cells and in many viruses. DNA codes genetic information for the transmission of inherited traits.

The chemical DNA was first discovered in 1869, but its role in genetic inheritance was not demonstrated until 1943. In 1953 James Watson and Francis Crick, aided by the work of biophysicists Rosalind Franklin and Maurice Wilkins, determined that the structure of DNA is a double-helix polymer, a spiral consisting of two DNA strands wound around each other. The breakthrough led to significant advances in scientists' understanding of DNA replication and hereditary control of cellular activities.

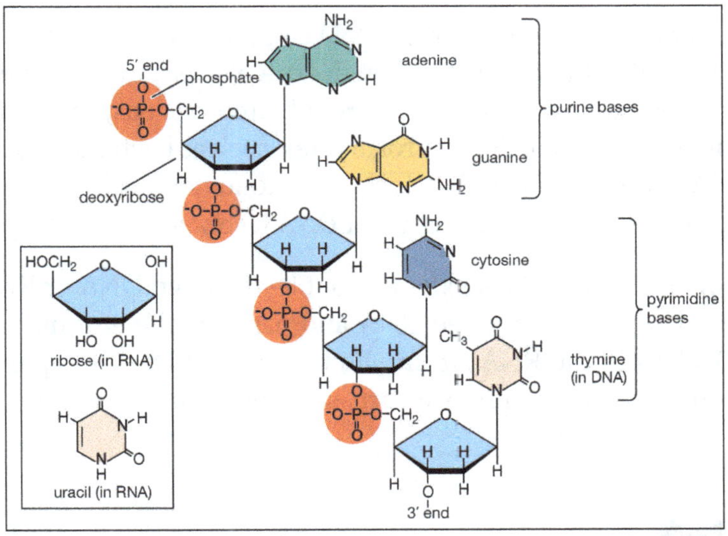

Polynucleotide chain of deoxyribonucleic acid (DNA)Portion of polynucleotide chain of deoxyribonucleic acid (DNA). The inset shows the corresponding pentose sugar and pyrimidine base in ribonucleic acid (RNA).

Each strand of a DNA molecule is composed of a long chain of monomer nucleotides. The nucleotides of DNA consist of a deoxyribose sugar molecule to which is attached a phosphate group and one of four nitrogenous bases: two purines (adenine and guanine) and two pyrimidines (cytosine and thymine). The nucleotides are joined together by covalent bonds between the phosphate of one nucleotide and the sugar of the next, forming a phosphate-sugar backbone from which the nitrogenous bases protrude. One strand is held to another by hydrogen bonds between the bases; the sequencing of this bonding is specific—i.e., adenine bonds only with thymine, and cytosine only with guanine.

The configuration of the DNA molecule is highly stable, allowing it to act as a template for the replication of new DNA molecules, as well as for the production (transcription) of the related RNA (ribonucleic acid) molecule. A segment of DNA that codes for the cell's synthesis of a specific protein is called a gene.

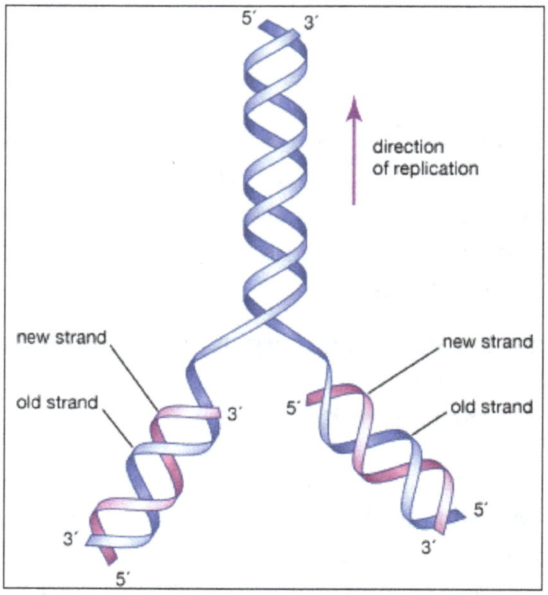

Initial proposal of DNA structure. The initial proposal of the structure of DNA by James Watson and Francis Crick, which was accompanied by a suggestion on the means of replication.

Within a cell, DNA is organized into dense protein-DNA complexes called chromosomes. In eukaryotes, the chromosomes are located in the nucleus, although DNA also is found in mitochondria and chloroplasts. In prokaryotes, which do not have a membrane-bound nucleus, the DNA is found as a single circular chromosome in the cytoplasm. Some prokaryotes, such as bacteria, and a few eukaryotes have extrachromosomal DNA known as plasmids, which are autonomous, self-replicating genetic material. Plasmids have been used extensively in recombinant DNA technology to study gene expression.

The genetic material of viruses may be single- or double-stranded DNA or RNA. Retroviruses carry their genetic material as single-stranded RNA and produce the enzyme reverse transcriptase, which can generate DNA from the RNA strand. Four-stranded DNA complexes known as G-quadruplexes have been observed in guanine-rich areas of the human genome.

Forms of DNA

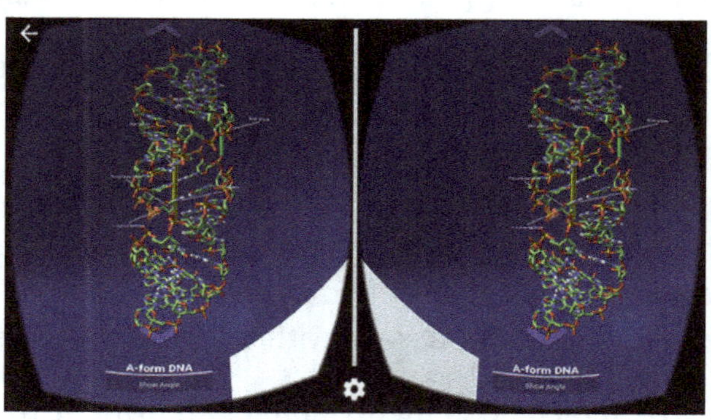

A-form DNA

A-DNA is a right-handed double helix made up of deoxyribonucleotides. It appears when the relative humidity of the environment is less than 75%, which means that it is rarely present in normal physiological condition. The two strands of A-DNA are anti-parallel with each other and not symmetrical. The molecule is asymmetrical is because the glycosidic bonds of a base pair are not diametrically opposite to each other. Therefore, major grooves and minor grooves can be observed in each turn. One turn of the helix consists of 11 base pairs with a length of 2.86nm. The backbone of A-DNA is formed by sugar phosphates that are linked continuous using phosphodiester bonds. All the nitrogenous bases are at the core centre of the helix. Hydrogen bonds between nitrogenous bases allow the molecule to exhibit the double helix structure. The helix width of A-DNA is 2.3nm. Overall, A-DNA is wider than the more commonly found B-DNA.

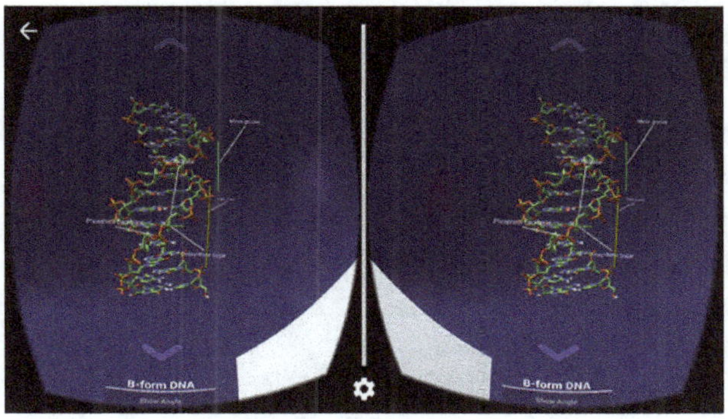

B-form DNA

B-form DNA is a right-handed double helix, which was discovered by Watson and Crick based on the X-ray diffraction patterns. It is the common form of DNA exists under normal physiological condition. The double strands of B-DNA run in opposite directions. The structure is asymmetrical with major grooves and minor grooves present alternatively. The molecule is asymmetrical is because the glycosidic bonds of a base pair are not diametrically opposite to each other. In one turn, there are 10 base pairs with a length of 3.4nm. The distance between adjacent deoxyribonucleotides is 0.34nm. Same as A-DNA, the backbone of B-DNA is formed by sugar phosphates that are linked continuous using phosphodiester bonds and the core region consists of nitrogenous bases. The two strands are held together by hydrogen bonds between nitrogenous bases. The helix width of B-DNA is 2nm. B-DNA is narrower than A-DNA.

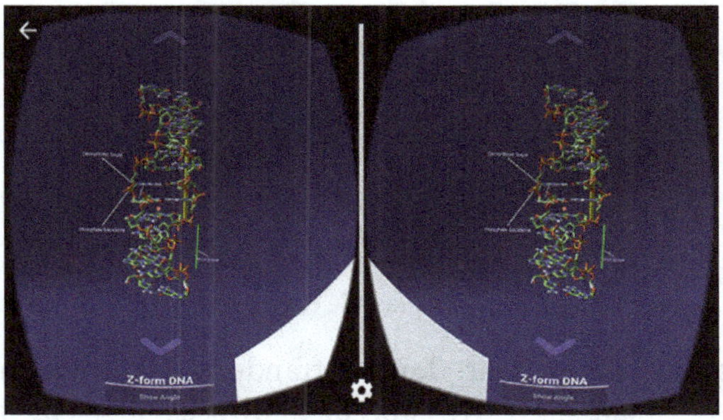

Z-form DNA

Z-form DNA is a left-handed double helix. It has a very different structure when compared with A-DNA and B-DNA. The zigzag appearance of backbone allows it to be distinguished from other forms of DNA. The helix width is 1.8nm, which is the narrowest among the three types. The structure consists of major and minor grooves. One turn of Z-DNA has 12 base pairs and the length is 4.56nm. The distance between two adjacent deoxyribonucleotides is 0.37nm. Same as the other forms of DNA, hydrogen bond is present to hold the two strands together. Z-DNA is difficult to be observed since it is unstable. It may take part in expression regulation of some genes or in genetic recombination. It can be found in bacteria, eukaryotes and viruses. In some viruses, they require Z-DNA binding proteins for pathogenesis.

DNA Structure

DNA or deoxyribonucleic acid is a helically twisted double chain polydeoxyribonucleotide macromolecule which constitutes the genetic material of all organisms with the exception of rhinoviruses. In prokaryotes it occurs in nucleoid and plasmids. This DNA is usually circular. In eukaryotes, most of the DNA is found in chromatin of nucleus.

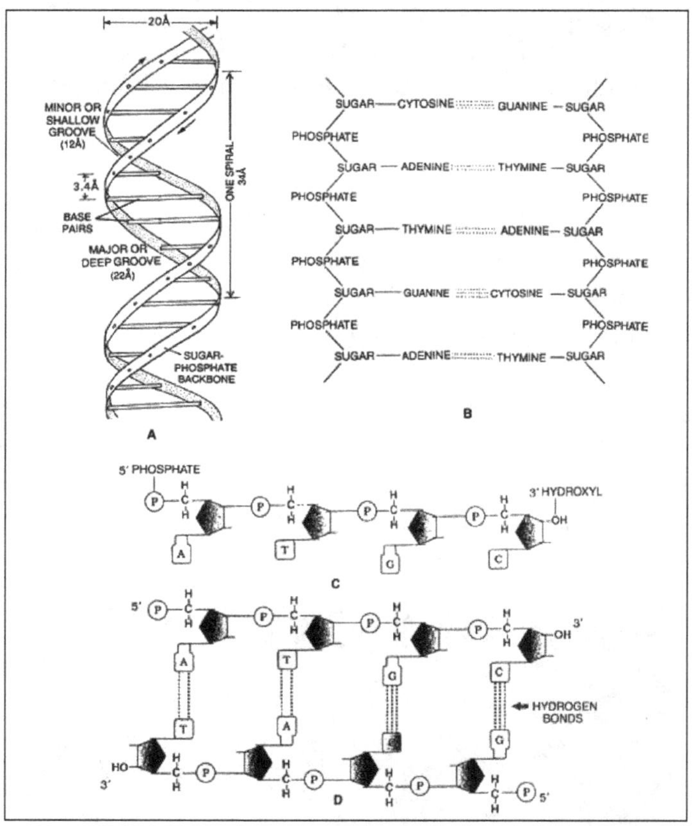

Structure of DNA: A. Coiling in double helix or duplex of DNA. B. Sequence of nucleoids in a part of the double helix of DNA. C. A single stranded polynucleotide chain. D. A double stranded polynucleotide chain.

It is linear. Smaller quantities of circular, double stranded DNA are found in mitochondria and plastids (organelle DNA). Small sized DNAs occur in viruses, ɸ x 174 bacteriophage has 5386 nucleotides. Bacteriophage lambda (Phage X) possesses 48502 base pairs (bp) while number of base pairs in Escherichia coli is 4.6 x 10^6. A single genome (haploid set of 23 chromosomes) has about 3.3 x 10^9 bp in human beings. Single-stranded DNA occurs as a genetic material in some viruses (e.g., phage ф x 174, coliphage fd, M_{13}). DNA is the largest macromolecule with a diameter of 2 nm (20Å or 2 x 10^{-9}m) and often having 3 lengths in millimetres.

It 15 negatively charged due to phosphate groups. It is a long chain polymer of generally several hundred thousands of deoxyribonucleotides. A DNA molecule has two un-branched complementary strands. They are spirally coiled. The two spiral strands of DNA are collectively called DNA duplex.

The two strands are not coiled upon each other but the whole double strand (DNA duplex) is coiled upon itself around a common axis like a rope stair case with solid steps twisted into a spiral. Due to spiral twisting, the DNA duplex comes to have two types of alternate grooves, major (22 Å) and minor (12 Å).

In B-DNA, one turn of the spiral has about 10 nucleotides on each strand of DNA. It occupies a distance of about 3.4 nm (34 Å or 3.4×10^{-9} m) so that adjacent nucleotides or their bases are separated by a space of about 0.34 nm (0.34×10^{-9} m or 3.4 Å).

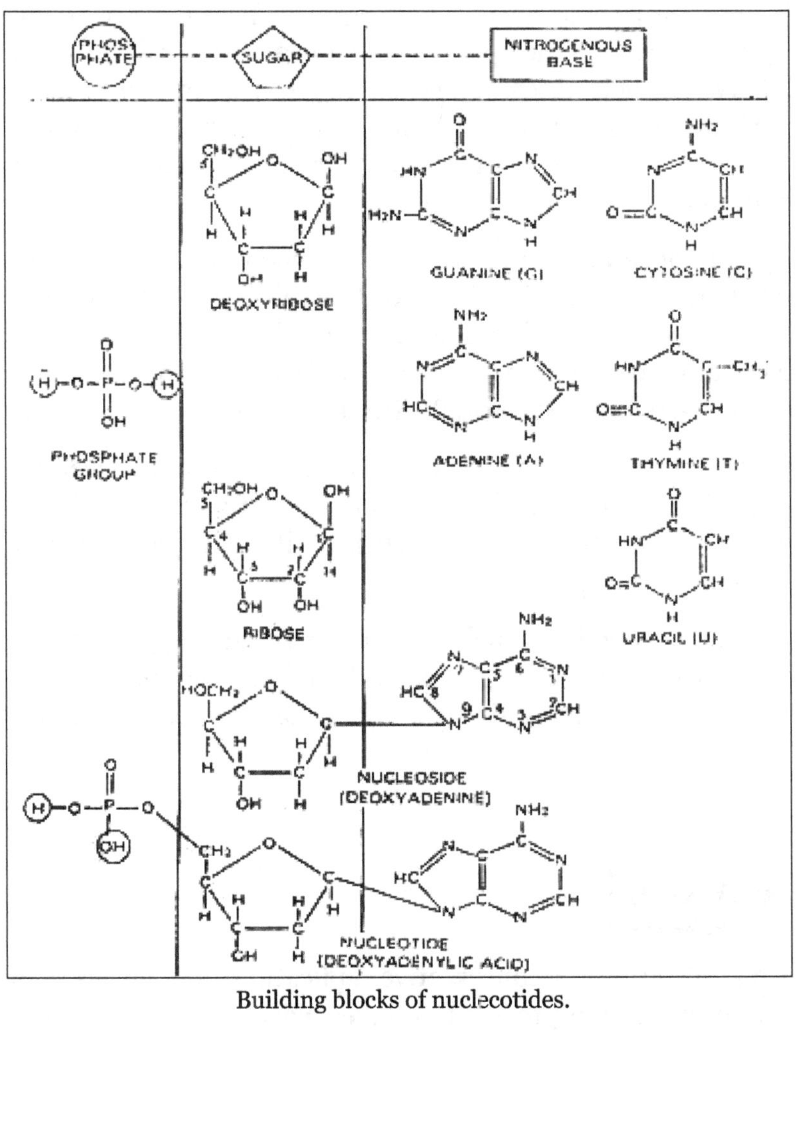

Building blocks of nuclecotides.

A deoxyribonucleotide of DNA is formed by cross-linking of three chemicals ortho- phosphoric acid (H_3PO_4), deoxyribose sugar ($C_5H_{10}O_4$) and a nitrogen base. Four types of nitrogen bases occur in DNA. They belong to two groups, purines (9-membered double rings with nitrogen at 1,3,7 and 9 positions) and pyrimidines (six membered rings with nitrogen at 1 and 3 positions). DNA has two types of purines (adenine or A and guanine or G) and two types of pyrimidines (cytosine or C and thymine or T).

Depending upon the type of nitrogen base, DNA has four kinds of deoxyribonucleotides —deoxy adenosine 5- monophosphate (d AMP), deoxy guaninosine 5-monophosphate (d GMP), deoxy thymidine 5-monophosphate (d TMP) and deoxy cytidine 5- monophosphate (d CMP).

The back bone of a DNA chain or strand is built up of alternate deoxyribose sugar and phosphoric acid groups. The phosphate group is connected to carbon 5′ of the sugar residue of its own nucleotide and carbon Y of the sugar residue of the next nucleotide by (3 '—5') phosphodiester bonds. -H of phosphate and -OH of sugar are eliminated as H_2o during each ester formation.

Phosphate group provides acidity to the nucleic acids because at least one of its side group is free to dissociate. Nitrogen bases lie at right angles to the longitudinal axis of DNA chains. They are attached to carbon atom 1 of the sugars by N-glycosidic bonds. Pyrimidine (C or T) is attached to deoxyribose by its N-atom at 1 position while a purine (A or G) does so by N-atom at 9 positions.

The two DNA chains are antiparallel that is, they run parallel but in opposite directions. In one chain the direction is 5→ Y while in the opposite one it is 3′ →5′. The two chains are held together by hydrogen bonds between their bases. Adenine (A), a purine of one chain lies exactly opposite thymine (T), a pyramidine of the other chain. Similarly, cytosine (C, a pyrimidine) lies opposite guanine (G a purine). This allows a sort of lock and key arrangement between large sized purine and small sized pyrimidine.

It is strengthened by the appearance of hydrogen bonds between the two. Three hydrogen bonds occur between cytosine and guanine (C = G) at positions 1'-1', 2′- 6′ and 6′-2′. There are two such hydrogen bonds between adenine and thymine (A=T) which are formed at positions 1'-3′ and 6′-4′. Hydrogen bonds occur between hydrogen of one base and oxygen or nitrogen of the other base. Since specific and different nitrogen bases occur on the two DNA chains, the latter are complementary.

Thus the sequence of say AAGCTCAG of one chain would have a complementary sequence of TTC-GAGTC on the other chain. In other words, the two DNA chains are not identical but complementary to each other. It is because of specific base pairing with a purine lying opposite a pyrimidine. This makes the two chains 2 nm thick.

A purine- purine base pair will make it thicker while a pyrimidine-pyrimidine base pair will make it narrower than 2 nm. Further, A and C or G and T do not pair because they fail to form hydrogen bonds between them. 5′ end of each chain bears phosphate radical while the 3′ end possesses a sugar residue (3'-OH).

DNA Replication

DNA replication, also known as semi-conservative replication, is the process by which DNA is essentially doubled. It is an important process that takes place within the dividing cell.

- This is a complex process that takes place during cell division, (interphase, S phase) whereby DNA makes copies (duplicates) before the cell divides through mitosis and meiosis.
- DNA replication is a semiconservative process where a parental strand (template) is used to synthesize a new complementary daughter strand using several protein elements which include enzymes and RNA molecules.
- DNA replication process uses DNA polymerase as the main enzyme for catalyzing the joining of deoxyribonucleoside 5′-triphosphates (dNTPs) forming a growing chain of DNA.
- Other proteins are also involved for initiation of the process and copying of DNA, along with proofreading capabilities to ensure the replication process takes place accurately.
- Therefore DNA replication is a process that produces identical helices of DNA from a single strand of the DNA molecule.
- DNA replication is an essential mechanism in enhancing cell growth, repair, and reproduction of an organism.

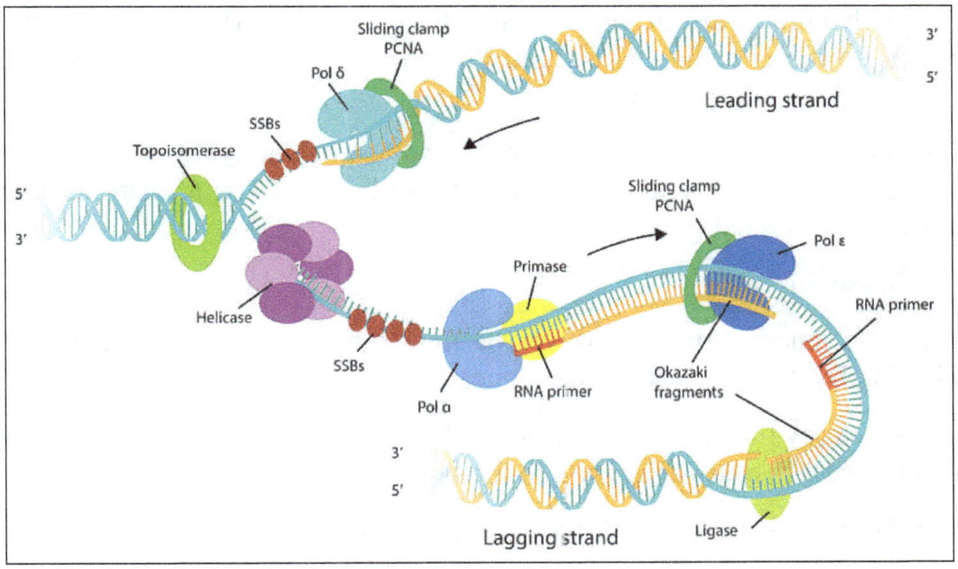

The Mechanism of DNA Replication

DNA replication takes place in three major steps.

- Opening of the double-stranded helical structure of DNA and separation of the strands.
- Priming of the template strands.
- Assembly of the newly formed DNA segments.
    - During the separation of DNA, the two strands uncoil at a specific site known as the origin. With the involvement of several enzymes and proteins, they prepare (prime) the strands for duplication.
    - At the end of the process, DNA polymerase enzyme starts to organize the assembly of the new DNA strands.

- These are the general steps of DNA replication for all cells but they may vary specifically, depending on the organism and cell type.
- Enzymes play a major role in DNA replication because they catalyze several important stages of the entire process.
- DNA replication is one of the most essential mechanisms of a cell's function and therefore intensive research has been done to understand its processes.
- The mechanism of DNA replication is well understood in *Escherichia coli,* which is also similar to that in eukaryotic cells.
- In *E.coli,* DNA replication is initiated at the oriClocus (oriC), to which DnaA protein binds while hydrolyzing of ATP takes place.

DNA Replication Enzymes and Proteins

- DNA Polymerase:
  - DNA polymerases are enzymes used for the synthesis of DNA by adding nucleotide one by one to the growing DNA chain. The enzyme incorporates complementary amino acids to the template strand.
  - DNA polymerase is found in both prokaryotic and eukaryotic cells. They both contain several different DNA polymerases responsible for different functions in DNA replication and DNA repair mechanisms.
- DNA Helicase Enzyme:
  - This is the enzyme that is involved in unwinding the double-helical structure of DNA allowing DNA replication to commence.
  - It uses energy that is released during ATP hydrolysis, to break the hydrogen bond between the DNA bases and separate the strands.
  - This forms two replication forks on each separated strand opening up in opposite directions.
  - At each replication fork, the parental DNA strand must unwind exposing new sections of single-stranded templates.
  - The helicase enzyme accurately unwinds the strands while maintaining the topography on the DNA molecule.
- DNA Primase Enzyme:
  - This is a type of RNA polymerase enzyme that is used to synthesize or generate RNA primers, which are short RNA molecules that act as templates for the initiation of DNA replication.
- DNA Ligase Enzyme:
  - This is the enzyme that joins DNA fragments together by forming phosphodiester bonds between nucleotides.

- Exonuclease:
    - These are a group of enzymes that remove nucleotide bases from the end of a DNA chain.
- Topoisomerase:
    - This is the enzyme that solves the problem of the topological stress caused during unwinding.
    - They cut one or both strands of the DNA allowing the strand to move around each other to release tension before it rejoins the ends.
    - And therefore, the enzyme catalysts the reversible breakage it causes by joining the broken strands.
    - Topoisomerase is also known as DNA gyrase in *E. coli*.
- Telomerase:
    - This is an enzyme found in eukaryotic cells that adds a specific sequence of DNA to the telomeres of chromosomes after they divide, stabilizing the chromosomes over time.

DNA Replication Steps/Stages

Initiation

- This is the stage where DNA replication is initiated.
- DNA synthesis is initiated within the template strand at a specific coding region site known as origins.
- The origin sites are targeted by the initiator proteins, which recruit additional proteins that help in the replication process to form a replication complex around the DNA origin.
- There are several origin sites on which DNA replication is initiated and they are all known as replication forks.
- The formed replication complex contains the DNA helicase enzyme whose function is to unwind the double helix, exposing the two strands, which act as templates for replication.
- The mechanism of DNA helicase enzyme is by hydrolyzing the ATP that is used to form the bonds between the nucleobases, thus breaking the bond that holds the two strands.
- Additionally, during initiation DNA primase enzyme synthesizes small RNA primers that kick-start the function of DNA polymerase.
- DNA polymerase enzyme functions by growing the new DNA daughter strand.

Elongation

- This is the phase where the DNA polymerase grows the new DNA daughter strand by attaching to the original unzipped template strand and the initiating short RNA primer.
- The DNA polymerase is able to synthesize a new strand that matches the template, by extending the primer via the addition of free nucleotides to the 3′ end.

- One of the templates reads in the 3′ to 5′ direction, and therefore, the DNA polymerase synthesizes the new strand in the 5′ to 3′ direction, which is known as the leading strand.
- Along the template strand, DNA primase synthesizes a short RNA primer at the beginning of the template in the 5′ to 3′ direction, which initiates the DNA polymerase to continue synthesizing new nucleotides, extending the new DNA strand.
- The other template (5′ to 3′) is elongated in an antiparallel direction, by the addition of short RNA primers which are filled with other joining fragments, forming the newly formed lagging strand. These short fragments are known as the Okazaki fragments.
- The synthesis of the lagging strand is discontinuous since the newly formed strand is disjointed.
- The RNA nucleotides from the short RNA primers must be removed and replaced by DNA nucleotides, which are then joined by the DNA ligase enzyme.

Termination

- After the synthesis and extension of both the continuous and discontinued stands, an enzyme knows as exonuclease removes all RNA primers from the original strands.
- The primers are replaced with the right nucleotide bases.
- While removing the primers, another type of exonuclease proofread the new stands, checking, removing, and replacing any errors formed during synthesis.
- DNA ligase enzyme joins the Okazaki fragments to form a single unified strand.
- The ends of the parent strand consist of a repetition of DNA sequences known as telomeres which act as protective caps at the ends of chromosomes preventing the fusion of nearby chromosomes.
- The telomeres are synthesized by a special type of DNA polymerase enzyme known as telomerase.
- It catalyzes the telomere sequences at the end of the DNA.
- On completion, the parent and complementary strand coil into a double helical shape, producing two DNA molecules each passing one strand from the parent molecule and one new strand.

Okazaki Fragments

The two DNA strands run in opposite or antiparallel directions, and therefore to continuously synthesize the two new strands at the replication fork requires that one strand is synthesized in the 5′ to 3′ direction while the other is synthesized in the opposite direction, 3′ to 5′. However, DNA polymerase can only catalyze the polymerization of the dNTPs only in the 5’to 3’direction. This means that the other opposite new strand is synthesized differently. But how?

By the joining of discontinuous small pieces of DNA that are synthesized backward from the direction of movements of the replication fork. These small pieces or fragments of the new DNA strand are known as the Okazaki Fragments. The Okazaki fragments are then joined by the action of DNA ligase, which forms an intact new DNA strand known as the lagging strand.

The lagging phase is not synthesized by the primer that initiates the synthesis of the leading strand. Instead, a short fragment of RNA serves as a primer (RNA primer) for the initiation of replication of the lagging strand. RNA primers are formed during the synthesis of RNA which is initiated de novo, and an enzyme known as primase synthesizes these short fragments of RNA, which are 3-10 nucleotides long and complementary to the lagging strand template at the replication fork. The Okazaki fragments are then synthesized by the extension of the RNA primers by DNA polymerase. However, the newly synthesized lagging strand is that it contains an RNA-DNA joint, defining the critical role of RNA in DNA replication.

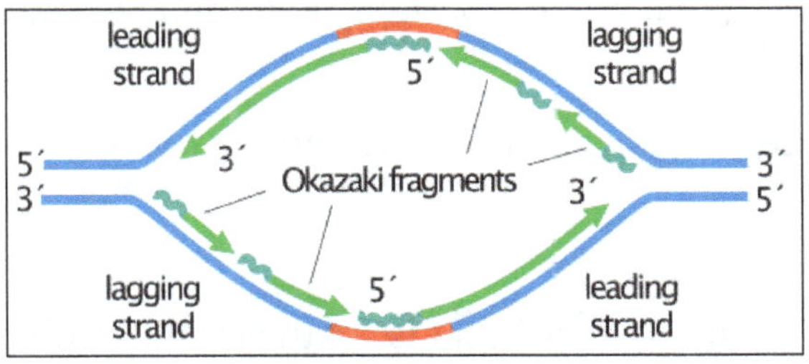

Okazaki fragments.

Replication Fork Formation and its Function

- The replication fork is the site of active DNA synthesis, where the DNA helix unwinds and single strands of the DNA replicates.
- Several sites of origin represent the replication forks.
- The replication fork is formed during DNA strand unwinding by the helicase enzyme which exposes the origin of replication. A short RNA primer is synthesized by primase and elongation done by DNA polymerase.
- The replication fork moves in the direction of the new strand synthesis. The new DNA strands are synthesized in two orientations, i.e 3′ to 5′ direction which is the leading strand, and the 5′ to 3′ orientation which is the lagging strand.
- The two sides of the new DNA strand (leading and lagging strand) are replicated in two opposite directions from the replication fork.
- Therefore the replication fork is bi-directional.

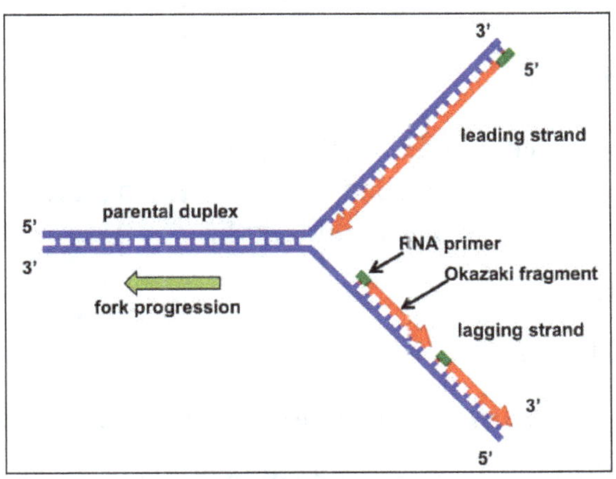

DNA Replication Fork.

Leading Strand

- The leading strand is the new DNA strand that is continuously synthesized by the DNA polymerase enzyme.
- It is the simplest strand that is synthesized during replication.
- The synthesis starts after the DNA strand has unzipped and separated. This generates a short piece of RNA known as a primer, by the DNA primase enzyme.
- The primer binds to the 3′ end (start) of the strand, thus initiating the synthesize of the new strand (leading strand).
- The synthesis of the leading strand is a continuous process.

The Lagging Strand

- This is the template strand (5′ to 3′) that is synthesized in a discontinuous manner by RNA primers.
- During the synthesis of the leading strand, it exposes small, short strands, or templates that are then used for the synthesis of the Okasaki fragments.
- The Okasaki fragments synthesize the lagging strand by the activity of DNA polymerase which adds the pieces of DNA (the Okasaki fragments) to the strand between the primers.
- The formation of the lagging strand is a discontinuous process because the newly formed strand (lagging strand) is the fragmentation of short DNA strands.

Importance of DNA Replication

- DNA replication takes place during cell division and it enables the multiplication and division of DNA by making two copies of the genome from a single parent genome.
- And therefore, its importance is in the creation of new and next copies of DNA giving rise to two daughter cells from a single parent cell.
- Each new cell is formed with its own genome.
- This enhances heredity via reproduction and cell division.

DNA Replication Stress

During DNA replication, the process and the DNA genome undergoes various stress arising from the mechanism. These stress a result installed replication and stalled replication fork formation. Several events contribute to these stresses, including:

- Unusual DNA structure.
- Mismatched ribonucleotides.
- Tensions arising from concurrent mechanisms of replication and transcription.
- Inadequate availability of important replication factors.

- Fragile sites on the replicating DNA strand.
- Overexpression or constitutive activation of oncogenes.
- Inaccessible chromatins.

Kinase regulatory proteins such as ATM (ATM serine/threonine kinase) and ATP are proteins that assist in alleviating replication stress. These proteins get recruited and activated by DNA damages.

Stalled replication forks may collapse if the regulatory proteins do not stabilize, and if and when this happens, initiation of repairing mechanisms to reassembling of the replication fork takes place. this helps to amend damages the damaged ends of DNA.

Prokaryotic DNA Replication

The prokaryotic chromosome is a circular molecule with a less extensive coiling structure than eukaryotic chromosomes. The eukaryotic chromosome is linear and highly coiled around proteins. While there are many similarities in the DNA replication process, these structural differences necessitating some differences in the DNA replication process in these two life forms.

How does the replication machinery know where to start? It turns out that there are specific nucleotide sequences called origins of replication where replication begins. *E. coli* has a single origin of replication on its one chromosome, as do most prokaryotes. The origin of replication is approximately 245 base pairs long and is rich in AT sequences. This sequence of base pairs is recognized by certain proteins that bind to this site. An enzyme called helicase unwinds the DNA by breaking the hydrogen bonds between the nitrogenous base pairs. ATP hydrolysis is required for this process because it requires energy. As the DNA opens up, Y-shaped structures called replication forks are formed. Two replication forks are formed at the origin of replication and these get extended bi-directionally as replication proceeds. Single-strand binding proteins coat the single strands of DNA near the replication fork to prevent the single-stranded DNA from winding back into a double helix.

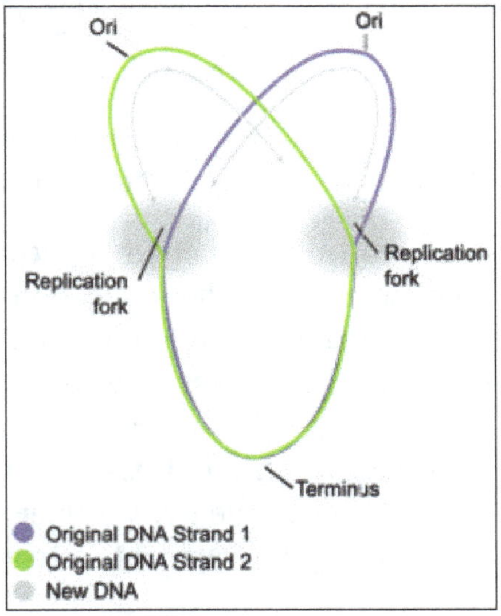

DNA replication in prokaryotes, which have one circular chromosome.

The next important enzyme is DNA polymerase III, also known as DNA pol III, which adds nucleotides one by one to the growing DNA chain. The addition of nucleotides requires energy; this energy is obtained from the nucleotides that have three phosphates attached to them. ATP structurally is an adenine nucleotide which has three phosphate groups attached; breaking off the third phosphate releases energy. In addition to ATP, there are also TTP, CTP, and GTP. Each of these is made up of the corresponding nucleotide with three phosphates attached. When the bond between the phosphates is broken, the energy released is used to form the phosphodiester bond between the incoming nucleotide and the existing chain.

In prokaryotes, three main types of polymerases are known: DNA pol I, DNA pol II, and DNA pol III. DNA pol III is the enzyme required for DNA synthesis; DNA pol I is used later in the process and DNA pol II is used primarily required for repair (this is another irritating example of naming that was done based on the order of discovery rather than an order that makes sense).

DNA polymerase is able to add nucleotides only in the 5′ to 3′ direction (a new DNA strand can be only extended in this direction). It requires a free 3′-OH group (located on the sugar) to which it can add the next nucleotide by forming a phosphodiester bond between the 3′-OH end and the 5′ phosphate of the next nucleotide. This essentially means that it cannot add nucleotides if a free 3′-OH group is not available. Then how does it add the first nucleotide? The problem is solved with the help of a primer that provides the free 3′-OH end. Another enzyme, RNA primase, synthesizes an RNA primer that is about five to ten nucleotides long and complementary to the DNA. RNA primase does not require a free 3′-OH group. Because this sequence primes the DNA synthesis, it is appropriately called the primer. DNA polymerase can now extend this RNA primer, adding nucleotides one by one that are complementary to the template strand.

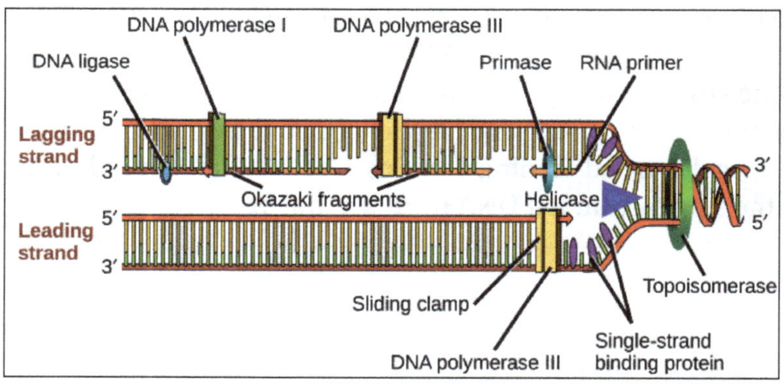

A replication fork is formed when helicase separates the DNA strands at the origin of replication.

The DNA tends to become more highly coiled ahead of the replication fork. Topoisomerase breaks and reforms DNA's phosphate backbone ahead of the replication fork, thereby relieving the pressure that results from this supercoiling. Single-strand binding proteins bind to the single-stranded DNA to prevent the helix from re-forming. Primase synthesizes an RNA primer. DNA polymerase III uses this primer to synthesize the daughter DNA strand. On the leading strand, DNA is synthesized continuously, whereas on the lagging strand, DNA is synthesized in short stretches called Okazaki fragments. DNA polymerase I replaces the RNA primer with DNA. DNA ligase seals the gaps between the Okazaki fragments, joining the fragments into a single DNA molecule.

The replication fork moves at the rate of 1000 nucleotides per second. DNA polymerase can only extend in the 5′ to 3′ direction, which poses a slight problem at the replication fork. As we know, the DNA double helix is anti-parallel; that is, one strand is in the 5′ to 3′ direction and the other is oriented in the 3′ to 5′ direction. One strand, which is complementary to the 3′ to 5′ parental DNA strand, is synthesized continuously towards the replication fork because the polymerase can add nucleotides in this direction. This continuously synthesized strand is known as the leading strand. The other strand, complementary to the 5′ to 3′ parental DNA, is extended away from the replication fork, in small fragments known as Okazaki fragments, each requiring a primer to start the synthesis. Okazaki fragments are named after the Japanese scientist who first discovered them. The strand with the Okazaki fragments is known as the lagging strand.

The leading strand can be extended by one primer alone, whereas the lagging strand needs a new primer for each of the short Okazaki fragments. The overall direction of the lagging strand will be 3′ to 5′, and that of the leading strand 5′ to 3′. A protein called the sliding clamp holds the DNA polymerase in place as it continues to add nucleotides. The sliding clamp is a ring-shaped protein that binds to the DNA and holds the polymerase in place. Topoisomerase prevents the over-winding of the DNA double helix ahead of the replication fork as the DNA is opening up; it does so by causing temporary nicks in the DNA helix and then resealing it. As synthesis proceeds, the RNA primers are replaced by DNA pol I, which breaks down the RNA and fills the gaps with DNA nucleotides. The nicks that remain between the newly synthesized DNA (that replaced the RNA primer) and the previously synthesized DNA are sealed by the enzyme DNA ligase that catalyzes the formation of phosphodiester linkage between the 3′-OH end of one nucleotide and the 5′ phosphate end of the other fragment.

Once the chromosome has been completely replicated, the two DNA copies move into two different cells during cell division. The process of DNA replication can be summarized as follows:

- DNA unwinds at the origin of replication.
- Helicase opens up the DNA-forming replication forks; these are extended in both directions.
- Single-strand binding proteins coat the DNA around the replication fork to prevent rewinding of the DNA.
- Topoisomerase binds at the region ahead of the replication fork to prevent supercoiling (over-winding).
- Primase synthesizes RNA primers complementary to the DNA strand.
- DNA polymerase III starts adding nucleotides to the 3′-OH (sugar) end of the primer.
- Elongation of both the lagging and the leading strand continues.
- RNA primers are removed and gaps are filled with DNA by DNA pol I.
- The gaps between the DNA fragments are sealed by DNA ligase.

Table: The enzymes involved in prokaryotic DNA replication and the functions of each.

Prokaryotic DNA Replication: Enzymes and their Function	
Enzyme/protein	Specific Function
DNA pol I	Exonuclease activity removes RNA primer and replaces with newly synthesized DNA.
DNA pol II	Repair function.
DNA pol III	Main enzyme that adds nucleotides in the 5′-3′ direction.

Helicase	Opens the DNA helix by breaking hydrogen bonds between the nitrogenous bases.
Ligase	Seals the gaps between the Okazaki fragments to create one continuous DNA strand.
Primase	Synthesizes RNA primers needed to start replication.
Sliding Clamp	Helps to hold the DNA polymerase in place when nucleotides are being added.
Topoisomerase	Helps relieve the stress on DNA when unwinding by causing breaks and then resealing the DNA.
Single-strand binding proteins (SSB)	Binds to single-stranded DNA to avoid DNA rewinding back.

DNA replication has been extremely well-studied in prokaryotes, primarily because of the small size of the genome and large number of variants available. *Escherichia coli* has 4.6 million base pairs in a single circular chromosome, and all of it gets replicated in approximately 42 minutes, starting from a single origin of replication and proceeding around the chromosome in both directions. This means that approximately 1000 nucleotides are added per second. The process is much more rapid than in eukaryotes.

Eukaryotic DNA Replication

Eukaryotic chromosome contains multiple origin of replication. However, similar to prokaryotic replication, in eukaryotes the replication happens to be bidirectional from each origin. Origins of Replication in eukaryotic chromosome are separated by tens to hundreds of kilo bases. In yeast, the initiation of DNA replication occurs at autonomously replicating sequences (ARS), which are conserved 11-bp sequences adjacent to easily unwound DNA. In mammalian genomes, replication origins do not exhibit sequence conservation, yet these sites all support the binding of a six subunits of origin recognition complex. A hexasubunit protein called origin recognition complex (ORC), binds to the replication origin. ORC in association with other proteins helps loading of hexameric helicases consisting of six homologous MCM proteins, which separate the parental strands, at an origin in an ATP dependent manner, with RPA proteins (eukaryotic homologue of bacterial SSB protein) binding to the resulting single-stranded DNA. The resulting pre-replication complex (pre-RC) is not competent to initiate replication until it has been activated by other factors that control progress through the cell cycle. Presumably, the separate control of pre-RC assembly and activation allows cells to select replication origins before MCM unwinds the template DNA and replication commences. Once DNA synthesis is initiated, the formation of pre-RC complexes is stopped, thereby ensuring that the DNA replication is occurred only once per cell cycle.

Initiation sites are uniformly distributed across the genome in early embryogenesis, when cell division is rapid. However, after cells have differentiated, the distribution of replication origins changes, possibly reflecting patterns of gene expression and/or alterations in DNA packaging in different cell types. Cytological observations indicate that the various chromosomal regions are not all replicated simultaneously. Rather, clusters of 20 to 80 adjacent replicons (replicating units; segments of DNA that are each served by an origin of replication) are activated simultaneously. New sets of replicons are activated until the entire chromosome has been replicated. DNA replication proceeds in each direction from the origin of replication until each replication fork collides with a fork from the adjacent replicon. Eukaryotes appear to lack termination sequences analogous to the Ter sites in *Escherichia coli*.

Primer synthesis and subsequent steps in replication of cellular DNA are discussed here. Chromosomal DNA replication and other physiological processes leading to the proliferation of the cells are tightly regulated, so that the appropriate numbers of cells constituting each tissue are produced during development and throughout the life of an organism.

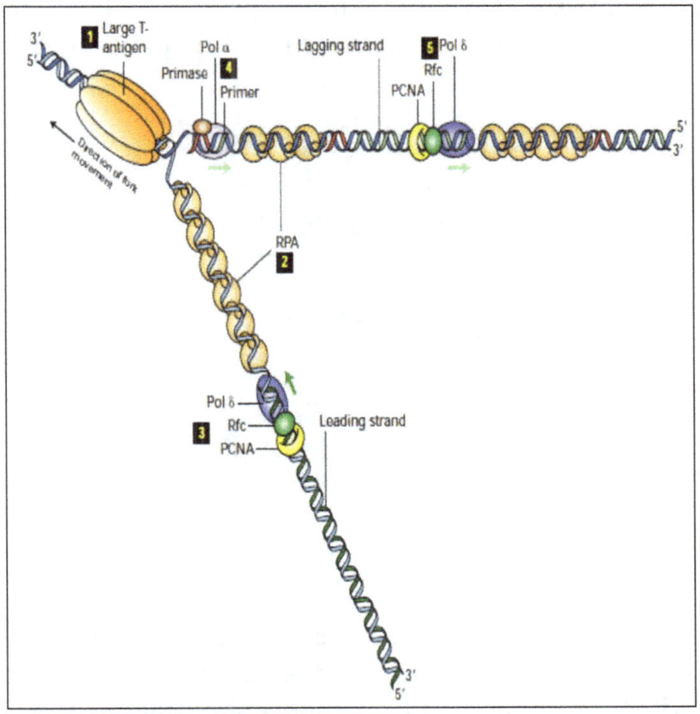

Model of an SV40 DNA replication fork and assembled proteins.

Hexameric T-antigen (1), a viral helicase protein unwinds the parental DNA strands. Multiple copies of protein RPA (2) binds to the single stranded region of parental template strands. A protein complex of DNA polymerase ε (Pol ε), PCNA, and Rfc (3) are responsible for the synthesis of leading strand. DNA polymerase α (Pol α) and primase (4) synthesizes lagging strand primers. The 3′-end of each primers are extended by DNA polymerase δ to synthesize each Okazaki fragment (5).

Several DNA Polymerases are Involved in Eukaryotic Replication

Nearly fifteen different DNA polymerases, till date, have been discovered in animal cell. Based on sequence homology, eukaryotic and prokaryotic polymerases are classified into six families: A, B, C, D, X, and Y. Although, several different proteins are actively participate in eukaryotic replication, three among them are very important in replication of nuclear DNA: polymerases α, δ, and ε, all of them belongs to B-family of polymerases. Like all DNA polymerases, DNA polymerase α replicates DNA by extending a primer in the 5'- 3' direction. However, this enzyme lacks exonuclease activity and hence cannot proofread its polymerization product. DNA polymerase α, which is moderately processive and can polymerize 100 nt at a time, remains in a complex with an enzyme called primase. Primase/α complex synthesizes small RNA primer on template strand and extend it to make nearly 20 nt long DNA. Even though the DNA polymerase α does not have any proofreading activity, these few nt are actually replaced along with the elimination of RNA primer. DNA polymerase δ, unlike DNA polymerase α, is not associated with a primase and contains proofreading activity

(3'-5' exonuclease activity). The processivity of this polymerase is dependent on its association with a sliding clamp protein called proliferating cell nuclear antigen (PCNA), which forms trimeric ring similar to Escherichia coli β2 sliding clamp although they rarely have any sequence similarity. In association with PCNA, DNA polymerase δ is required for lagging strand DNA synthesis. A replication factor C is required to load PCNA on to the template strand near primer, which displaces DNA polymerase α from the strand. This process is called Template switching.

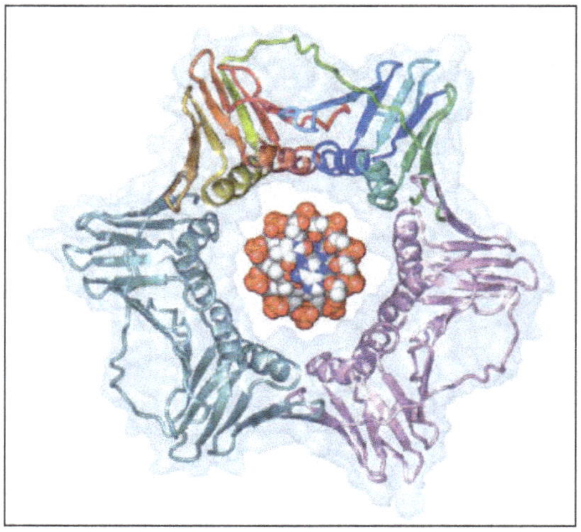

Crystal structure of PCNA.

The three subunits, which form a threefold symmetric ring, are drawn in ribbon form embedded in their semi-transparent surface diagram. One of the subunits is colored in rainbow order from its N-terminus (blue) to its C-terminus (red), another is pink, and the third is light green. A space filling model of B-DNA viewed along its helix axis has been drawn in the center of the PCNA ring (PDB ID 1AXC).

A heterotetrameric nuclear enzyme DNA polymerase ε is the most enigmatic participant in DNA replication. Even in absence of PCNA, Pol ε is highly processive. It has a 3'- 5' exonuclease activity which instead of replacing mononucleotide, degrades single-stranded DNA to 6- or 7-residue, as does that of pol δ. Although pol ε is necessary for the viability of yeast and its essential function can be carried out by only the noncatalytic C-terminal half of its 256-kD catalytic subunit, which is unique among B-family DNA polymerases. This suggests that the C-terminal half of the pol ε catalytic subunit is required for the assembly of the replication complex. Nevertheless, Thomas Kunkel has shown that pol ε is probably the leading strand replicase, although it may also contribute to lagging strand synthesis. Moreover, pol ε may also participate in leading strand synthesis.

Table: Some Important Eukaryotic DNA Polymerases and their Properties.

DNA Polymerases	Exonuclease (3'-5')	Requires PCNA	Processivity	Primase Association
α	No	No	Moderate	Yes
δ	Yes	Yes	High	No
ε	Yes	No	High	No

Reports show that there are some DNA polymerases like DNA polymerase γ, are organelle specific. DNA polymerase γ carries out polymerization of deoxyribonucleotides in mitochondria. Likewise, chloroplast in plant cell contains a specific DNA polymerase for replication.

Retrovial DNA Synthesis

Reports show that the host cell genome can be integrated by genomes of a number of animal viruses. Retroviruses, which are enveloped viruses, contain two identical RNA strands in their genome. In retroviruses, RNA strand acts as a template for the formation of a DNA molecule. This mechanism, on the contrary of most common transcription of DNA into RNA, involves the reverse flow of genetic information. Retroviral genome contains a viral enzyme called reverse transcriptase initially copies the viral RNA genome into complementary single stranded DNA to the viral RNA, resulting in the generation of a RNA-DNA hybrid. The viral RNA strand is then nucleolytically degraded by an RNase H (an RNase activity that hydrolyzes the RNA of an RNA–DNA hybrid helix). Same enzyme then catalyzes the synthesis of a complementary DNA strand using the existing DNA strand. The resulting double-stranded DNA is integrated into the chromosome of the infected cell, thus making a way to escape from host defense machinery. The integrated proviral DNA, is finally transcribed in to RNA by the host's transcription machinery. Viral genome specific RNA is either translated into viral proteins or packaged within viral coat proteins resulting in the formation of infectious progeny virus that are released from the host-cell by budding out from the membrane. Majority of the retroviruses do not kill their host cells. Once infected, cells can replicate, and divide producing daughter cells containing integrated proviral genome. These daughter cells continue transcribing the proviral genome and bud progeny virions. RT has been a particularly useful tool in genetic engineering because it can transcribe mRNAs to complementary strands of DNA (cDNA). Messenger RNA-derived cDNAs can be used, for example, to express eukaryotic structural genes in E. coli. Since E. coli lacks the machinery to splice out introns, the use of genomic DNA to express a eukaryotic structural gene in E. coli would require the prior excision of its introns—a technically difficult feat.

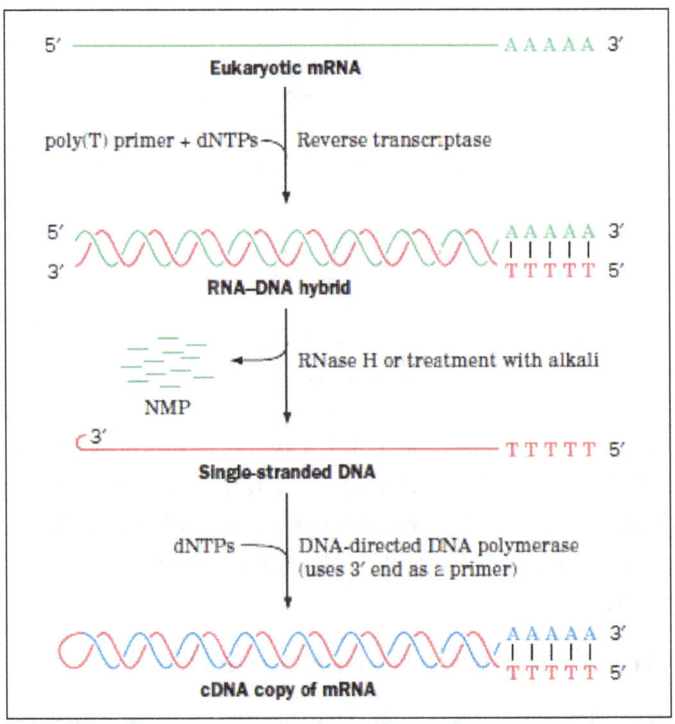

Process of Retroviral Complementary DNA synthesis: Retroviral RNA is used as a template for complementary DNA (cDNA) synthesis. RNase H removes viral RNA in RNA-DNA hybrid and then DNA directed DNA polymerase synthesizes another stand of DNA forming double helix.

Removal of RNA Primers in Eukaryotes

In mammalian systems where the DNA polymerase does not have 5'-3' exonuclease activity (like DNA polymerase I in E. coli), Okazaki fragments are connected by a two-step process. RNA primer of the preceding fragment is displaced by the synthesis of Okazaki fragments in the form of "flap". In the first step RNase H1 cleaves major portion of the RNA, leaving only several 5'-ribonucleotides adjacent to the DNA. An enzyme called "flap endonuclease I" (FEN) removes rest of the bases of flap by endonuclease action. It has been seen in eukaryotic system that pol α extends the RNA primer by ~15 nt of DNA before pol δ displace it from template DNA. The lack of proofreading ability in pol α results in an erroneous primer extension which is not observed in the polymerization reaction of pol δ. However, PCNA recruits FEN1 which provides what is, in effect, pol α's proofreading function: It is also an endonuclease that excises mismatch-containing oligonucleotides up to 15 nt long from the 5' end of an annealed DNA strand. Moreover, FEN1 can make several such excisions in succession to remove more distant mismatches. The excised segment is later replaced by pol δ as it synthesizes the succeeding Okazaki fragment.

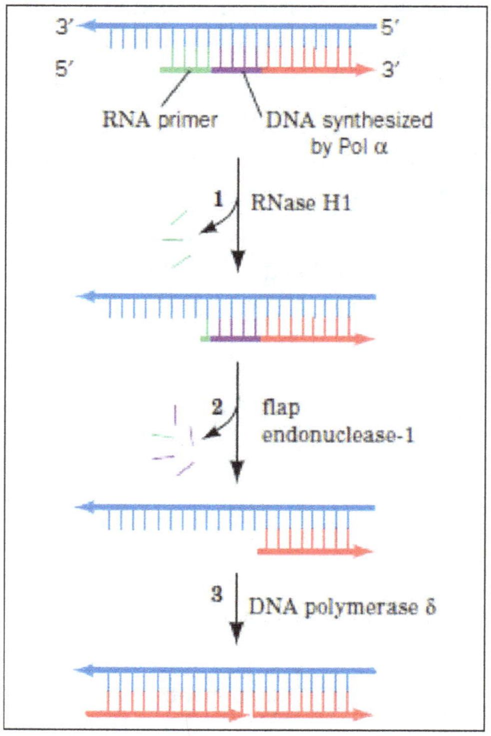

Elimination of RNA primers in eukaryotes: (1) RNase H1 removes all but the 5'-ribonucleotide of the RNA primer. (2) FEN1, a 5'-3' endonuclease, then removes the remaining ribonucleotide along with a segment of adjoining DNA if it contains mismatches. (3) The excised nucleotides are replaced as DNA polymerase δ completes the synthesis of the next Okazaki fragment. The nick is eventually sealed by DNA ligase.

Telomerase Solves the End-replication Problem

The replication machinery always have a problem replicating the end of a linear chromosome, specifically, DNA polymerase is unable to synthesize the extreme 5' end of the lagging strand. Even if an RNA primer were paired with the 3' end of the DNA template, it could not be replaced with

DNA (recall that DNA polymerase operates only in the 5'-3' direction; it can only extend an existing primer, and the primer must be bound to its complementary strand). Consequently, in the absence of a mechanism for completing the lagging strand, linear chromosomes would be shortened at both ends by at least the length of an RNA primer with each round of replication. Telomeres Are Built from an RNA Template. The ends of eukaryotic chromosomes, the telomeres, have an unusual structure. Telomeric DNA consists of 1000 or more tandem repeats of a short G-rich sequence (TTGGGG in the protozoan Tetrahymena and TTAGGG in humans) on the 3'-ending strand of each chromosome end. telomeric DNA is synthesized and maintained by an enzyme named telomerase, which is a ribonucleoprotein (a complex of protein and RNA). The RNA component (451 nt in humans) includes a segment that is complementary to the repeating telomeric sequence and acts as a template for a reaction in which nucleotides are added to the 3' end of the DNA. Telomerase translocates a repetitive G-rich sequence to the new 3' end of the DNA strand, thereby adding multiple telomeric sequences to the DNA. The sequences complementary to the telomeric-G-rich strand is apparently synthesized by the normal cellular machinery for lagging strand synthesis, leaving a 100- to 300-nt single-stranded overhang on the G-rich strand. Telomerase functions similarly to reverse transcriptase; in fact, its highly conserved catalytic subunit, called TERT, is homologous to reverse transcriptase. In association with finger, palm and thumb domain, TERT includes an additional N-terminal RNA-binding domain, which is not present in normal polymerases. The absence of telomerase, which allows the gradual truncation of chromosomes with each round of DNA replication, contributes to the normal senescence of cells. Conversely, enhanced telomerase activity permits the uncontrolled replication and cell growth that occur in cancer.

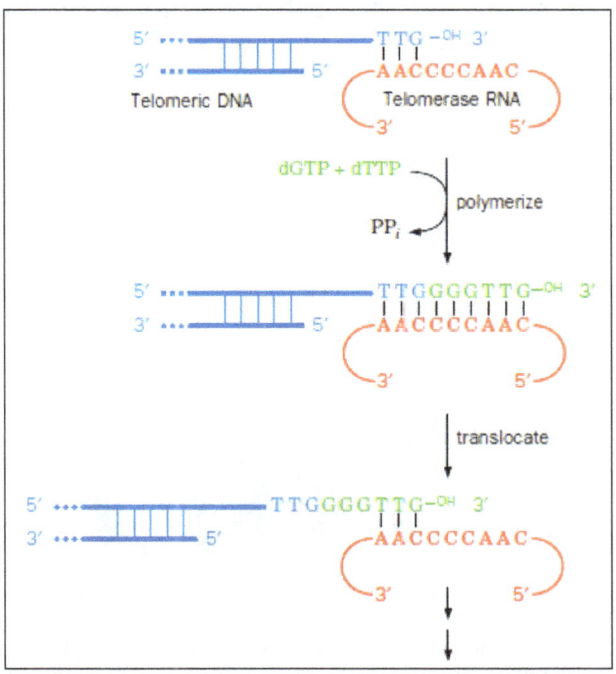

Mechanism for the synthesis of telomeric DNA by Tetrahymena telomerase. The red strand contains the G-rich repetitive sequence. The telomere's 5'-ending strand is later extended by normal lagging strand synthesis.

Cells from knockout mice that cannot produce the telomerase-associated RNA exhibit no telomerase activity, and their telomeres shorten successively with each cell generation. Such mice can breed and reproduce normally for three generations before the long telomere repeats become substantially eroded. Then, the absence of telomere DNA results in adverse effects, including fusion of chromosome termini and chromosomal loss. By the fourth generation, the reproductive potential of these knockout mice declines, and they cannot produce offspring after the sixth generation.

The human genes expressing the telomerase protein and the telomerase-associated RNA are active in germ cells but are turned off in most adult tissues, in which cells replicate only rarely. However, these genes are reactivated in most human cancer cells, where telomerase is required for the multiple cell divisions necessary to form a tumor. This phenomenon has stimulated a search for inhibitors of human telomerase as potential therapeutic agents for treating cancer. While telomerase prevents telomere shortening in most eukaryotes, some organisms use alternative strategies. Drosophila species maintain telomere lengths by the regulated insertion of non-LTR retrotransposons into telomeres. This is one of the few instances in which a mobile element has a specific function in its host organism.

G-quartets are Formed in Telomere

The G-quartet is a cyclic hydrogen bonded array of four guanine nucleotides that can be formed by G-rich sequences in both DNA and RNA. Several G-quartets can stack upon each other to form quadruple helical structures. G-rich polynucleotides are essentially difficult to work with due to their propensity to aggregate via Hoogsteen-type base pairing to form cyclic tetramers known as G-quartets. Indeed, the G-rich overhanging strands of telomeres fold back on themselves to form a hairpin, two of which associate in an antiparallel fashion to form stable complexes of stacked G-quartets. Such structures presumably serve as binding sites for capping proteins, which may help regulate telomere length and prevent activation of DNA repair mechanisms that recognize the ends of broken DNA molecules.

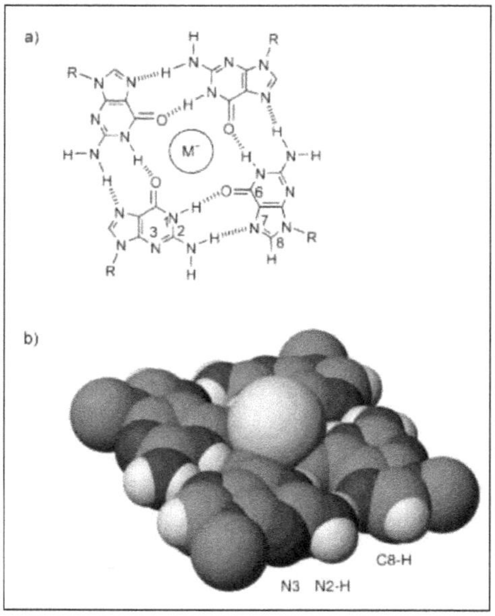

a) The G-quartet; b) space-filling model from the crystal structure of a typical G-quartet with a K+ ion bound above the planar assembly. The riboses have been removed for clarity.

Nucleosomes Reassembly Happens behind the Replication Fork

In contrast to prokaryotic DNA, eukaryotic DNA is packaged in nucleosomes. Some alteration of this structure is probably necessary for initiation, but once replication is under way, nucleosomes do not seem to impede the progress of DNA polymerases. Experiments with labeled histones indicate that nucleosomes just ahead of the replication fork disassemble and the freed histones, either individually or as dimers or tetramers, immediately reassociate with the emerging daughter duplexes. The parental histones randomly associate with the leading and lagging duplexes. DNA replication (which occurs in the nucleus) is coordinated with histone protein synthesis in the cytosol so that new histones are available in the required amounts.

DNA Damage

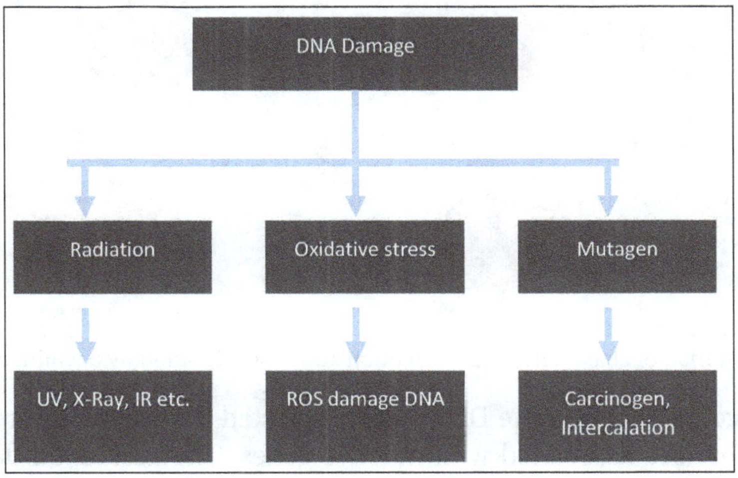

The proofreading function of DNA polymerases is essential for the fidelity of DNA replication and thus for the inheritance of correct genetic information to the descendants during cell division. Yet errors in DNA polymerization have been seen occasionally and may alter the nucleotide sequences of genes if not corrected. Chemical agents either naturally present in the cell or in the external environment of the cell can also alter DNA sequence. Severity of the damage in DNA determines whether the damage portions can be repaired transmit the intact genetic information or can't be repaired leading to the loss of genetic information and sometimes cell death. Sometime altered genetic information is passed on to the daughter cells because of imperfect restoration of gene sequences producing mutations. Genetic changes are usually notable only when they occur in germline cells of multi-cellular organisms, so that the change is transmitted to all the cells of the organism's offspring. Damage to the DNA of a somatic cell, in contrast, rarely has an effect beyond that cell unless the mutation contributes to a malignant transformation.

Spontaneous Alterations in DNA Structure by Hydrolytic Cleavage Reaction

DNA, a complex organic molecule, is a highly stable material, which is required for the storage of genetic information. DNA is susceptible to spontaneous changes that would lead to mutation if left unrepaired even under normal condition. There are three kinds of bonds in DNA with the potential for hydrolytic cleavage: Phosphodiester bonds, N-glycosyl bonds and bonds linking exocyclic amine groups to bases. In DNA strand, spontaneous phosphodiester bond cleavage is a very rare occurrence and probably does not make any significant contribution to DNA damage. However, an abasic site is formed upon N-glycosidic bond cleavage, which is also known as AP (apurinic or apyrimidinic) site. Reports show that DNA of each human cell loses about 18000 purine bases (adenine and guanine) every day because their N-glycosyl linkages to sugar (deoxyribose) hydrolyze by a spontaneous reaction called depurination. In addition, reactive compounds generated from cellular metabolism can cause the deamination of DNA bases. The nature of the four bases of DNA makes the distinction between undamaged and damaged bases very clear. For example every possible deamination event in DNA yields an "unnatural" base, which can be directly recognized and removed by host repair system.

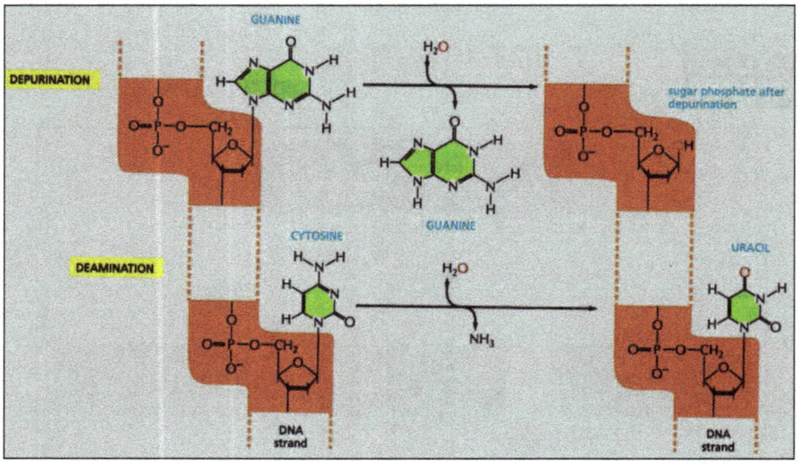

Generation of abasic site by depurination and base alteration by deamination.

A special situation occurs in vertebrate DNA, where selected C nucleotides are methylated at specific CG sequences that are associated with inactive genes. The accidental deamination of these methylated C nucleotides produces the natural nucleotide T in a mismatched base pair with a G on the opposite DNA strand. This is another example of spontaneous base alteration in DNA.

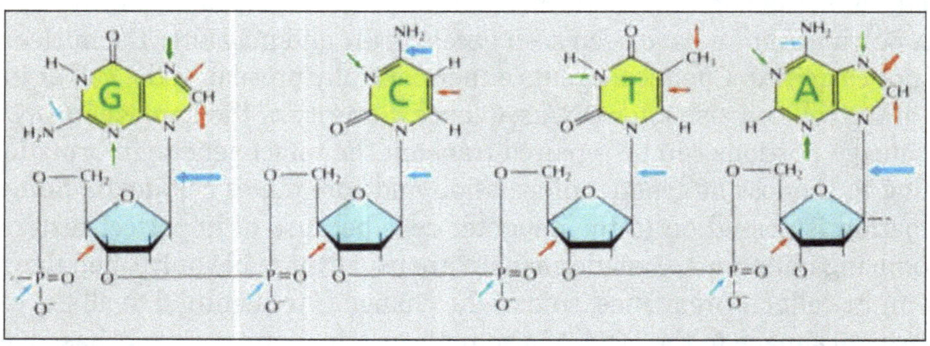

Spontaneous alterations cause DNA damage. The red arrows indicate the target of oxidative damage. Blue and green arrows indicate the site of hydrolytic damage and alkylation.

DNA Damage by Radiation

Different high energy electromagnetic radiations in the environment including ultraviolet light, ionizing radiation, and certain chemical agents can physically damage DNA. For example, UV radiation (200–300 nm) promotes the formation of a cyclobutane pyrimidine dimer (CPD) between adjacent thymine residues on the same DNA strand to form an intra-strand thymine dimer. Similar cytosine and thymine–cytosine dimers also form but less frequently. Such pyrimidine dimers locally distort DNA's base-paired structure, interfering with transcription and replication. This kind of damage accounts for about 75% of UV induced DNA damage. The cyclobutane ring is generated by forming one bond between C-5 atoms and another between C-6 atoms, on adjacent pyrimidine ring. Structural studies show: (1) B-DNA can accommodate a single T-T dimer, forcing the helical axis to bend by about 30°toward the major groove; (2) the dimer's 3′-thymine can form normal Watson-Crick base pair with complementary adenine; and (3) the interaction between the 5′-thymine and its complementary adenine will be weaker than the normal because a single hydrogen bond can be formed. Ionizing radiation also damages DNA either through its direct action on the DNA molecule or indirectly by inducing the formation of free radicals, particularly the hydroxyl radical, in the surrounding aqueous medium. This can lead to strand breakage. The DNA damage produced by chemical mutagens, substances that induce mutations, falls into two major classes:

1. Point mutations, in which one base pair is replaced by another one. These are further classified as:
   (a) Transitions, in which one purine (or pyrimidine) is replaced by another.
   (b) Transversions, in which a purine is replaced by a pyrimidine or vice versa.
2. Insertion/deletion mutations, in which one or more nucleotide pairs are inserted in or deleted from DNA.

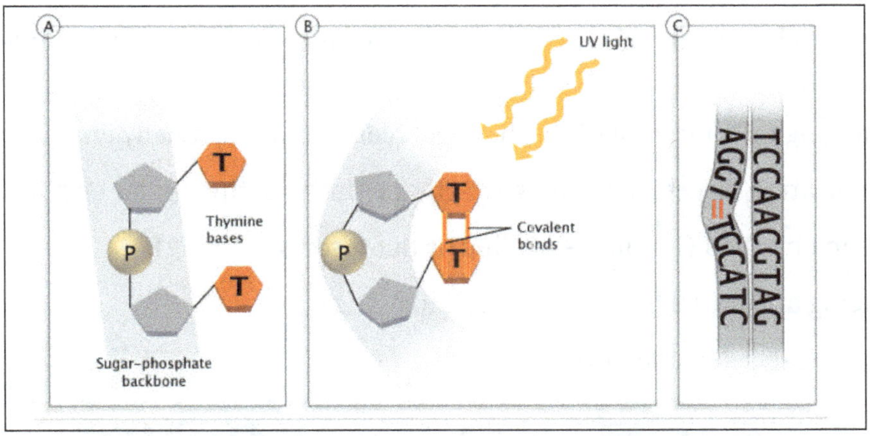

Distortion of DNA by UV Radiation. DNA structure before (A) and after (B) UV radiation which lead to the distortion of DNA strand (C).

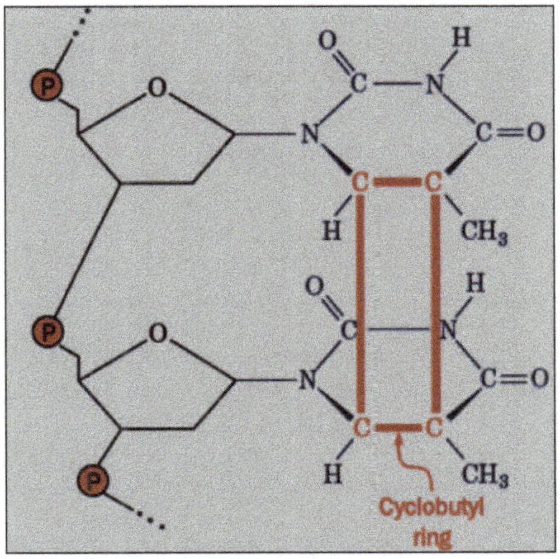

The cyclobutylthymine dimer. Upon UV irradiation of two adjacent thymine residues on a DNA strand forms dimer. Long covalent bonds (~1.6-Å) joining the thymine rings (red) are much shorter than the normal 3.4-Å spacing between stacked rings in B-DNA, which distorts DNA locally.

Detection of Complex DNA Damage

The In Situ Detection of Complex DNA Damage and the Co-localization Concept

Some of the first DNA damage detection techniques were based on electrophoresis or flow cytometry. They use the collective signal from thousands of molecules and they can be performed in vitro, but not in situ. On the other hand, cytogenetics and in situ immunofluorescence (like γ-H2AX foci assay) provide the ability of damage detection at the site they occur. The in situ detection of complex DNA damage relies on the colocalization concept, additionally to all of the "DNA damage in situ detection" principles. Colocalization is, by definition, the presence of two or more fluorophores at the same physical location. In other words, colocalization can be described as the spatio-temporal overlap of two or more dyes in a multichannel image. The biological meaning is that two or more (macro-) molecules of different kinds, labeled with fluorophores, and exhibiting colocalized signal are connected to the same subcellular structure. This proximity—when not spontaneous—implies potential biological interaction between the labeled molecules.

The experimental orchestration of complex DNA damage detection in situ includes the following five modules, as they are also presented in figure:

- The damaging factor (irradiation, chemical reagent, or genetically engineered clones).
- The visualization target (the damage itself or a participating repair enzyme).
- The labeling method (immuno-labeling or fluorescent labeling).
- The imaging apparatus (static or live cell imaging).
- The image analysis (static or tracking).

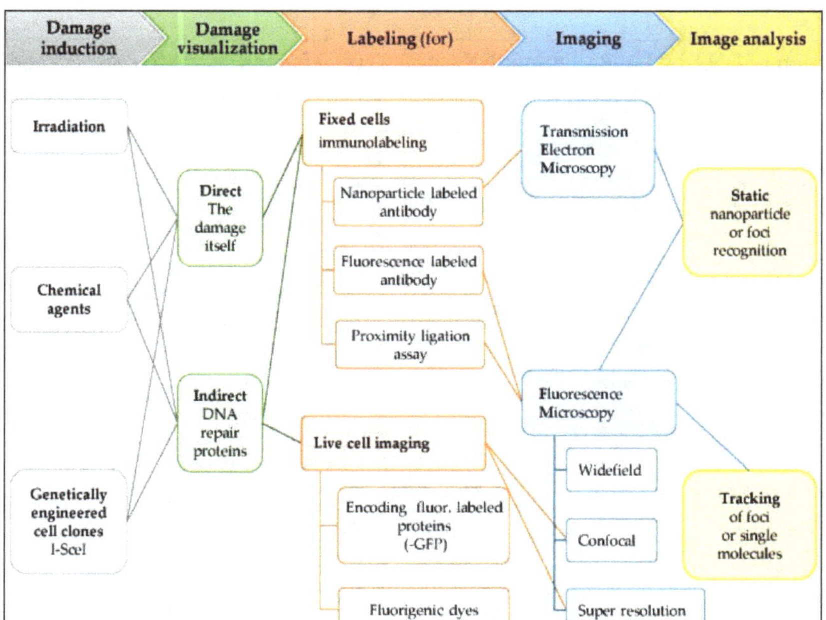

Synopsis of the common steps among clustered DNA damage in situ detection techniques. Synergistic action of involved enzymes is confirmed via detection of co-localizing signal, produced by the labeled proteins of interest.

Damage Induction

Since the interest for clustered/complex DNA damage emerged from radiation biology, the most common damaging cause is radiation, and usually IR. Sophisticated irradiations—targeting a sub-cellular area—can be performed utilizing a laser microbeam, or even an ion microbeam. In addition to the exposure of the biological material to irradiation, occasionally for comparison reasons, the latter is exposed to chemical agents or is genetically engineered.

Damage Visualization: Direct or Indirect

The most common way to visualize (complex) DNA lesions is via labeling DNA repair enzymes. The predominant technique is immunofluorescence, where antibody technology is used to target the object to be detected. The most widely used enzyme using antibody technology is the phosphorylated histone H2AX, i.e., γH2AX, $\gamma-$H2AX, or gamma H2AX. There is also the option of labeling the damage itself. Suitably developed antibodies have the ability to bind onto specific DNA modifications, e.g., anti- 8-oxodG antibody for the detection of base lesions.

Labeling Techniques

Immuno-labeling—Fixed Cells

- In Situ Immunofluorescence-Fluorophore Conjugated Antibodies: The predominant method for the in situ clustered DNA damage detection is immunofluorescence. Damaged cells, after repair for a specific period of time, are incubated with antibodies against the targets of interest; the antibodies used are conjugated with fluorophores. Clustered DNA damage genome regions are revealed when antibodies against DNA repair enzymes of different DDR (DNA damage response) pathways are used.
- Transmission Electron Microscopy (TEM) Analysis: Nanoparticle Conjugated Antibodies Colocalization analysis via TEM follows the same concept as in situ immunofluorescence assay. The telling difference between the two methods is the labeling molecule; here, antibodies are conjugated with nanoparticles. The exploration of colocalization events is achieved by the use of nanoparticles of dissimilar size.
- Proximity Ligation Assay: In Situ PLA Protein colocalization can also be visualized by the one-and-only fluorescent dye, unlike traditional colocalization techniques, which utilize one dye for every different target-enzyme. This assay is called proximity ligation assay (PLA) and as its name proclaims, fluorescent signal is produced only when the molecules of interest are found to be located in close proximity. A fluorescent macromolecule is then synthesized, producing enough fluorescence to be visualized under a conventional fluorescence microscope. PLA assay has the ability to make the proximity of two single molecules visible.

Live Cell Imaging

Fluorescence labeling of target proteins in live cells facilitates visualizing their intracellular distribution and interactions, as well as their time evolution (dynamics). Here, the main methods with regard to DNA repair are presented:

- Encoding Fluorescence Labeled Proteins: The principle behind live cell imaging is to induce cells to express the desired proteins (here, DDR-associated ones) that in addition to their properties, are also fluorescent. To this end, a (set of) suitable plasmid(s) has to be transfected into the cells, carrying the modified gene(s) of the protein(s) of interest. As a result, and upon DNA damage, the cells produce the DDR proteins that conserve their initial properties, but they are also fluorescent, thus allowing their visualization via light microscopy. Such proteins are called recombinant fusion proteins labeled with fluorescent proteins (FP) or autofluorescent proteins (AFP).

- Fluorogenic Dyes: Fluorogenic or fluorigenic dyes are—as their name proclaims—not fluorescent themselves, but become fluorescent in a complex with the molecule of interest. They are based on antibody technology to identify the proper molecule and to form a fluorescent complex with it. However, the substrate that they are added to (i.e., cells) needs to be genetically engineered. If, for example, the target of interest is a protein, then a tagging gene needs to be added next to the gene sequence of the protein of interest. Thus, the target protein is expressed along with its tag. This tag is recognized by a fluorogen, and then labeling is achieved.

Imaging: Microscopy and Image Analysis

Microscopy: TEM and Fluorescence Microscopy

TEM microscopy, although it offers an impressive resolving power down to a few nanometres, is not very popular for the study of DNA damage or repair, due to its "inherent" disadvantages; TEM requires the presence of high vacuum and a specimen slicing thinner than cell thickness, making the observation of live cells impossible. Moreover, specimen preparation protocols are quite demanding in terms of experimentalist's technical skills. Another limiting factor is the TEM apparatus cost (~100k USD). On the other hand, fluorescence microscopy is suitable for the observation of both fixed and live cells:

- Conventional (Widefield) Fluorescence Microscopy: Conventional microscopy, i.e., traditional microscopy (MS), is characterized by the illumination of the entire specimen simultaneously. It is also referred to as widefield MS (WFM), to be differentiated from confocal MS where the specimen is illuminated partially and sequentially. WFM was the first fluorescence MS system, and nowadays is the most widely used technique, as it has become cost efficient (beginning from a few k USD). WFM resolving power is theoretically diffraction limited (~200 nm axial and ~500 nm lateral), but in practice, it is much weaker.
- Confocal Microscopy: Scanning Laser or Spinning Disk In confocal microscopy (CFM), the illumination is accomplished by lasers. Since the lasers emit in one wavelength, no excitation filter is needed. Optical sectioning of the specimen is achieved by CFM. Scanning laser CFM was the first confocal adaptation. Later, spinning disk technology came up. In spinning disk, the single pinhole is replaced by two or more pinholes on the same aperture (light source), allowing more than one voxel to be simultaneously illuminated, thus accelerating the image acquiring process. CFM offers better resolution than WFM, but is still diffraction limited; its cost begins from ~100k USD.
- Super Resolution Microscopy (SRM): Beyond the Diffraction Limit SRM relies on the invention of new materials, the metamaterials, with negative refracting index, i.e., negative

permittivity and magnetic permeability. Lenses of this type of metamaterials are called superlenses. A superlens can propagate not only the far-field part of emitted light, but also the near-field (evanescence waves). Evanescence is the part of emitted signal that contains information about the object details that are smaller than the diffraction limit. Key SRM applications for DNA damage detection can be found in. SRM resolution is ~100 nm axial and ~300 nm lateral, while it costs several hundreds of k USD.

Image Analysis: Co-localization Coefficients, Parameters and Methods

Common co-localization approaches like Pearson correlation coefficient, Overlap coefficient, r, Overlap coefficient, r^2, k_1 and k_2, M_1 and M_2 overlap coefficient, Costes' automatic threshold, Van Steensel's CCF, or Cytofluorogram creation, correlate the pixel intensity between the fluorescence channels. Such approaches provide results with regard to a 'given' image, and not to a 'given' cell. However, the core of complex DNA damage analysis is the cell nucleus and especially the DSB-detection fluorescence focus. Some approaches, based on the DSB focus topology, are discussed here.

Defining an Extra Type of Foci a quite simple method: To investigate the degree of co-localization in a given cell population—is to introduce an extra focus type—let us call it "colocalization focus"—in addition to DSB foci. Colocalization foci conserve all the detection characteristics of the given DSB focus type, along with an intensity threshold criterion for the second channel (i.e., the non-DSB foci channel when clustered DNA damage is to be detected). By determining the ratio:

$$y = \frac{\text{number of colocalization foci per cell}}{\text{number of DSB foci per cell}}$$

the co-localization degree is estimated.

R_{col} A second quantitative colocalization parameter is the colocalization ratio R_{col}, introduced by Martin et al. and expressed as:

$$R_{col} = \frac{2I_{53BP1}}{I_{\gamma H2AX} + I_{53BP1}}$$

Where I_{53BP1} is the average per pixel intensity of the 53BP1 channel for a given γH2AX focus and $I_{\gamma H2AX}$ is the analogous entity for the γH2AX channel. Although R_{col} was introduced to estimate the degree of colocalization between the two DSB repair proteins γH2AX and 53PB1, it could be also used for the clustered DNA damage detection, in terms of the detection of non-DSB repair enzymes in the vicinity of a DSB. Therefore, R_{col} could have the general formula of:

$$R_{col} = \frac{2I_{\text{non DBS repair enzyme}}}{I_{\text{DBS}} + I_{\text{non DBS repair enzyme}}}$$

Where I_{nonDSB} repair enzyme and I_{DSB} would be the averages per pixel intensity for the non-DSB and DSB repair enzyme channel, correspondingly, for a given DSB focus.

P_{clc} Another co-localization parameter is the P_{clc} *parameter,* and compares the intensity of the non-DSB protein fluorescence signal between the DSB foci area and the rest of cell nucleus; i.e., the cell nucleus area (volume) excluding DSB foci area (volume). P_{clc} is described by the expression:

$$P_{clc} = \frac{\text{average mean Intensity of nonDSB channel over DSB foci volume}}{\text{average mean Intensity of nonDSB channel over the rest of the nucleus}}$$

Where red color indicates the fluorescence signal of the antibody against the non DSB repair protein and green color corresponds to DSB foci marker and blue is for the DAPI staining of nucleus. When P_{clc} tends towards zero, reversed colocalization is implied, whilst when P_{clc}~1 random staining occurs, and for P_{clc} >1 true colocalization takes place. This parameter, like R_{col}, conserves DSB topology and is very sensitive: It detects colocalization even under high background signal conditions. Pclc has already been incorporated to one foci software so far, namely to the JQuantPro developed by Dr. Pavel Lobachevsky , which is a later version of JCountPro software . Its concept is quite simple, making it workable for users of almost any other software.

DNA Sequencing for Genome-wide Nucleotide-resolution DNA Damage Identification

DNA sequencing has just entered the arena of DDR detection, although it was a widely used technique in other applications. Until recently, the short-read technologies, though precise, had the ability to accurately detect only genome point-mutations like base deletions or substitutions, due to their inability to span larger DNA lengths. In case of more complex type of abnormalities, they fail to do so, due to their intrinsic limitations. The detection of more complex chromosomal aberrations (e.g., genomic rearrangements or structural variations) requires—in addition to short-read technologies—the assistance of long-read technologies, for precise characterization. This adaptation offers confidence in the detection of alterations that may span kb of DNA. Long-read sequencing technologies are already applied clinically towards the aim of personalized treatments. They recently started to attract interest in the detection of DNA damage long-term consequences, to find genomic rearrangements such as deletions, translocations, or inversions with high accuracy, low cost, and in a relatively short time. Sequencing adaptations suitable for DNA damage induction demand a site-specific DSB labeling step. Such techniques are NGR, BLISS (UMIs), BLESS, and i-BLESS:

- NGR detects and labels the nick-gaps, induced during DNA replication.
- BLISS can detect DSBs induced by endonucleases, using their corresponding unique molecular identifiers (UMIs).
- BLESS is able to detect DSBs at nucleotide resolution. It is independent of proteins that bind to DNA or single-stranded DNA, which are both sources of bias. BLESS' innovation is that it uses an amplification step for the fragments created by two DSBs.
- i-BLESS is a BLESS adaptation that is suitable for very small and fragile cell genomes (yeast), while it achieves the incredible detection accuracy of one DSB per 10^5 cells.

A very recent adaptation, qDSB-Seq, combines both long- and short-sequencing techniques in order to quantify absolute numbers of DSBs. In order to quantify the DSBs of interest, spike-ins

DSBs are induced to the genome, using site-specific endonucleases. The presence of spike-ins DSB offers a measure for DSB number counting calibration.

To conclude, DNA sequencing nowadays offers high throughput and it can substantially give added value in the analysis of the induction of DSBs and their consequences, concerning chromosomal instability. Up to now, the detection of DNA damage induction has been limited to DSB detection that can be traced in pre-known genome loci. A promising field that is currently under intense investigation is the identification of the consequences of DSBs on the genome after repair mechanisms have taken place. To this end, long-read technologies (e.g., Nanopore sequencing) have developed structural variation analysis pipelines that utilize neural networks and optimized machine learning algorithms. The use of such methods can reduce the so-far existing technological limitations that led to the inability of whole genome analysis in the field of DSB-induced genomic alterations.

Biological Response to Clustered DNA Damage and its Significance

The potential importance of clustered DNA damage is highlighted through evidence of associated biological effects, such as mutagenesis, carcinogenesis, lethality, and reduced reparability. As discussed, strong theoretical (MC simulations) and experimental evidence suggests an increment of the complexity of DNA damage and therefore repair resistance with increasing LET. Significant improvements have been made recently towards the identification of key parameters relating to the efficient detection of complex DNA damage. However, the molecular mechanisms and the basis of the underlying repair pathway choices that eventually lead to the observed biological consequences from DNA damage clustering are not yet fully understood.

Table: Summary of in vitro studies on biological responses to clustered DNA damage induction.

Cell Type	Radiation Type	Biological Response
HeLa and oropharyngeal squamous cell carcinoma (UMSCC74A and UMSCC6) cells	High-LET α-particles (121 keV/μm) or protons (12 keV/μm), versus low-LET protons (1 keV/μm) and X-rays	The signaling and repair of complex DNA damage, particularly induced by high-LET IR is coordinated through the specific induction of $H2B_{ub}$ catalyzed by MSL2 and RNF20/40, a mechanism that contributes to decreased cell survival after irradiation.
Human peripheral blood lymphocytes	Mixed beam of alpha-particles (^{241}Am source, 0.223 Gy/min, LET: 90.9 keV/μm) and X-rays (190 kV, 4.0 mA, 0-2 Gy)	Induced DNA damage was above the level predicted by assuming additivity. The activation levels of DDR proteins and mRNA levels of the studied genes were highest in cells exposed to mixed beams. The repair of damage occurs with a delay.
Human dermal fibroblasts	High-LET IR with carbon ions (9.5 MeV/n; LET 190 keV/μm; calculated mean dose: 1.52 Gy) or calcium ions (7.7 MeV/n; LET 1800 keV/μm; calculated mean dose: 14.4 Gy), versus low-LET IR with 6-MV photons (10 Gy)	High-LET-IR induced clustered DNA damage and triggered profound changes in chromatin structure along particle trajectories. DSBs exhibited delayed repair despite cooperative activity of 53BP1, pATM, pKap-1.
Normal human skin fibroblasts	^{60}Co γ-rays (LET=0.3 keV/μm), accelerated ^{11}B (E = 8.1 MeV/nucleon, LET=138 keV/μm) ions	It has been found that heavy charged particles induce clustered DNA damage in the genome of cells that can lead not only to gene mutations, but also to large deletions.

HeLa Kyoto cells	Pulsed UV laser (micro-irradiation, IR)	Recruitment and dissociation of 70 DNA repair proteins to laser-induced complex DNA lesions.
Human uveal melanoma (92–1) cells	Carbon ions (LET: 80 keV/μm) and iron ions (LET: 400 keV/μm) at different doses, versus X-rays (LET: 4 keV/μm)	Heavy ions were more effective at inducing senescence than X-rays. Less-efficient repair was observed when DNA damage was induced by heavy ions compared to X-rays and most of the irreparable damage was complex of SSBs and DSBs, while DNA damage induced by X-rays was mostly repaired in 24 h. Results suggest that DNA damage induced by heavy ion is complex and difficult to repair, thus presenting as persistent DNA damage, and pushes the cell into senescence.
Human dermal fibroblasts, NFFh-TERT foreskin fibroblasts	Low-LET irradiation with 6 MV photons, versus high-LET irradiation with carbon ions (9.5 MeV/n; LET = 190 keV/μm)	High-LET irradiation caused localized energy deposition within the particle tracks and generated highly clustered DNA lesions with multiple DSBs in close proximity. Huge DSB clusters predominantly localized in condensed heterochromatin. High-LET irradiation-induced clearly higher DSB yields than low-LET irradiation, and large fractions of these heterochromatic DSBs remained unrepaired.
Human osteosarcoma cell line (U2-OS)	X-rays (250 keV, 16 mA; LET: 2 keV/μm), versus heavy ions: ^{238}U ions (LET: 15,000 keV/μm), ^{207}Pb ions (LET: 13,500 keV/μm), ^{197}Au ions (LET: 13,000 keV/μm), ^{119}Sn ions (LET: 7,880 keV/μm), ^{59}Ni (LET: 3,430 keV/μm), ^{48}Ti (LET: 2,180 keV/μm), ^{14}N ions (LET: 400 keV/μm), and ^{12}C ions (LET: 170 keV/μm)	DSB complexity plays a critical role in the decision for DSB end-resection in G1-cells. CtIP, MRE11, and EXO1 are required for the resection of complex DSBs in G. Repair of complex DSBs relies on resection independent of the cell cycle stage.
Human cells (fibroblasts, HBECs)	1 Gy of Si (LET: 44 keV/μm) or Fe (LET: 150 keV/μm) ions	Direct visualization of clustered DNA lesions at the single-cell level using 53BP1, XRCC1, and hOGG1 as surrogate markers for DSBs, SSBs, and base damage, respectively, reveals that most complex DNA damage is not repaired in human cells. Unrepaired clustered DNA lesions result in the generation of a spectrum of chromosome aberrations. Checkpoint release before the completion of clustered DNA damage repair is a major cause of chromosomal aberrations.
Human Lung Adenocarcinoma (A5490) cells	^{12}C ions (62 MeV, LET: 290 keV/μm), versus ^{60}Co γ-rays (1–3 Gy)	Carbon ions were three times more cytotoxic than γ-rays. The observed decrease in number of γ-H2AX foci 4 h after γ-rays irradiation indicates repair of damage and is supported by nearly 100% survival, whilst the decrease in γ-H2AX foci after carbon ion irradiation was not indicative of repair.

HF12 primary male human fibroblast cells	^{238}Pu α-particles (range, ~20 μm; peak energy, 3.26 MeV; LET=121.4 keV/μm)	Many α-particle-induced mutations are large deletions. Rejoining at microhomologies characterizes large deletion junctions. Intra- and interchromosomal insertions and inversions occur at the sites of some large deletions. Novel fragments found in complex rearrangements derive from other sites of radiation damage in the same cell.

The Role of Delayed Repair

Complex DNA lesions including DSB and non-DSB clusters are a major challenge for the DNA repair mechanisms and the fate of a cell. The biological significance of clustered DNA lesions formed by multiple ionizations relates to the inability of cells to process them efficiently compared to isolated DNA damages, and the outcome in case of erroneous repair can vary from cell apoptosis to mutations and chromosomal instability. It has been shown that clustered damaged sites (DSB and non-DSB clusters) compromise the base excision repair (BER) pathway leading to the lifetime extension of the lesions within the cluster, compared to isolated ones; thus, the chances that the lesions persist to replication and induce mutation are drastically increased. Chromosomal abnormalities can originate from the repair of DNA damage via the less accurate pathways of non-homologous (NHEJ) or alternative non-homologous end joining (alt-EJ).

Complex DNA damage is known to have a slow rate of repair and to require the activation of DDR via the ATM pathway. the non-DSB clustered lesions generated in mammalian cells by IR show a considerably longer lifetime compared to isolated DNA lesions. Recently, mixed beam (alpha particles and X-rays) irradiation of human peripheral blood lymphocytes revealed a lagging repair of the induced damage, presumably contributing to increased damage misrepair, particularly of complex DSBs. Large 53BP1 foci formation presumably at sites of clustered DNA damage, to the detriment of small foci generated in dispersed, simple damage areas, indicates that DDR network is mainly engaged by the alpha-particles-induced complex damage, so that the low-LET radiation-induced damage is detained. Such results mechanistically demonstrate how cells cope with concurrent clustered and simple DNA damage, but also show that health effects of mixed beams may be greater than expected due to additivity of the individual beam components.

Repair of base damage within a heavily clustered damage site is likely to lead to the generation of additional DSBs, either through mistimed endonuclease action at base damage on complementary DNA strands or through interaction of base damage with the replication machinery. Such DSBs generated as repair intermediates may be surrounded by unrepaired base damages and become an unrepairable, complex DSB. Complex DSBs either induced directly by IR or by the processing of non-DSB clustered lesions are repaired by slow kinetics or left unrepaired and cause cell death or pass mitosis. In surviving cells, large deletions, translocations, and chromosomal aberrations are observed. An additional factor contributing to the increased mutagenicity of complex DNA damage is DNA repair deficiencies. Previous evidence suggests that deficiencies in repair enzymes like DNA-PK, APE1, and others significantly elevate the accumulation of clustered DNA lesions and genomic instability. Last but not least, one should not disregard the environmental stress fueling DNA damage in cells and tissues through either direct induction of DNA lesions (radiation, chemicals) and epigenetic changes like DNA methylation of key repair gene promoters negatively impacting repair efficiency. Whilst the delayed or incorrect repair of clustered DNA lesions in human results in the generation of mutations and genetic instability in normal tissue, an "optimistic" idea could be that it could also positively serve to cell killing of tumor cells.

Double-strand Break Clustering

As already stated, damage complexity is usually determined by the existence of extra lesions in the vicinity of a DSB. Another level of complexity that can compromise lesions' processing by chromatin destabilization in the area surrounding a cluster is DSB clustering. The formation of DSB clusters is induced by the exposure of cells to IR (particularly of high-LET radiation). Studying their processing will allow us to elucidate whether the non-random distributions of breaks produced by high-LET particle tracks have any effect on their repair and biological effectiveness.

Extensive mathematical modelling provides theoretical support for the biological severity of DSB clustering. However, the majority of experimental techniques used to detect IR-induced DNA damage does not provide a lot of information about the precise number and spatial arrangement of the lesions within one or two helical turns of the DNA, thus leading to an uncertainty over the biological responses. Towards this direction, revolutionary advances have been made in methods allowing the generation of enzymatic DSBs at random or in well-defined sites in the genome (by the use of *I-SceI* meganuclease), which might throw some light on several assumptions such as the exact processing of DSB clusters and the mechanisms underpinning their severity as DNA lesions. *I-SceI* is a rare cutting homing endonuclease that is normally absent from the human and mouse genomes. *I-SceI* recognition sequences can be artificially introduced in a genome and the subsequent expression of *I-SceI* generates a DSB. To study the biological consequences of DSB clustering in a cogent manner, cell lines harboring DNA sequences, at which single DSBs and DSB clusters of known constitution are enzymatically generated, can be constructed. Efficient activation of the DDR has been demonstrated using this model system, as well as a significantly increased potential of DSB clusters, as compared to single DSBs, to kill cells and generate PARP1-dependent chromosomal translocations. Increasing formation of chromosomal translocations is also observed at numbers and complexities, with increasing DSB-clustering. Such findings indicate that DSB repair relying on first-line DSB-processing pathways (NHEJ and HRR) is compromised within DSB clusters, presumably through the associated chromatin destabilization, consequently increasing the contribution of Alt-EJ, which has high tendency to generate chromosomal rearrangements. These observations offer a mechanistic explanation for the increased efficacy of high-LET radiation.

Carcinogenesis Associated with Clustered DNA Damage

Amongst all types of DNA damage generated by IR, DSBs are considered as the most critical since they can affect cellular fate by leading to cell death or carcinogenesis when unrepaired or mis-repaired. One of the main etiological factors in the initiation of mutagenesis and the promotion of carcinogenesis is the lower fidelity or deficiencies of cellular repair mechanisms, which are responsible for removing or bypassing the damaged sites and restoring the initial sequence after exposure to IR. The formation of chromosomal translocations is considered as the key event that might come out of DSB processing. Translocations are well known to cause cell death and carcinogenesis. Recent work also demonstrates that translocations might not be directly induced by IR, but they can be generated due to DSB processing that takes place in damaged cells by repair pathways that have evolved to process this type of lesions.

Considering the threats DSBs pose to cells, it comes as no surprise that the biological consequences of complex DNA damage (either DSB and non-DSB lesions, or DSB clusters) can range from point mutations and loss of genetic material to cellular lethality due to repair failure. Singleton et al. demonstrated that irradiation-induced clustered DNA damage leads to complex genetic changes in human cells and that damage clustering is reflected in the complexity of mutations. At the sequence level, genetic changes, such as intra- and interchromosomal insertions and inversions, in association with some large deletions, were observed as a result of heavy clustering of DNA damage from α-particle irradiation. Such complex genetic changes are clear indicators of genetic instability and potentially of carcinogenesis. It has also been suggested that non-DSB clusters, if unrepaired, can lead to the formation of mutations and chromosome abnormalities.

It is generally accepted that DSBs induced by high-LET IR are rejoined less efficiently than those induced by low-LET radiations, leaving a higher rate of non-re-joined DNA breaks. High-LET radiation-induced complex DNA damage is more potently carcinogenic, generating tumors with significantly higher frequency and shorter latency compared to tumors generated by low-LET, as low LET radiation induces DNA DSBs that are rapidly repaired. In terms of genomic instability, photon and particle radiation of high-LET exhibits a variety of dynamic chromosomal aberrations, including chromosomal rearrangements, such as chromosome breaks, dicentrics, translocations, and deletion mutations. Through the years, scientific interest has been raised in the biological effects of IR induced DNA damage on human health and particularly the effects on astronauts during periods of space travel. The potential carcinogenetic effect of space radiation was recently reviewed based on the induction of clustered DNA damage, chromosomal aberrations, micronuclei, and mutations as the main triggers of tumor initiation and progression.

The strong linkage between the induction of clustered DNA damage and its deficient repair, the constant DDR triggering and activation of immune system has also been considered as a possible pathway leading to carcinogenesis. Maladaptive inflammation is a main pathogenic mechanism driving chronic diseases, and a key factor leading to carcinogenesis. Chronic inflammation can result in DDR down regulation, due to the release of inflammatory mediators and reactive oxygen species (ROS) driving genomic instability. Thus, cancer susceptibility is likely to depend on specific genetic predisposition to the type and duration of this response. Numerous experimental studies show inflammation to be a key process underlying the carcinogenic potential of high-LET radiation.

Table: High-linear energy transfer (LET) radiation triggering immune responses.

Biological System/Cell Type	Radiation Type	Immune Response
Peripheral blood mononuclear cells (PBMCs) of head and neck (HNSCC) cancer patients	Intensity modulated radiotherapy (IMRT), (51–74 Gy total dose, 1.6–3 Gy dose/fraction)	Expression of the FXDR, SESN1, GADD45, DDB2, and MDM2 radiation-response genes were altered in the PBMCs of patients after RT. All changes were long-lasting, detectable one month after RT. Local tumor irradiation induces systemic changes in the level of immune and inflammation-related plasma proteins. RT induces changes in the immune phenotype of PBMCs of HNSCC patients.

Murine CT26 colorectal cancer cells	8 Gy proton beams at 1.09 keV/μm (low), 2.58 keV/μm (medium) and 7.7 keV/μm (high) LET.	Increase in percentage expression of immune markers (OX40L, CD40, ICAM-1, and MHC-I) in high-LET irradiated cells. High-LET proton radiation can be used to stimulate better immunogenic phenotype in tumor cells compared to low LET proton radiation.
Human cancer cell lines: TE2, KYSE70, A549, NCI-H460 and WiDr	Carbon ions (290 MeV/n, LET 30 keV/μm)	Carbon-ion beams significantly increased HMGB1 (a damage-associated molecular pattern—DAMP) levels in the culture supernatants.
NCI-H446 (lung tumor cells)	Carbon ions (290 MeV/n, LET 13 keV/μm)	Cyto- and chemokine response release by tumor cells after irradiation. (TNF-α)
Tumor-bearing mice (C3H/He, Balb/c nude mice)	Carbon ion irradiation (290 MeV/n, LET=77 keV/μm)	Increased cytotoxic T-lymphocytes (CTL)-associated lysis of isolated tumor splenocytes after carbon ion irradiation treatment with supplementary intratumoral dendritic cell (DC) injection.
Tumors of mouse squamous cell carcinoma (NR-S1) cells inoculated in the legs of C3H/HeSlc mice	Carbon ions (290 MeV/n, 6-cm spread-out Bragg peak, 6 Gy)	Even when exposed to the same equivalent doses, carbon ion therapy might activate the immune system to a greater extent than conventional RT.
NR-S1 and SCCVII (squamous cell carcinoma), NFSa, #8520 (fibrosarcoma)	Carbon ions (290 MeV/n, LET 50 keV/μm)	Significant C-ion induced upregulation of stress-responsive and cell-communication genes common to different tumor types.
Rat skin	^{56}Fe ions (1.01 GeV/n)	^{56}Fe ion radiation significantly induced inflammation-related genes, including many in the categories of 'immune response', 'response to stress', 'signal transduction', and 'response to biotic stress', that contribute to carcinogenesis.
Highly aggressive HT1080 human fibrosarcoma and LM8 mouse osteosarcoma cells	Carbon ion beams (290 MeV/n), versus X-rays	When compared with photon irradiation, carbon ion exposure reduced the number of distant lung metastasis in carcinoma models in immunocompetent mice.

Chemical Agents Generate Mutations

Oxidative Damage

Cellular respiration involves the generation of reactive oxygen species (ROS) produced in electron transport chain, which probably do not cause any damage to nuclear DNA under normal physiological conditions. The reason behind this facts includes- (i) respiratory chain does not normally release these reactive oxygen species, (ii) cells contain ROS scavenging enzymes like superoxide dismutase which converts superoxide radical into molecular oxygen and hydrogen peroxide and catalase which converts hydrogen peroxides to oxygen and water, (iii) superoxide or hydroxyl radicals if released are so reactive that they react with nearby biomolecules before acting on DNA. However, ROS produced by other mechanisms can cause considerable damage of nuclear DNA. The primary reason for this is the generation of hydroxyl radical from hydrogen peroxide which is a normal metabolic product of many biochemical pathway including one which converts guanine to uric acid. Hydrogen peroxides may escape the action of catalase and peroxidase and can be converted to a hydroxyl radical by the following reaction:

$$Fe^{2+} + H_2O_2 \rightarrow Fe^{3+} + HO^{\bullet} + OH^-$$

This reaction is called Fenton reaction which was named in the honour of its discoverer Henry J. H. Fenton. Hydroxyl radical generated in fenton reaction can cause many different kind of DNA damages including oxidation of guanine to 8-oxo-guanine and thymine to thymine glycol etc. More than 100 different oxidative modifications leading to DNA damage have been reported till date. Deficiecy in the repair of these modified bases results in transversion mutation. When a DNA strand with such kind of modified nucleotides is replicated, the oxoG can base-pair with either C or A, causing a G.C to T.A transversion. Thymin glycol has a cytotoxic effect on cell due to its inhibitory effect on DNA replication.

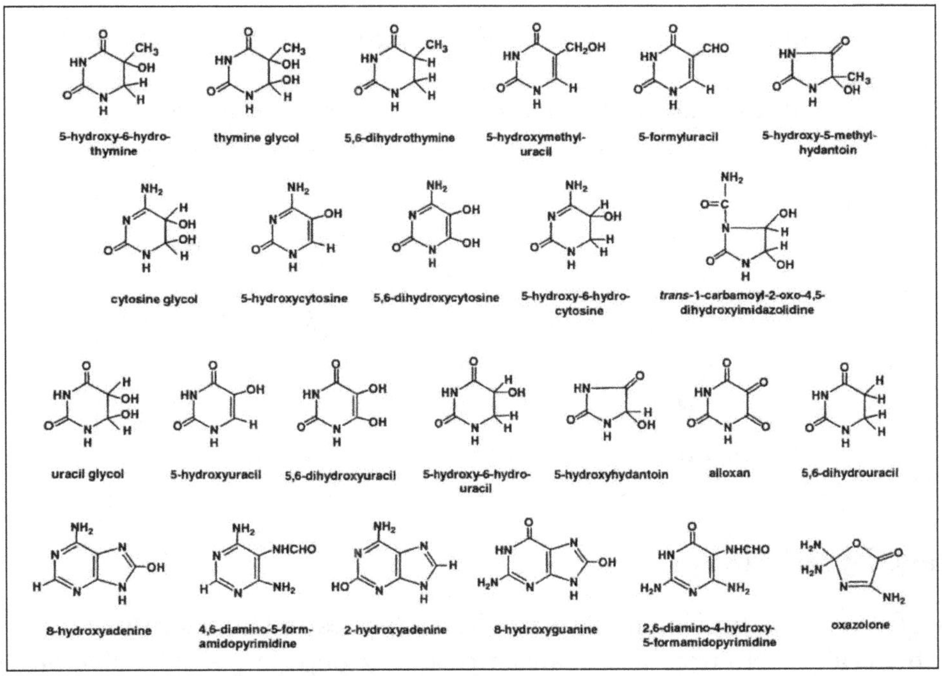

Products generated as a result of interaction of DNA bases with reactive oxygen and free radical species.

Alteration of Bases due to Deamination of DNA Bases

Oxidative deamination of aromatic primary amines is one of the reasons of point mutations. Cytosine is thereby converted to uracil and adenine to the guanine like hypoxanthine (which forms two of guanine's three hydrogen bonds with cytosine). Hence, deamination of DNA results in both A.T to G.C and G.C to A.T transitions. One example of such DNA deaminating agent is nitrous acid (HNO_2), which is formed from nitrates used as a preservatives in processed meats such as bacon, sausage hot dogs etc. Despite its potential mutagenic activity, nitrite (the conjugate base of nitrous acid) is still used as a preservative because it also prevents the growth of Clostridium botulinum, the causative agent of botulism. Bisulfite (HSO^{3-}) an additive that is sometime present in wine beer, fruit juices and dried fruits also significantly increases the rate of cytosine deamination without effecting purine or 5- methylcytosine. A few cytosine bases in eukaryotic DNA are converted to the modified base 5-methyl cytosine. This modified base is located in the so called CpG islands, which are small segment of DNA often present in regulatory elements termed promoters, the upstream sequence of transcription start site. The rate of spontaneous deamination of 5-methyl cytosine bases in CpG islands is even greater than that for cytosine. The product in this case in thymine resulting in the conversion of a C-G base pair to a T-A base pair.

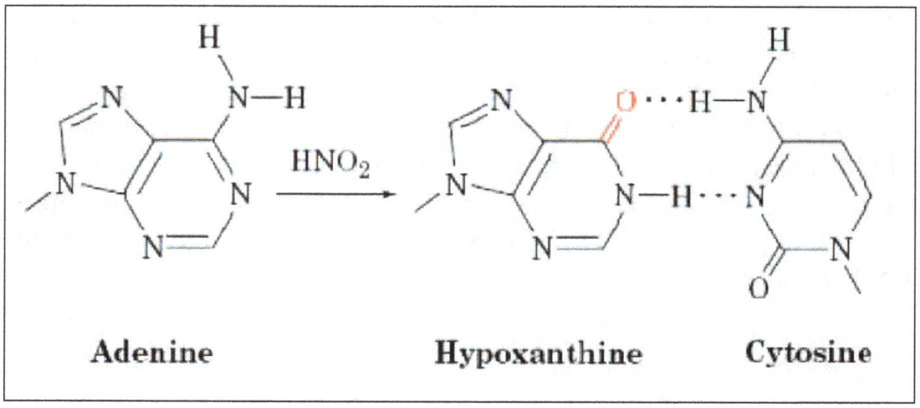

Deamination of different DNA bases Cause mutations in DNA.

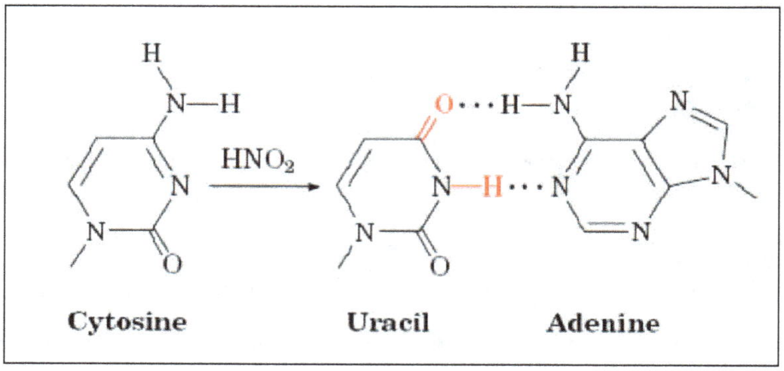

DNA Damage by Alkylation

The backbone of DNA consists of highly electron-rich molecules and therefore are target of attack electron seeking chemicals or electrophiles. Transfer of alkyl group e.g methyl, ethyl, propyl etc. to the electron-rich atoms of DNA by a highly reactive group of electrophiles leads to damage DNA by alkylation. Alkylation takes place at: (i) nitrogen and oxygen atom external to the base ring system (ii) non-bridging oxygen atom of the phosphate group and (iii) nitrogen atom in the base ring except those linked to ribose sugar. The example of such alkylating agents are dimethyl sulfate (which is frequently used in research laboratories as a mutagenic agent for random mutagenesis), nitrogen mustard, ethylnitrosourea, N-methyl-N'-nitro-N-nitrosoguanidine (MNNG). The exposure of DNA to MNNG yields, among other products, O 6 -methylguanine residues, which can base-pair with either C or T.

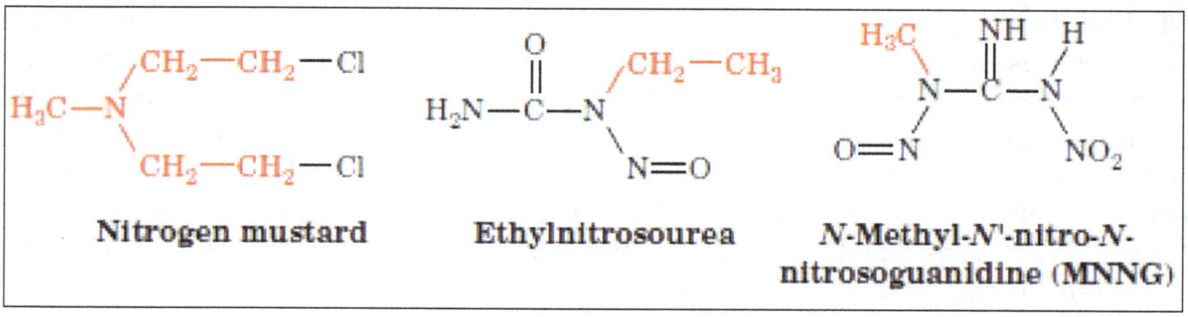

Structure of alkylating agents.

S-adenosylmethionine, a metabolic methylating agent, occasionally nonenzymatically methylates a base to form derivatives such as 3-methyladenine and 7-methylguanine residues. However, base methylations also have normal physiological functions. The alkylation of the N7 position of a purine nucleotide renders its glycosidic bond susceptible to hydrolysis, leading to loss of the base. The resulting gap in the sequence is filled in by an error-prone enzymatic repair system. Transversions arise when the missing purine is replaced by a pyrimidine.

In the environment, different hazardous materials in soot is a mixture of polycyclic aromatic hydrocarbon (PAHs), formed by the incomplete combustion of wood or coal used as a fuel. Tobacco smoke and charbroiled meats also contain similar types of PAHs. The common structure of different types of PAHs is two or more fused aromatic rings. These hydrophobic hydrocarbons are not able to damage DNA unless they are metabolically activated. Metabolic activation pathway requires the action of several different enzymes which includes cytochrome P450, epoxide hydrolase etc. However, during this enzyme-mediated activation process DNA can be alkylated by PAHs.

Insertion/Deletion Mutations are Generated by Intercalating Agents

Intercalating agents are reported to have the capacity to mutate DNA. Insertion/deletion mutations may arise from the treatment of DNA with intercalating agents such as acridine orange, proflavin, ethidium bromide etc. Intercalation of such a molecule in a DNA strand increases the distance two times between two consecutive base pairs. The replication of the distorted DNA occasionally results in the insertion or deletion of one or more nucleotides in the newly synthesized polynucleotide. (Insertions and deletions of large segments generally arise from aberrant crossover events).

A large number of reports regarding mutagenesis is that mutations, as a result of polymerase errors, spontaneous modification, or chemical damage to DNA, occur at random. John Cairns explained this paradigm and demonstrated that bacteria, which are unable to digest lactose, preferentially acquired the mutations they needed in order to use lactose when it was the only nutrient available. This observation, which suggests that bacteria can "direct" mutations that benefit them, more likely reflects a nonspecific adaptive response in which the overall rate of mutation (whether it is useful or non-useful) increases when the cells are under metabolic stress. The activation of error-prone DNA repair and recombination systems that are relatively inactive in normally growing cells, render the cell in a hypermutable state.

DNA Repair

Maintenance of chromosomal integrity requires the repair of the damage part of the DNA before it is transferred to daughter cells after cell division. Cells with defective DNA repair system are unable to repair the damage site of the DNA, which results in permanent alteration of DNA bases leading to gene mutation. Consequences of mutation of essential gene of the cell are lethal for its survival. The presence of great variety of repair mechanisms in simple organisms like E. coli identifies its great physiological importance. These systems involves the action of a single enzyme which can eliminate the wrongly incorporated nucleotides with the correct one or a series of enzymes acting at the damaged site to excise and repair the damage segment of DNA.

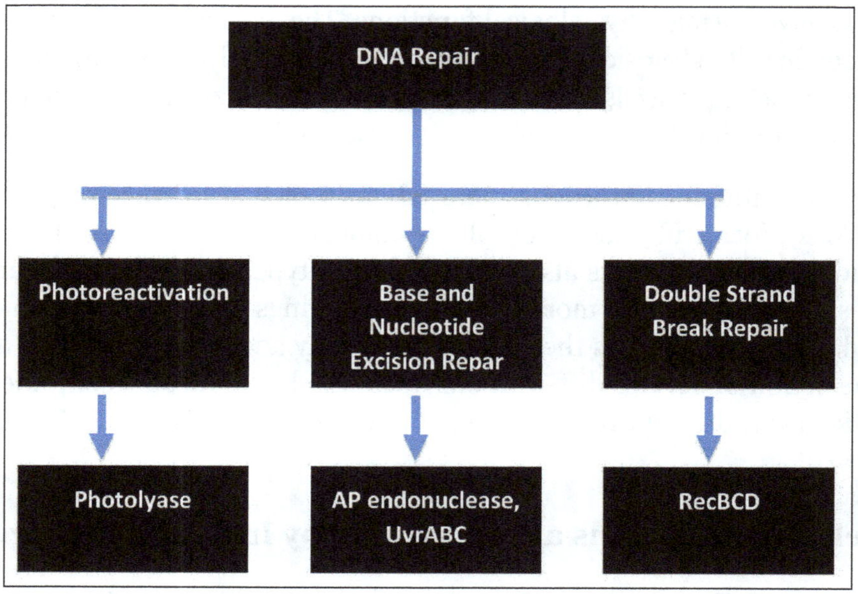

Direct Reversal of DNA Damage

Damage on DNA at a particular site can be repaired by a single catalytic reaction. Reports show that enzyme identifies and reverse certain types of damage on DNA. One of such kind of DNA damage by UV irradiation is the formation of pyrimidine dimers which can be repaired to their monomeric forms by a process called photoreactivation. In that process, an enzyme known as DNA photolyase catalyse the reversal reaction to restore the original DNA sequence. This protein with molecular weight of 55- to 65-kD is active in its monomeric form and prevalent in many prokaryotes and eukaryotes but not in humans. Two prosthetic groups are present in photolyase: a light-absorbing cofactor and FADH-. The mechanism of photoreactivation in the E. coli enzyme involves the excitation of light absorbing cofactor (N^5,N^{10} -methenyltetrahydrofolate) by absorbing UV–visible light (300–500 nm), followed by transfer of the excitation energy to the FADH- . Finally, an electron is transferred from FADH to the pyrimidine dimer, thereby splitting it. FADH at the end is reduced by resulting pyrimidine anion to regenerate the enzyme.

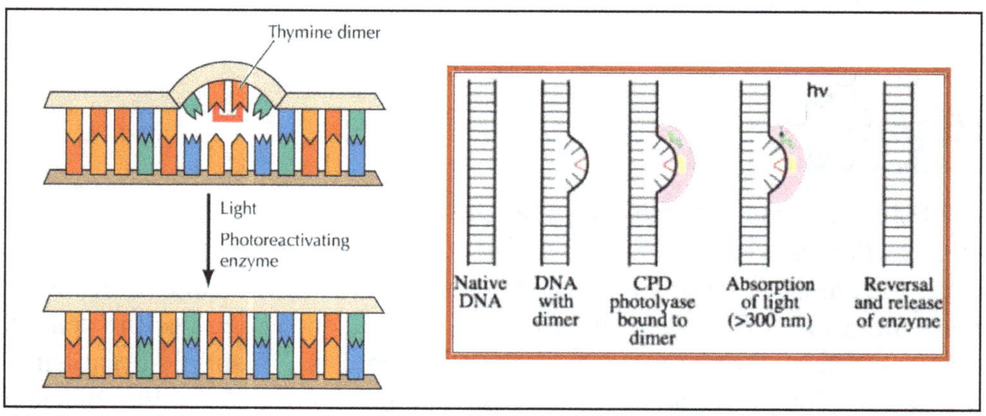

Mechanism of photoreactivation and DNA repair. Model in the left side show how thymine dimmer can be repaired in one catalytic step by the enzyme photolyase. Right side model shows the steps of photoreactivation and catalysis.

Replication Error can be Corrected by DNA Mismatch Repair

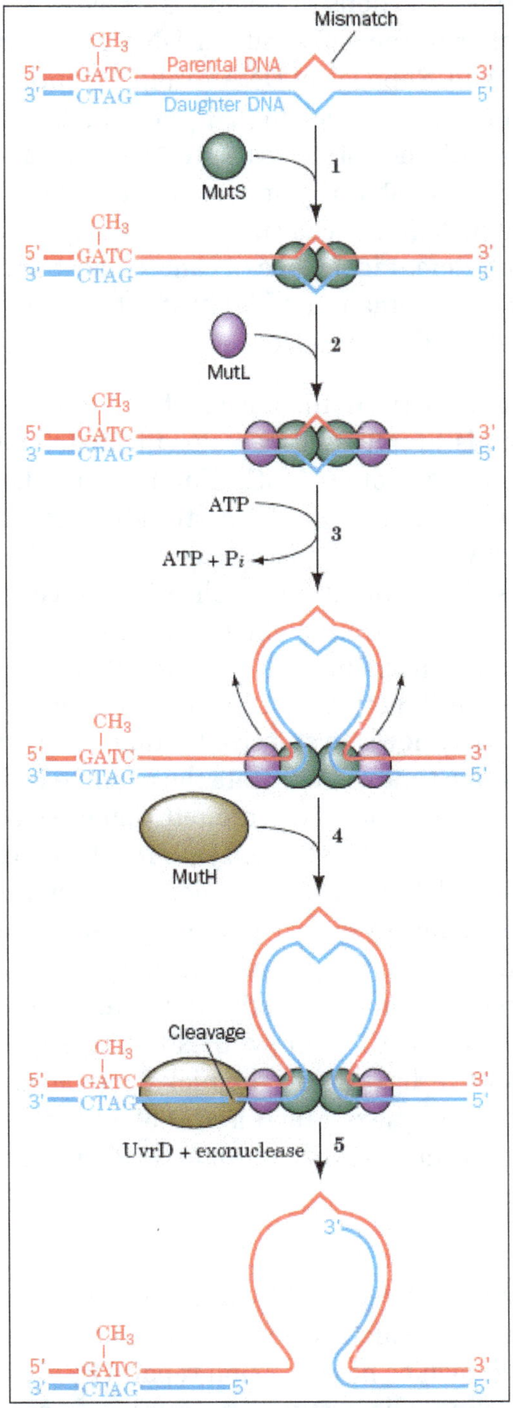

DNA mismatch repair system in E. coli. (1) Binding of MutS. (2) Binding of MutL. (3) The mismatch-containing DNA loops out. (4) Binding of MutH endonuclease. (5) Mismatched base is excised in unmethylated DNA strand.

Any misincorporation of nucleotides during replication which elude the proofreading activity of DNA polymerases can be replaced with a correct nucleotide by a repair mechanism known as mismatch repair (MMR). One per 10^5 nucleotides is mistakenly incorporated during replication by DNA polymerase, the proofreading activity of which increases the fidelity of DNA replication by 100 fold with the removal of mispaired nucleotides before further extension of the DNA strand. Moreover, up to four nucleotide insertions or deletions which is arised from the slippage of one DNA strand relative to the other in the active site of DNA polymerase, can also be corrected by DNA mismatch repair mechanism. Slippage of DNA strand over other happens at particular region of DNA which contain the sequence motif rich in $[CA]_n$ or $[A]_n$. All these different kind of incidents which leads to misincrporation of nucleotides can produce permanent mutation if not repaired. Therefore, cells with impaired mismatch repair system will have high mutation rate and if that happens in essential gene will be lethal to cell survival. Reports show that defects in human mismatch repair system result in frequent incidence of cancer, most notably hereditary nonpolyposis colorectal cancer syndrome. Several organs are highly affected in this syndrome which may be the most common inherited predisposition to cancer.

In E. coli, three different proteins carry out the mismatch repair process. A protein known as MutS in its homodimmeric form binds to unpaired bases or mismatch base pair of DNA, followed by binding of another homodimmeric protein MutL. The resulting heterotetramer complex ($MutS_2MutL_2$) translocates in both direction of the DNA strands in an ATP-dependent process which cause the double stranded DNA to form a mismatch-containing loop that is closed by the MutS2MutL2 complex. Correcting the errors in the replication process, MMR system has to distinguish parental strand from daughter strand, which contains the misincorporated nucleotide. In E. coli, mismatch repair system can distinguish the two DNA strands because newly replicated DNA remains hemimethylated until methyltransferases have had sufficient time to methylate the daughter strand. After replication newly systhesized strand is not methylated by deoxyadenine methylase (Dam) which take some time to methylate the adenine residue in GATC sequence. MutS2MutL2 complex when encounters a nearby hemimethylated GATC palindrome (since parental strand is already methylated), recruits another enzyme MutH, an endonuclease, which makes a nick at the 5'-end of the unmethylated GATC sequence. The hemimethylated GATC site may be located on either side of the mismatch over nearly 1000 bp distant from it. A helicase UvrD, which is a member of nucleotide excision repair system, unwinds and seperates the nicked part of the daughter strand from the parental strand. Defective daughter strand is then digested completely by an exonuclease. DNA polymerase III then synthesizs that segment of daughter strand with correct nuleotides and finnaly DNA ligase seals the nick. Homologs of MutS and MutL are found in eukaryotic cells, however, homologue of MutH is absent and therefore eukaryotic cell must use some other cue than methylation status to differentiate the daughter and parental strands.

Base Excision Repair

Not all types of DNA damage can be repaired by a single step, some repair pathway essentially involve a series of enzymes, each of which performs a specific job in the complete repair process. One of such repair process is known as base excision repair (BER), which is operative in an identical mechanism in almost all living cells. As the name implies, base excision repair pathway starts with removal of the damaged base by a variety of DNA glycosylases, which are present in the cell. DNA glycisylase cleaves the glycosidic bond linking deoxyribose to altered nucleotides resulting in the generation of a deoxyribose residue without any attached base, which is called apurinic or apyrimidinic sites (AP) or abasic sites. This situation can also be occasionally found after the spontaneous hydrolysis of glycosidic bonds. An enzyme, AP endonuclease, then cleaves off one side of deoxyribose residue, which along with some adjacent residues are removed by some exonucleases possibly associated with DNA polymerase. The gap generated by exonuclease action finally filled by DNA polymerase and the nick is sealed by DNA ligase. BER, which is found to be the most frequent type of DNA repair system, contains a glycosylase that identifies oxidized guanine nucleotide, 8-oxoguanine. Another example of such enzyme is uracil–DNA glycosylase, which excises uracil residues, which arise from cytosine deamination as well as the occasional misincorporation of uracil instead of thymine into DNA. Abasic sites in mammalian DNA have cytotoxic effect because mammalian topoisomerase I is trapped irreversibly in its covalent complex with DNA. In addition to that, since the sugar moiety at abasic site is devoid of a glycosidic bond, which can convert it to its linear form whose aldehyde group is very reactive and can cross-link to other cell components. This rationalizes why AP sites are tightly bound to UDG in solution. AP endonuclease enhances UDG activity, the following enzyme in BER pathway; however, two enzymes do not interact in the absence of DNA. This further suggests that an abasic site is always bound to UDG thereby protecting the cell from the AP site's cytotoxic effects. It is very likely that other DNA damage-specific glycosylases function in an identical pathway.

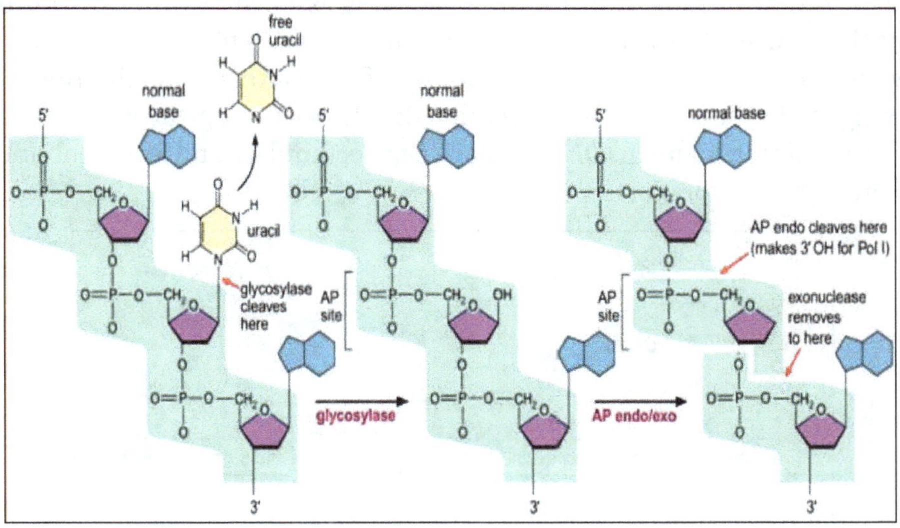

Base excision repair system in E. coli. The enzyme glycosylase catalyse the cleavage of N-glycosidic bond eliminating defective base. Another enzyme AP endonuclease catalyse the removal of sugar at the AP site.

Nucleotide Excision Repair

Exposure to mutagens can damage DNA and result in bulky lesions that distort the double-helix structure or impede proper transcription. Damaged DNA can be detected and repaired in a process called nucleotide excision repair (NER). NER employs a set of specialized proteins that first scan DNA to detect a damaged region. Next, NER proteins separate the strands and excise the damaged area. Finally, they coordinate the replacement with new, matching nucleotides.

DNA Distortion and Damage

Cells are regularly exposed to mutagens—factors in the environment which can damage DNA and generate mutations. UV radiation is one of the most common mutagens and is estimated to introduce a significant number of changes to DNA. These include bends or kinks in the structure which can block DNA replication or transcription. If these errors are not fixed, the damage can cause mutations which in turn can result in cancer or disease depending on which sequences are disrupted.

Identification and Repair of Damaged Regions

Nucleotide excision repair relies on specific protein complexes to recognize damaged regions of DNA and flag them for removal and repair. In prokaryotes, the process involves three proteins—UvrA, UvrB, and UvrC. The first two proteins work together as a complex, traveling along the DNA strands to detect any physical aberrations.

Once identified, the strands at the damaged location are separated, and endonuclease enzymes such as UvrC cut and excise the affected region. DNA polymerase fills the gap with new nucleotides, and then the DNA ligase enzyme seals the edges between the new and old DNA.

Mutations in NER have Severe Consequences

In prokaryotes, the NER complex consists of the three Uvr proteins, but in eukaryotes, more than a dozen proteins operate to regulate DNA repair. In humans, mutations in the NER pathway can cause diseases such as xeroderma pigmentosum (XP), which is associated with a 2000-fold increase in the incidence of skin cancer. Individuals suffering from XP are highly sensitive to UV exposure and can develop severe skin burns after just a few minutes of exposure to sunlight. Additionally, XP patients can show signs of premature aging and often develop neurological abnormalities. Without a properly working repair mechanism, DNA damage can accumulate and lead to abnormal cell death or potentially cancerous tumors.

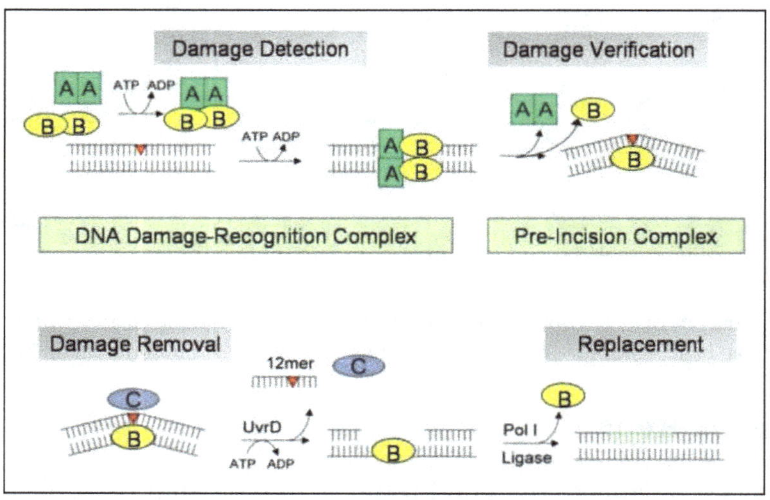

Nucleiotide excision repair system in E. coli.

In bacteria like E. coli, NER is an ATP-dependent process where the actions of UvrA, UvrB, and UvrC proteins (the products of the uvrA, uvrB, and uvrC genes) are required to repair the helix distortion in DNA. This pathway is often referred to as the UvrABC endonuclease repair pathway, which is responsible for the cleavage of the damaged part of the DNA strand at the seventh and at the third or fourth phosphodiester bonds from the lesion's 5' and 3' sides, respectively. The DNA segment of 11- or 12-nt oligonucleotide long containing damage part is excised by UvrABC endonucleases and then displaced by a helicase UvrD. In this mechanism, a heterotrimeric complex consisting of UvrA2 and UvrB recognizes a pyrimidine dimmer or other damage sites of the DNA. UvrA dissociates from the complex after damage site recognition with the expense of ATP. Endonuclease UvrC then joins UvrC and make an 5'-incision at a distance of 7 nuceotides from the damage site and another 3'- incision which is three to four nuceotides away from the damaged site. This step also requires energy from ATP hydrolysis. A helicase UvrD then helps unwind and release the single strand between two cuts. The gap is then filled by the actions of DNA polymerase I and DNA ligase seals the nick at the end.

A homodimer of UvrA in association with a homodimer of UvrB identify the DNA lesion. UvrA falls off from UvrB after damage site recognition. UvrC protein binds to UvrB and make two incisions on both sides of damaged base generating a 12 nt long segment of DNA, which is removed by UvrD and rest of the job done by DNA pol I and ligase.

Defective NER causes genetic disorder like Xeroderma Pigmentosum and Cockayne Syndrome. A segment of 30 nucleotides long DNA can be excised out by eukaryotic NER system, which contains nearly 16 different proteins. The proteins are conserved in eukayotic system, however they don't show any sequence similarity with prokaryotic protein although they are doing the same kind of functions. This suggests that the two NER systems evolved in convergent pathway. Malfunction of any of these enzymes in human pathway have been identified through mutations that are manifested as two genetic diseases. Xeroderma pigmentosum (XP) is one of such diseases mainly inherited from ancestors to descendents and is associated with the impaired NER system in skin cells where the skin cells are unable to repair UV-induced DNA lesions. Individuals with autosomal recessive condition to XP are extremely sensitive to sunlight. Skin during infancy changes such as excessive freckling, dryness, and keratoses (a type of skin tumor, together with eye damage, such as opacification and ulceration of the cornea. Moreover, cells with impaired NER system often develop fatal skin cancers at nearly 2000-fold greater rate than normal and internal cancers at a 10- to 20-fold increased rate. Reports show that many individuals with XP also are prone to develop many physiological dysfunction including neurological degeneration and developmental deficits.

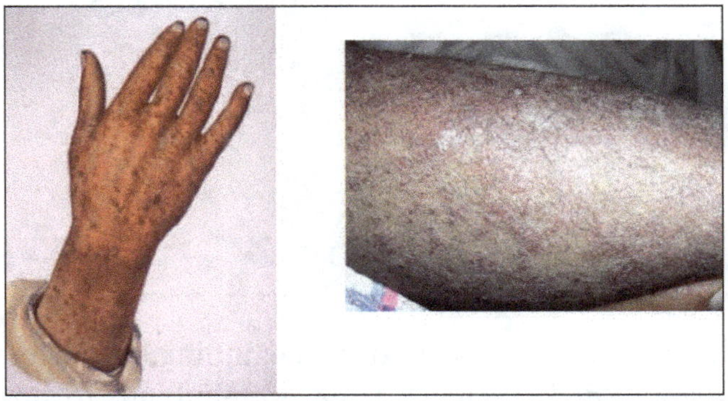

Individuals with impaired NER develop Xeroderma pigmentosum. Above pictures show the appearance of the skin of an individual suffering from Xeroderma pigmentosum.

Cockayne syndrome (CS) is another inherited disease, which is also associated with defective NER. This disease is developed from defects in same three genes that are defective in XP as well as in two additional genes. Individuals with CS develop hypersensitiveity to UV radiation and exhibit stunted growth. Other effects include neurological dysfunction as a result of neuron demyelination, and the appearance of premature aging but, intriguingly, have a normal incidence of skin cancer. The defective proteins in CS normally recognize a halted RNA polymerase at a damaged or distorted DNA template. To resume halted transcription process the stalled RNA polymerase should be removed so that NER system can repair the damaged part of the DNA, which causes the helix distortion. Individuals with CS cannot repair which causes the cell to go through programmed cell death or apoptosis. The death of cells, which are transcriptionally active may account for the developmental symptoms of Cockayne syndrome.

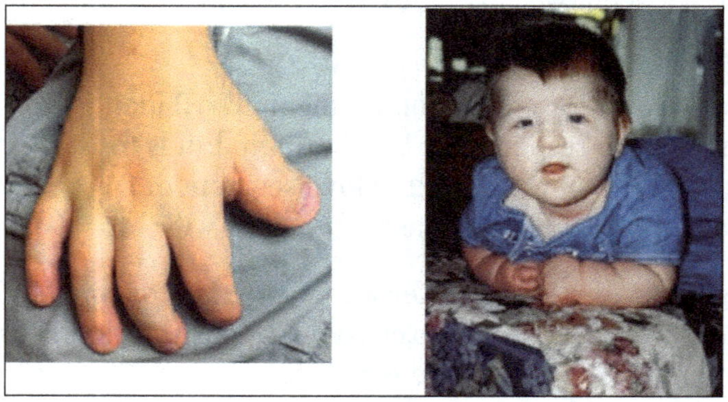

Above pictures show the appearance of the skin of a child suffering from Cockayne syndrome.

Double Strand Break Repair

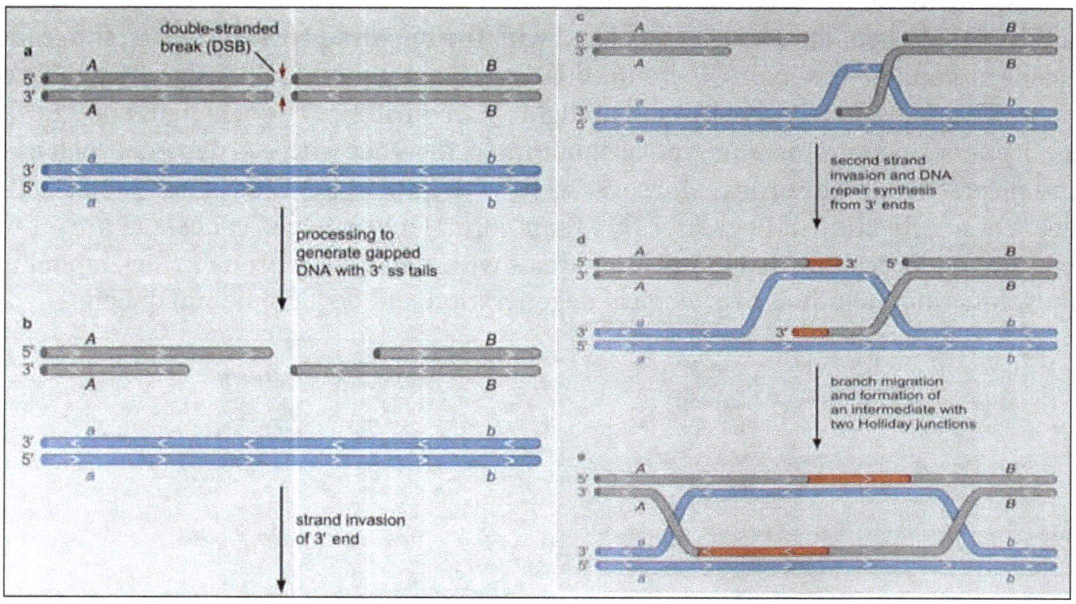

Steps of Double strand break repair pathway. At the beginning exonuclease processing generates gapped DNA with a single-stranded overhang. RecA protein binds to single-stranded region and help promote the strand invasion. DNA repair synthesis starts from 3′-end. Branch migration occurs and generates intermediates with two cross over points.

DNA damage often times causes double strand break, which should be repaired otherwise replication machinery will be stalled at the site of double strand break. The repair system, which fix this break and make a normal continuous DNA, requires the action of a multi-enzyme complex, RecBCD protein in E. coli. This complex is a 330-kD heterotrimeric product of genes recB, recC, and recD. The proteins have both helicase and nuclease activities and binds to double strand break site of DNA and then with its helicase function unwinds dsDNA with the expense of ATP. Its nuclease activity degrades the unwound single strands and move forward until they encounter a Chi site on DNA from its 3' end. Chi site is very specific sequence of GCTGGTGG occurs after every five kilobases in E. coli genome. Once RecBCD reaches chi sequence it stops trimming from 3'-end of the trands and increases the rate of cleavage from 5'-end of the strands, thereby generating a 3'-end single-strand overhang to which recombination inducing protein RecA binds to form a filamentous structure. This fact also explains why regions containing Chi sequences have elevated rates of recombination. RecA, a recombination inducing protein, is recruited by RecB protein and is also nucleated to the 3′-overhang of the DNA strand. RecA in association with 3'-overhang of DNA strand scan both the strands of an undamaged DNA to identify the complementary sequence to that overhang and then direct to exchange strand forming a branch point. Using both stands of undamaged strands as template the gaps generated by the action of RecBCD are filled by DNA polymerase followed by the action of DNA ligase, which seals the nick.

SOS Response

SOS Response is mutagenic in E. coli. DNA damaging agents in E. coli induce a complex system of cellular changes, which is known as the SOS response. In this condition, cell division is stopped and cell's ability to repair DNA damage is increased. A repressor protein called LexA and a DNA binding protein RecA, which is also a key player in homologous recombination, regulates the activity of the SOS response. RecA protein is also involved in DNA repair after replication. Under normal physiological growth condition, LexA represses SOS gene expression. However, when DNA is not completely replicated due to DNA damage the resulting single strands bind to RecA to form a filamentous complex that activates LexA to cleave and thereby inactivate itself. Once the repression is removed due to the damage signal mediated by RecA different SOS genes start to express. In bacteria there are nearly forty different kinds of SOS genes can be expressed, which include recA and lexA as well as the nucleotide excision repair genes like uvrA and uvrB. Again in the absence of any DNA damaging agents, the DNA.RecA complex is no longer present, so the expression of the SOS genes is repressed by the newly synthesized LexA. The SOS repair system is error-prone and consequently mutagenic.

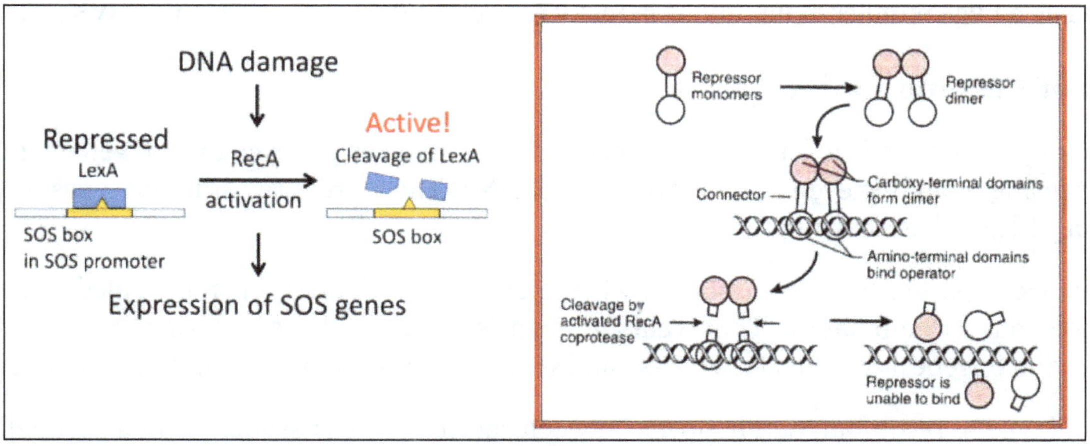

Expression of SOS genes.

Figure above shows the overall process of SOS response. LexA binds to the SOS box and suppress the expression of SOS genes. However, different radiation induces and activates RecA protein, which facilitates the autocatalysis of LexA and thereby turn on the SOS response genes. Picture in left side shows that a dimeric repressor LexA activated under normal condition and binds to the operator site of the SOS genes through its amino terminal domain.

DNA Recombination

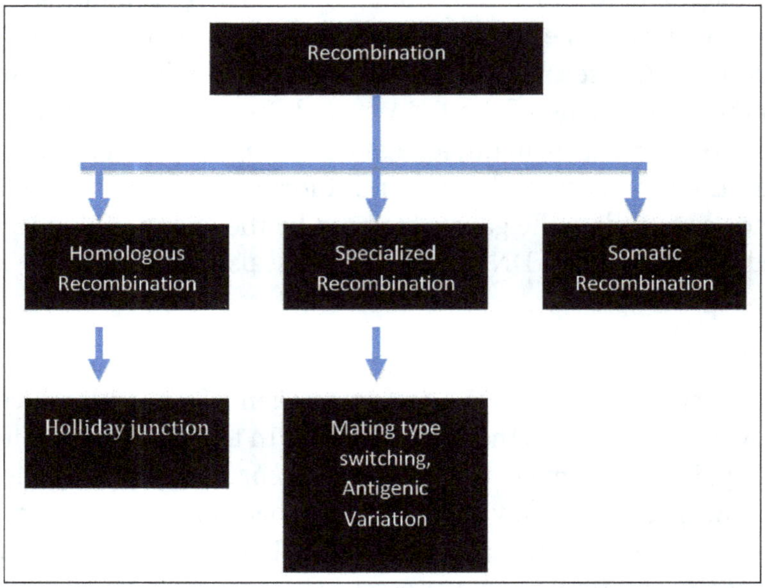

Genetic recombination of the homologous part of the chromosome is an essential process for genetic diversity and eventually for evolution. When homologous chromosomes are aligned, pairs of genes are exchanged by crossing over in higher organisms. Bacteria although lacks duplicate chromosomes, also have an elaborate mechanism for recombining genetic information. Addition to that, recombination is also seen when a foreign DNA is integrated in a host's chromosome. If genetic material would not exchange at the homologous site in the chromosomes, the content of each chromosome would be fixed irreversibly in its particular alleles. If mutation occurs in a particular gene, which cannot be exchangeable with a wild type copy of the same gene in other chromosome, cells will accumulate a huge number of mutations in many essential genes and will die in a very short time.

Types of Recombination

Genetic recombination is a very precise process, which involves the exchange of genetic material without losing or adding even a single base pair of DNA. There are three different types of recombination pathway have been reported.

1. Homologous part of the chromosomes, when recombine with each other are called generalized or homologous recombination. This occurs at meiosis in eukaryotes e.g. during spermatogenesis of males and oogenesis of females. Four-strand stage of meiosis allows recombination to happen.

2. Second type of recombination happens between specific pairs of sequences at a particular site, mainly observed in prokaryotes, and is known as specialized recombination or site specific recombination. This was first discovered in bacteria where phage genome was integrated in bacterial chromosome. In the recombination event, the crossing over happens at a short stretch bacterial DNA which shares homology with phage DNA.

3. In special circumstances, gene expression can be controlled by gene rearrangement through a different kind of recombination, known as somatic recombination.

Homologous Recombination

In this type of recombination event, two duplexes are recombined at their homologous site. During meiosis homologous chromosomes approach each other and pair at multiple region forming bivalents. This process is called chromosome pairing or synapsis. Recombination allows the interaction of the duplex DNA of one sister chromatid with the duplex DNA of a sister chromatid from another chromosome. At the synaptic point, chromosomes can be held together at discrete sites called chiasmata where the cross over event occurs. The recombination event is initiated by a double strand break by an endonuclease in the recipient DNA duplex. Spo11 is an endonuclease, which produce a double strand break in a mechanism related to topoisomerase. Following endonucleolytic cleavage, exonuclease in association with a DNA helicase trims one strands of either side of the break generating a duplex with a 3'–overhang in a process known as 5'-end resection. Homologous region of the donor DNA duplex is then invaded by one of the free 3'-overhang in a process called single-strand invasion, which generates a displacement loop (D loop) in the hetero-duplex region of the chromosome. Within the D-loop, there is cross over point known as recombinant joint, where a single strand of duplex DNA crosses from one duplex to the other one facilitating the displacement of one strand of donor duplex. One interesting feature of recombinant joint is its mobility along the duplex DNA, which is called the branch migration. Branch migration can allow the cross over point to move in either direction.

D-loop eventually grows large enough to correspond to the length of the gap on recipient chromatid. DNA synthesis occurs to fill the gap. Cross-strands then flank each sides of the gap generating a Holliday junction. Strand exchange, which leads to form joint molecule, should be resolved into two separate duplex DNA requiring additional pair of nick.

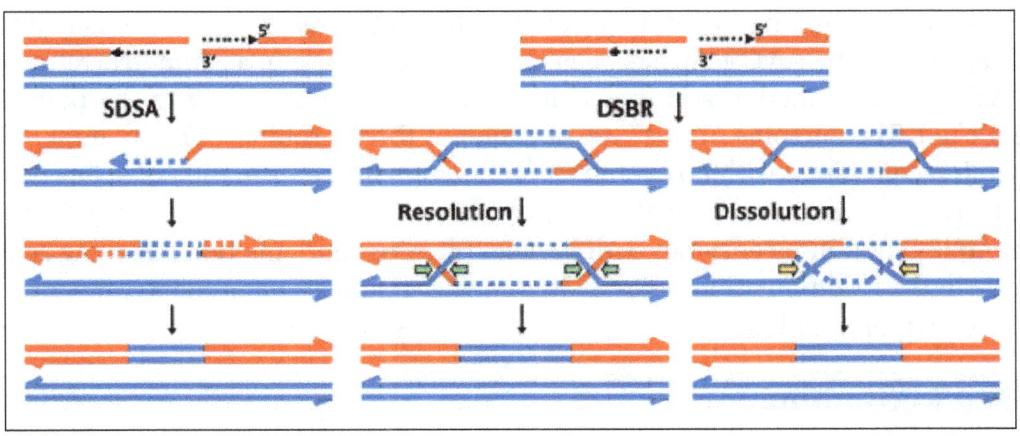

Synthesis dependent strand invasion (SDSA) and Double strand break repair (DSBR) model. A double strand break by an endonuclease (Spo11) initiates the recombination event. A single stranded 3'-OH tail is formed following nuclease degradation. Strand invasion results in the formation of a D-loop. DNA synthesis occurs to repair the break.

Function of RecA Protein

RecA was the first protein discovered in Escherichia coli, which plays a central role in promoting the strand transfer during recombination. Rad 51 is another protein, which serves the same function in eukaryotes. Bacterial strains with the mutation in recA gene, accumulates duplex DNA with double strand break, thus failing to form normal synaptonemal complexes. RecA protein in bacteria serves two important functions; it activates the protease activity of LexA protein that in turn produces the SOS response and it also has the ability to facilitate base pairing between a single stranded DNA with its complementary sequence in a duplex DNA in a process known as single-strand invasion. In other words, RecA protein by promoting single-strand invasion initiates the recombination process. In bacteria, RecA works on the substrate, which is produced by the processing of a duplex DNA with double-strand break by RecBCD complex. All the proteins belong to RecA family associate into a long presynaptic filament with single-stranded DNA. Six monomers of RecA are found per turn of the filament. DNA duplex is then incorporated forming a kind of triple stranded structure. Physical exchange of material occurs following synapsis. The promotion of strand exchange activity of RecA is associated with ATPase activity.

Resolution of Holliday Junction

The extent of recombination can be determined by the resolution of the Holliday junction, most critical step in recombination. The length of the generation of hybrid DNA is determined by branch migration from the strand exchange site. In Escherichia coli, ruv gene product has been reported to stabilize and resolve the Holliday junction. RuvA protein recognizes and binds to the structure of the Holliday junction. Each of the strands at the cross over point is bound to two tetramers of RecA from both sides. Ruv protein, a hexameric protein, is a helicase with ATPase activity, which provide the energy for branch migration with expense of ATP. RuvB protein binds to each DNA duplex upstream to the cross over point and in association with RuvA, it can cause the branch migration at a rate of 10-20bp/sec. Another protein known as RecG also found to have similar kind of function like RuvAB complex. In the process of resolution of the Holliday junction, RecA proteins are displaced from the DNA by RuvAB complex. The recombination activity in E. coli is completely abolished if the cells have the mutations in both RuvAB and RecG encoding gene, since these two proteins have the overlapping function at the branch point to resolute the Holliday junction. An endonulease, encoded by ruvC gene has been shown to recognize an asymmetric tetra-nucleotide sequence ATTG at the junction and direct the resolution corresponding to the pair of strands, which are nicked. The outcome of this type of action of RuvC protein results in either the formation of a patch recombinant with no overall recombination or the formation of splice recombinant involving the recombination between flanking regions. All of these incidents suggest that a "resolvasome" complex containing enzymes with branch migration and junction resolving activity plays a crucial role in the recombination process. It is conceivable that mammalian cells may contain a similar kind of complex and mechanism in the recombination process.

Specialized Recombination

In specialized recombination, two specific sequences of nearly 15-50 nucleotides long are involved in the reaction. Two specific sites are homologous in some cases; however, in some other cases they are nonhomologous. Enzymes known as recombinanse promotes site-specific recombination. Nearly hundred different kinds of recombinases have been reported till date. One example of site-specific recombination is the integration of phage DNA in the bacterial chromosome. Underline recombination process in the phage integration involves recombinase known as integrase, which is prevalent in different phages including λ phage. Bacterophages can exist in two different lifestyles; lytic and lysogenic. Phage DNA in the lysogenic state is integrated in the bacterial genome by site-specific recombination (which is called prophase). Virus particles are released into their lytic lifestyle by an excision mechanism as an independent, circular molecule in the infected bacterium.

Bacterial and phages DNA have specific loci known as attachment (att) site, where recombination occurs for the integration or excision to happen. Bacterial attachment site is called attB, whereas that in phage is known as attP. The attachment sites in bacterial and phage DNA contain a characteristics core sequence, which is flanked by accessory sequence in both sides known as arms. Recombination during integration or excision occurs in the core sequence regions and the arm regions promote the recombination to happen. Once viral DNA is integrated into bacterial genome, two new arms now surround the viral DNA in bacterial chromosome, which are known as attL and attR. Therefore, attP and attB sites are recognized for the recombination to promote integration of viral DNA into the bacterial chromosome, whereas, the excision of viral DNA from bacterial genome requires the recognition of attL and attR sites. So, site-specific recombination has a unique directional character, which is controlled by identity of two recombining sites. There are specific sequence requirement region of att sites. Site attB, ~23 bp long segment including a 4 bp core region, is much smaller than attP, which is nearly 240 bp long. Difference in size between attB and attP suggests that their roles in recombination process are different. Site-specific recombination involves breakage and reunion of DNA fragments in an intermediate stage of recombination process where a core sequence on attB is nicked. Cleavages in both attB and attP are staggered by 7 bp followed by a cross-wise end joining occurs. This kind of recombination process also involves sequential DNA exchange.

A series of proteins is involved to promote the recombination event, which is reversible in nature. An integrase enzyme, encoded by phage gene int, facilitates the integration between attB and attP with the help of a bacterial protein called integration host factor (IHF). Mechanism of integrase action is similar to that of type I topoisomerase. Unit of two enzymes binds to either sides of recombination site and promotes transfer reaction through a synaptic complex formation. A tyrosine residue in the catalytic site of integrase is actively involved in cleavage and rejoining of the DNA strands. Excision in the recombination process is facilitated by a protein, which is encoded by gene xis. Proteins Int and IHF are essentially involved in both the processes; however, association of Xis to Int and IHF proteins inhibits integration and controls the directionality of recombination process.

Somatic Recombination

Somatic recombination is a type of gene rearrangement by which cells of the adaptive immune system physically cut out small regions of DNA and then paste the remaining pieces of DNA back together in an error-prone way. Because there are many ways to cut and paste the DNA, and because repairing a cut is imperfect and allows for the insertion of random DNA bases, an almost infinite number of DNA sequences can be stochastically generated. This, in turn, leads to the production of an equally infinite number of proteins. These proteins, like the antibodies produced by B-cells and T cell receptors expressed on T-cells, are unique across individual adaptive immune cells. For example, each individual B-cell (called a clone) produces a single type of antibody capable of recognizing a single set of antigens, and this antibody is different from that produced by any other B cell clone in the body.

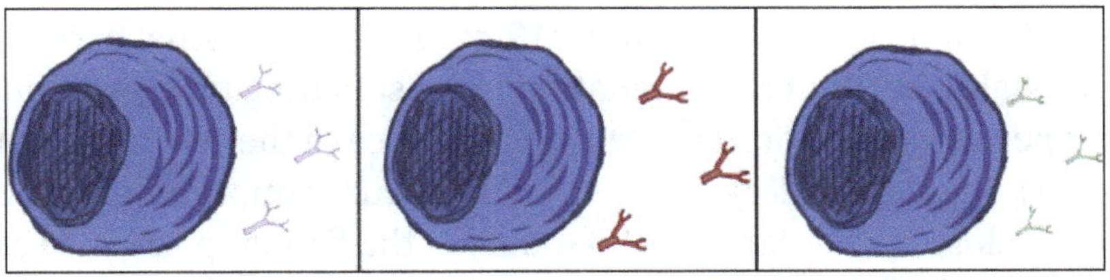

The production of unique antibodies by different antibody-secreting B-cells.

However, taken together, all of the unique antibodies produced by ever B cell clone in our body can collectively recognize an innumerable quantity of antigens present on invading pathogens. Individually, though, each adaptive immune cell produces proteins that only recognize a small handful of molecules present on particular pathogens. A similar process (T cell receptor recombination) occurs for the antigen-binding genes in T-cells that encode the T cell receptor, meaning that each T cell clone also produces a unique receptor that recognizes specific antigens. Importantly, the specificity of this response is the key difference that separates cells of the adaptive immune system from those of the innate immune system. Somatic recombination is different than somatic hypermutation (SHM), a process that takes place as part of an immune response, after a B cell has encountered antigen.

Yeast Mating Type Switching

Conversion of yeast Saccharomyces cerevisiae haploid and diploid condition takes place by mating and sporulation. Haploid cells can fuse together to give a diploid cell. Likewise, meiosis of diploid cells gives haploid spores. Two different kind of mating types are possible; one is a and another is α. Mating is only possible between haploid a and α to generate diploid cell of type a/α. However, diploid cells can regenerate haploid spores of either type. Yeast mating type locus MAT determines genetic behavior depending on the kind of allele present in that locus i.e. it is either MATa or MATα. Cells carrying a allele at MAT locus called a cell type which secretes a-type pheromone, whereas, those carrying α allele at MAT locus called α cell type responsible for the secretion of α-pheromone. A cell of a particular mating type carries receptor for opposite type of pheromone. Mating process is initiated by the interaction of one type of pheromone to the receptor of other cell type. Mutation, which abolishes any one of these interactions, prevents the mating process. Remarkable ability of yeast strain is that they can switch their mating type.

Presence of a dominant allele HO is the reason of yeast mating type switching. This fact suggests that all cells express only one mating type although they have all the information needed to be either MATa or MATα. Mating type cassette model explains how yeast cells switches its mating type. Reports show that additional two loci are involved to facilitate switching cell's mating type. One of them is HMLα, which is required for switching to produce cells of MATα type, whereas, HMRa is required to give MATa type cells. Both HML and HMR loci have silent cassette, whereas an active cassette of either a or α is present at MAT locus. Although, all cassettes carry information, which codes for a particular mating type, however, MAT locus containing an active cassette of a particular mating type is expressed. When the active cassette at the MAT locus is replaced by information of silent cassette present in either HML or HMR loci, only then switching occurs.

Gene Conversion

Gene conversion Initiates switching in yeast mating type in which MAT locus, a recipient site, acquires the sequence of donor type either from HMR or HML. Mutational experiments carried out by researchers have established that recombination in yeast mating type switching is unidirectional. In each mating type cassette, a common flanked sequence in the central region differs either in a or α (called Ya or Yα). Right boundary of Y in the MAT locus is important for the switching event. A double strand break close to the right of Y boundary is the first step in gene conversion process. An endonuclease encoded by HO locus, called HO endonuclease, makes a staggered cut on double stranded DNA right boundary of Y region in the MAT locus. The cleavage generates a sticky end with a single stranded overhang of 4 bases. HO endonuclease does not act on mutant MAT locus which cannot switch. It is also inactive on outside region of the MAT locus (HML or HMR). Recombination process occurs through a synthesis dependent strand annealing mechanism.

Recombination is Involved in Antigenic Variation

Trypanosoma, a single celled parasite, is the cause of human diseases like sleeping sickness and Chagas disease. This organism can escape human immune system by a process called antigenic variation which is the ability of these organisms to alter their surface antigen (variant surface glycoprotein). Human immune system target variant surface glycoprotein (VSG) of Trypanosoma cell. However, once antibody is produced against a given VSG, Trypanosoma cells are able to switch the expression of one to many hundreds of VSG genes present in their genome. A recombination event controls the switching event where a silent VSG gene is moved to an expression site (ES) which is transcriptionally active. The presence of open chromatin in transcriptionally active ES makes it a hotspot of recombination. The frequency of VSG recombination is higher than the rate of antibody production of the human cell, which is the reason why Trypanosoma evades human immune response.

References

- DNA, science: britannica.com, Retrieved 11, July 2020
- Different-form-of-DNA, vr-biomolecules: bch.cuhk.edu.hk, Retrieved 13, March 2020
- DNA-types-structure-and-function-of-DNA, dna-1301: biologydiscussion.com, Retrieved 02, August 2020
- DNA-replication: microbenotes.com, Retrieved 25, January 2020
- DNA-replication-in-prokaryotes, mhccmajorsbio: openoregon.pressbooks.pub, Retrieved 21, June 2020
- Nucleotide-excision-repair, science-education-10792: jove.com, Retrieved 09, February 2020
- Somatic-recombination, somatic-recombination-is-a-type-in-an-error-prone-way: immunobites.com, Retrieved 24, April 2020

3

Understanding RNA

RNA stands for ribonucleic acid. It is an important macromolecule which is present in all cells and is genetic material of some viruses. RNA is responsible for protein synthesis and carries instructions from DNA. It is of three types; rRNA, mRNA and tRNA. The aim of this chapter is to explore these types of RNAs. These topics are crucial for a complete understanding of the subject.

Ribonucleic acid or RNA is one of the three major biological macromolecules that are essential for all known forms of life (along with DNA and proteins). A central tenet of molecular biology states that the flow of genetic information in a cell is from DNA through RNA to proteins: "DNA makes RNA makes protein". Proteins are the workhorses of the cell; they play leading roles in the cell as enzymes, as structural components, and in cell signaling, to name just a few. DNA(deoxy-ribonucleic acid) is considered the "blueprint" of the cell; it carries all of the genetic information required for the cell to grow, to take in nutrients, and to propagate. RNA–in this role–is the "DNA photocopy" of the cell. When the cell needs to produce a certain protein, it activates the protein's gene–the portion of DNA that codes for that protein–and produces multiple copies of that piece of DNA in the form of messenger RNA, or mRNA. The multiple copies of mRNA are then used to translate the genetic code into protein through the action of the cell's protein manufacturing machinery, the ribosomes. Thus, RNA expands the quantity of a given protein that can be made at one time from one given gene, and it provides an important control point for regulating when and how much protein gets made.

Ribosomal RNA

The major RNA types occurring in cells include rRNA, mRNA and transfer RNA (tRNA) which are processed from longer precursor molecules to final RNA products. The initial precursor molecules called primary transcripts or preRNA correspond to the DNA length being transcribed in a transcriptional unit. Nascent ribosomal RNA (rRNA) molecules are the core entities that partake the cooperating role with protein-synthesising organelles, ribosomes, that divulges to translate the messenger RNA (mRNA) sequence information into protein in the cytoplasm. The precursor molecules, transiently coupled to the proteins are covered into smaller functional units through serial processing steps that include truncation and editing reactions. RNA processing is concordant with the functioning of variety of small RNAs (typically 90-300 nucleotides in length) as well their associated proteins.

rRNA molecules are synthesised in specialised environs of the denser nuclear region of the cell nucleus called nucleolus, encoding rRNA encoding genes. The encoded rRNAs differ in sizes as large and small rRNA molecules, thus contributing to formation of sub-units and units of ribosomes. The functional integration of the large and small rRNAs with ribosomal proteins forming the large and small sub-units of the ribosome are designated as per their sedimentation rate aligned as Svedberg units (S) in a centrifugal force (e.g. 50S, 30S, 28S, 15S). Ribosomal proteins are synthesised in the cytoplasm and transported for the nucleolar based sub-assembly, with their final assembly accomplished in the cytoplasm.

The ribosomal RNAs form extensive secondary structures, playing a crucial role in recognition of the conserved areas of mRNAs and tRNAs. There is radical gap in the number as well as sequences of rRNA genes between prokaryotes and eukaryotes as exemplified by 7 copies of rRNA genes in E.coli synthesising 15,000 ribosomes per cell while in eukaryotes with define nuclear boundaries, anywhere from 50 to 5000 sets of rRNA genes resulting in the synthesis of as many as 10 million ribosomes per single cell. The sequence difference in the prokaryotic versus eukaryotic rRNA genes reflects the evolutionary architectural conformations. The 16S rRNA gene relates to the conserved genes existent in all bacteria, and an allied form in eukaryotes, with certain portions undergoing rapid genetic changes and serving a distinguishing characteristic of different speciation under the same genus. The ability of catalyzing the peptidyl transferase reaction during protein synthesis, is an evolutionary implication of rRNA, thus serving as both repository of nucleic acids and as catalyst, as a contribution to molecular evolution.

Synthesis and Processing of Ribosomal RNAs

Ribosomal RNAs populates as the most replete universal noncoding entities in living organisms. Millions of ribosomes in eukaryotes consists of several molecules of rRNA together with associated with proteins called ribonucleoproteins (RNPs). The DNA transcripts termed as rDNA that generate the milieu of such vast rRNAs are customarily repeated hundreds of times. rDNA often clustered at scattered loci of the genome e.g. in humans being dispersed at 5 different chromosomes. The interphase stages of cell contain rDNA cluttered forming irregularly shaped nuclear structures, nucleoli, playing a central role in ribosome biogenesis. The granular nucleolus mostly composed of ribosomal sub-units, with fibrillar materials that are indicateve of rDNA templates and nascent rRNA transcripts.

Synthesising the rRNA Precursors

The breakthrough of understanding the genes in action for manufacturing the rRNA repository was through the use of Electron Micrographic images, where the fibrillar cores of the growing oocytes nucleoli were described by Oscar Miller in late 1960s. Some of the facts that were established after these studies indicate the following:

i) Tandem arrangement of the repeated rRNA genes situated alongside a single rDNA molecule.

ii) Formation of the fibrillar structures called Christmas trees, by the 'in-action' elongating rRNA precursor transcripts along the DNA, with terminal RNA polymerase I molecule in process of synthesising the transcript. The fibrillar length increments extending lengthwise through (end to end) of the connected Christmas trees. The size demarcation for the shorter trees indicate shorter fibrillar RNA molecules with fewer nucleotides tethered to polymerase bound to the DNA closer to transcription initiation site, while the longer fibrils indicate the completion of the transcriptional processes. The rDNA length between the shortest and longest RNA fibrils accord for a single transcription unit, where the promoter is placed upstream of the transcription initiation sequence site.

iii) RNA fibrils are composed to be associated proteins and RNA particles, clumped to form the pre-rRNA processing units that allow them to be assembled as ribosomal sub-units.

iv) The rDNA in-between transcriptional units is devoid of nascent RNA chains, and are not transcribed, termed non-transcribed spacers (e.g. present between randomly repeated genes for tRNAs and histones).

rRNA Precursor Processing

Eukaryotes are endowed with four distinct ribosomal RNAs, three in the large subunit comprising 28S, 5.8S and 5S RNA molecules and one (18S RNA molecule) in the small sub-unit. These ribosomal RNAs constitute the 40S and 60S subunits devised through varied action of nucleases from a long polycistronic pre-ribosomal RNA (28S, 5.8S and 18S) transcript. The only exception here is with 5S rRNA that is synthesised from a divergent RNA precursor outside the nucleolus, where a multifarious progression of processing is accompanied for the gradual release of the mature RNAs from its precursor, synchronous with the assembly of the 79 ribosomal proteins.

The pre-rRNA is peculiar for its composition of highly methylated residues and pseudouridine residues in the transcript length, whose function is still unclear. The human pre-rRNA before its first cleavage acquires over 100 methyl groups added to its ribose residues and around 95 percent of uridine residues have been chemically converted into pseudouridines. The modifications are persistent post-transcriptionally when the nucleotides have been incorporated into the nascent RNA, where the altered nucleotides are clustered as conserved regions in the RNA molecule. These altered regions are saved in the processed transcripts while the rest portions are discarded from the final products during processing. The functional integrity of such modification of the pre-rRNA transcripts might correspond to prevention of unwanted cleavages of pre-rRNA by enzymes, promoting rRNA three dimensional structural folding, and promote rRNA interactions with other molecules.

The biogenesis and processing of the pre-rRNA transcripts have been followed by radioactive methionine labelling 'pulse-chase' experiments that point to the heavily loaded 13,000 nucleotide long 45S precursor RNA molecule, that is further cleaved into smaller fragments, processed to generate around 700 nucleotide long 28S, 18S and 5.8S rRNA combined molecular pool. The information generated depicts the presence of 32S RNA, as a transitory precursor to mature 28S and 5.8S RNA molecules while the other precursor molecules of the 45S generates the 18S RNA molecule whence transported to the cytoplasm.

Extensive collection of trans-acting factors chaperone this process, including small nucleolar ribonucleoparticles. The process of maturation and synthesis of the RNA molecules have renewed interest as per long-lasting questions fuelled by the associative genetic diseases revealing mutations in genes encoding both ribosomal proteins and ribosome biogenesis factors. The intricate details of concurrent scheme of pre-ribosomal RNA maturation would furnish perspective anticancer therapeutic regimes that target the mechanisms of ribosome synthesis.

Role of snoRNAs in rRNA Processing

The processing of rRNAs is tightly coupled to the presence of small nucleolar RNAs (snoRNAs), that are packaged with particular snoRNPs (Small nucleolar proteins). The dual association of more than two dozen snoRNPs before completion of transcription, initiated by the U3 snoRNA is an example of the requirement of this huge rRNA processing machinery. The other steps of rRNA processing is controlled by 'exosome', an RNA degrading enzyme that inhere nearly a dozen different exonucleases. Several low abundance snoRNAs categorised based on their function and nucleotide sequence similarity are grouped as the box C/D snoRNA (responsible for methylationon of ribose residues in pre-rRNAs) and as the box H/ACA snoRNA (determining the conversion of uridine into pseudouridines). The structure of nucleotides modified by these two groups of snoRNAs.

The snoRNA groups comprising of long complementary stretches of nucleotide sequences to sections of the rRNA transcript, forming the uniques example of double stranded hybrids or an snoRNA-rRNA duplex. The snoRNA guides the methane or pseudouridylase enzyme within the snoRNPs to modify the particular nucleotides in pre-rRNA. These snoRNAs account for roughly 200 molecules per pre-rRNA transcript that is ribose-methylated or pseudouridylated. The ribsosome sub-unit assembly is closely coupled in the nucleolus region, where two types of proteins i.e. ribosomal proteins (remaining in sub-units) and accessory proteins (acting as transitory processing proteins with rRNA intermediates), are a part of the enzymatic machinery resulting in the structural rearrangements as part of the ribosome biogenesis.

Role of Ribosomal Proteins in rRNA Processing

The explicit detail of the ribosomal proteins in pre-rRNA processing and ribosome biogenesis remains evasive to extensive details, though they have been implicated for the assembly, processing as well as nuclear export of the pre-ribosomal particles. Ribosomal proteins or RPs comprising common structural domains are indicative of the their mechanistic interactions with rRNAs and other RPs. Functionally acting as chaperones, RPs assist in an ordered fashion for the required folding of the nascent pre-rRNAs. Their depletion from the cellular pool confers blockage of differential prerRNA processing steps e.g. initiation of the cleavage at sites A0,1 and E in humans, strictly directed by the assembly of fractions of the small subunit RPs (RPSs) making contacts with the 5' part of the 18S domain. Orthologous counterparts of RPs are found as primary and secondary binders in the prokaryotic small subunit assembly. Conducive work of RPSs before being incorporated in pre-ribosomes and 3' structure region of the 18S rRNA, is the continued efficient processing at ITS1 sites - 1, E, or 3 culminating in nuclear export. Kinetic studies indicate that the assembly of 40S structure (both in prokaryotes and eukaryotes) is a sequential event with the body domain enclosing the central domains + 5'-end of 18S rRNA, subsequent to the head domain comprising major 3'-end domain of the 18S rRNA are coordinating with primary and secondary interacting proteins, RPs. Similar mechanistic detailing have been demarcated for the large subunit 60S assembly in yeast, where rRNA folding and RPLs interaction steps induce a hierarchical structural assemblage.

The RPLs categorised as early acting have complementary association with 5' branch domains of the 25S and 5.8S rRNAs stabilises a pre-ribosomal structure facilitating the A2 and A3 cleavages. The clamp structure hence formed favours the 27S-B pre-rRNA formation, that are acted upon by middle-acting RPLs allowing the cleavage at the C2 site in the ITS1. The consequent adherence at the 3' mature 25S rRNA sequence region of middle- and late-acting RPLs at is coupled to 7S pre-rRNA processing. Final assembly from the pre-60S particles to the large 60S subunit is continued by structural modulation or a semi-circular movement of the the 5S RNP complex (consisting of 5S rRNA, Rpl5p, and Rpl11p) resulting in protuberance form of the 60S subunit. This crucial step is tightly monitored and chaperoned by late-acting RPLs that result in processing and folding of the 6S precursor to mature 5.8S rRNA molecule.

Synthesis and Processing of 5S rRNA

The 5S rRNA, is presented in both prokaryotes as well as eukaryotes, conveying around 120 nucleotides in length. In eukaryotes, the 5S rRNA molecule is encoded outside the nucleolus by a large number of identical genes, organised as tandem arrays somewhat similar to precursor 45s pre-rRNA, with alternate presence of transcribed portions and non-transcribed spacers. The transcription here is done by RNA polymerase III, where mostly the 3' end of the 5S pre-rRNA is truncated of extra nucleotides during processing. The mature 5S rRNA molecule is then transferred back to the nucleolus to join the ribosome biogenesis counterpart rRNA molecules. RNA polymerase III categorises as a unique polymerase as it can bind the promoter region located internally through the transcribed portion of the gene, not requiring any of the flanking gene regions for transcription initiation.

The ribosomes in eukaryotes are composed of two subunits, that are assembled through an intricate association of ribosomal proteins (RPs) with 4 distinct ribosomal RNAs (rRNAs). The 40S small subunit is comprised of 18S rRNA assembled with around 33 RPs (RPSs), while the large 60S subunit enclosing the 5S, 5.8S and 28S rRNAs affiliated with 46 RPs (RPNs). The assembly of such detailed ribosomal units expend the chief module utilisation of the cellular energy and extend the usage of extensive nuclear domains while acquiring its eventual maturity in the cytoplasm. Ribosome biogenesis is an organised process that is actuated in the nucleolar bodies where three of the rRNAs (18S, 5.8S and 28S rRNAs) are produced as longer precursor transcripts emanating from the head to tail tandem repeats of rDNA clusters (300-400 in diploid human cell), called nucleolar organiser regions (NORs) while the precursor to the 5S rRNA is synthesised from multiple genes positioned in the close juxtaposition around the nucleolus.

The eukaryotic primary rRNA transcript architecture reveals the mature 18S, 5.8S, and 25S/28S rRNAs are interrupted by the internal transcribed spacers (ITS)- 1 and 2, flanked at the 5' and 3' sites by external transcribed spacers i.e. 5' ETS and 3' - ETS. The co-transcriptional interconnection of nascent primary transcripts with ribosomal proteins (RPs), various pre-ribosomal factors (PRFs), small ribonucleoprotein particles (snoRNPs) conform as large RNPs series, wherein preRNA folding and modifications are executed accompanying the RP assembly. The process is constituting the sequential elimination of the transcribed spacers through intricate complexation of endonucleolytic and exonucleolytic truncations. The pathways leading to production of the two ribosomal subunits are diversified after the cleavage of internal transcribed spacer (ITS1). The maturation pathway of the pre60S particles, form the granular nucleolar component, whereas pre-40S particles swiftly transmitted to the cytoplasm.

The proteomic investigations of maturation and production pathways of preribosomes, reveal that rRNA processing include a highly dynamic configuration as per its protein compositions, with progressive inducement of RPs, model association / dissociation of PRFs, as either individual factors or as multi-protein modules. The processes are usually undertaken in the nucleolus, though many maturation events are associated in the nucleoplasm. Pre-ribosome once formed are transported to the cytoplasm, where terminal maturation events to generate mature ribosomal sub-units occur. There are varied pathways for pre-rRNA processing which are pre-requisite of ribosome biogenesis, which are attributes of unicellular or multicellular eukaryotes.

Pre-rRNA Processing Pathways

The maturation pathway in mammalian cells for the nascent 18S pre-rRNA to 18S rRNA is initiated with the constitutive endonucleolytic processing at the 5'-ETS and exonucleolytic truncation within the ITS1 sequences. These steps, for the production of 5.8S rRNA are proceeded by tailoring the 5' end using alternate processing pathways yielding the long and short form of 5.8S rRNA. The ITS2 elimination is instituted through endonucleolytic cleavages by allowing the 3' to 5' and 5' to 3' exonuclease activity to correspondingly accomplish the maturity at 3' end for the 5.8S rRNA and 5' end for the 25S/28S rRNA.

The onset by cleavage at the 5'-ETS frames the 41S pre-rRNA, while the 30S preRNA is indicative of the ITS1 cleavage at site 2 preceding the 5'-ETS cleavage (Figure 2). A similar pre-cleavage at site E prior to the cleavage at site 2 proceeds in the formation of the 36S pre-rRNA in mouse cell lines. These are the indicators of kinetic variability of the systemised early cleavages observed in eukaryotes, is often acknowledged to the existence of different rRNA processing pathways, thus creating a possibility of a plethora of structural and functional heterogeneity among ribosomes. The presence of alternative processes also indicate the regulation and robust control over the maturation pathways for rRNAS. The radical differences in the truncation kinetics are linked to the time-line detection and concurrent manifestation of post-transcriptional processing as well as co-transcriptional cleavage of the pre-rRNAs, well evidenced by the visualisation of rDNA transcription units after chromatin spreading, popularly called electron micrograph Christmas tree displays. These are also a clear image of variable length nascent transcripts, their corresponding sequence distances from the promoter regions, having terminal globular structures at their 5' ends. These structures have been recollected as processome, where RNPS are ramified in early maturation steps of the small ribosomoal sub-units. The terminal globular structures disappear from the unprocessed transcripts immediately after the polymerase commands the 5' region of the 25S coding sequence suggestive of the co-transcriptional processing and truncation of 5'-ETS and ITS1 cleavages at site A2 in yeast, while in higher eukaryotes, similar chromatin spreads shows no co-transcriptional processing of pre40S particles.

Maturation of the 18S rRNA

1. Elimination of the 5' - ETS - The endonucleolytic cleavage at simultaneous sites A0/A0 and A1/1 for the coordinated removal of the 5'- ETS is required for formation of 18S rRNA. The uncoupling of cleavage generated by endoribonucleases at sites A0 and 1 leads to generation of 43S and 26S precursors. The enzymes, or the factors required for the processing have been identified as U3, snoRNA30/U17 and U14 snoRNAs and SSU processome as well as riboproteins (RPSs). This cleavages facilitate processing machinery for further maturation.

2. Processing of the ITS1 - The 3' end of the 18S rRNA is accomplished through either endonucleolytic alone or both ends and exo-nucleolytic cleavages at sites A2/ site 2 and D/E of the ITS1 junctions by the respective endo/ or exo-nucleases.

The pre-rRNAs thus cleaved in the ITS1 and further trimmed by $3' \rightarrow 5'$ exonucleolytic activities. The terminal nucleotide removal from the 3' end of the 21S pre-rRNA results in transitory 21S-C (domain C) intermediates, as similar to the 18S-E (domain E) precursor that is gradually shortened in the nucleus and cytoplasm. For final processing and maturation, there is definitive role of exonucleases as exosome units to ensure the trimming of the precursors to final rRNA products.

The 3′ →5′ exonuclease are probably involved in the processing of the 18S-E pre-rRNA. The 18S -E pre-rRNA after nuclear export is further cleaved at site 3 by PIN domain containing protein NOB1 yielding the mature 18S rRNA. The previous trimming processes as well as modification likewise of oligouridylation in the cytoplasm stimulates the recruitment of exonucleases or NOB1 proteins to the specified sites for processing. Thus, all such processing steps contribute to the directional and site specific tailoring of the pre-rRNA molecules to processed mature rRNAs, that are inclusive of RNA cyclases, RNAses (endo and exo-nucleases), RNPs that allow in varied organisms analogous or alternate (slightly variant) pathways for rRNA processing.

Maturation of the 5.8S rRNAs

Maturation of the 5.8S rRNA 5'-end is coupled with the 3'-end formation of the 25S/28S rRNA. Alternative pathways of processing results in different 5'-ends of 5.8S rRNAs. This is exemplified by the major 5.8SS (S-for short) is a resultant of the 5′ →3′ exonucleolytic clipping of the 27S-A3 pre-rRNA down to site B1S by two exoribonucleases: Rat1p and Rrp17. The exoribonuclease Rat1p is inducted within the 27S-A2 intermediates, precedent to A3 cleavage. The sequence of events co-ordinating the processing developments appearing at the 5' and 3' ends of 5.8S rRNA, here necessitates a string of factors referred to as 'A3 cluster' or 'Nop7psubcomplex' that structurally bind to the 3'-ITS2 of the 5.8S rRNA as well 5' end of the 25S rRNA. The 5'-end for the minor 5.8SL entity (long form) is modified through endonucleolytic activity at site B1L of the 27S-A2 RNA, or 27S-A3. Further trimming using endonucleolytic enzymes at the C2 site of the ITS2 is initiated after maturation of both 5' - 5.8S rRNA end and 3' - 25S rRNA end. The resultant is the 7S pre-rRNA corresponding to the mature 5.8S rRNA entity with extended 3- end ~140-nucleotide extension. This 7S pre-rRNA precursor unit processing continues through a complex string of exonuclease activity events that includes, instigation by nuclear exosome 3′→5′ directional transformation of the 7S intermediates from site C2, assisted through untwining of secondary structures or displacement of bound proteins by RNA helicases. The exosome (Dob1p/Mtr4p) is hindered sterically around ~30 nucleotides from the 5.8S rRNA 3′ end, generating the 5.8S+30 (nucleotide) transitory intermediates. Concomitantly the alternate Rrp6p exonuclease collaterally degrades the ITS2 sequence leaving behind 8 nucleotides, resulting in the 6S intermediates. The conclusion of the above events is played by Mg2+-dependent exonuclease Ngl2p trimming the 3′ end of the cytoplasmic pool of 6S pre-rRNA, producing the mature 3′ end of the 5.8S rRNA.

The above processing scheme is conserved across evolution, where in humans XRN2 (yeast Rat1p) work upon the downstream products generated at the sites 2 and E (E16) required to tailor the 5' end of the 32S rRNA i.e. cleavage at its ITS2 (site 4 in humans) giving rise to 12S and 28.5S species, that serve as precursor entities to mature 5.8S and 28S rRNAs, respectively. The 28.5S rRNA 5' end is trimmed the same XRN2 to generate 28S rRNA while the 5.8S rRNA is generated through 3' processing of the 12S precursor through multi-step mechanistics. The mammalian cells display shorter 7S pre-rRNA precursors whose processing is carried out by exonuclease, ISG20-L2. The adequate maturation of the 7S-pre rRNA is achieved using several subunits and associated factors. In humans and mice, RRP6 depletion results in 5.8S+40 intermediates, similar to 5.8S+30S species as described for yeast. The final processing is through a similar trimming of few nucleotides of the 6S transitory precursor followed by by the 3′ →5′ exoribonuclease ERI1, as reported in mouse cells. The overall processes of processing in yeast and mammals are in parallel, though the finer details of the contributing factors are scientifically deciphered through continued scientific investigations.

Messenger RNA

The major function inscribed to genes in a cell's DNA is to represent the amino acid sequence of the protein products, thus the RNA molecules derivative of these genes are called messenger RNA (mRNA) molecules. The final product of a minority of genes, however, is the RNA, which are the enzymatic and structural units catering for a variety of cellular processes.

There are small nuclear RNA (snRNA) molecules directing the splicing of pre-mRNA to form mRNA, ribososmal RNA (rRNA) forming the core of ribosomes, and transfer RNA (tRNA) forming the adaptors that select amino acids and hold them in position on a ribosome for incorporation into protein. A transcription unit consisting of each transcribed DNA segment, in eukaryotes carry a single gene information thus coding for either a single RNA molecule or single protein or a group of related proteins obtained from alternatively spliced mRNA. In prokaryotes, however, adjacent set of genes is transcribed as a unit, thus making mRNA there carrying information for several distinct proteins at a time. Most of the RNA population is composed of rRNA, whereas mRNA comprises of only 15% of the typical eukaryotic cells while tRNA is least in population of all such RNAs.

RNA Processing Events and Mechanisms

RNA processing is functionally the structural and chemical modifications of newly synthesized RNA molecules. In the appropriate cell type and at the correct developmental stage, ribonucleic acid (RNA) polymerase transcribes an RNA copy of a gene, the primary transcript. However, the primary transcript may contain many more nucleotides than are needed to create the intended protein. In addition, the primary transcript is vulnerable to breakdown by RNA-degrading enzymes.

Before the primary transcript can be used to guide protein synthesis, it must be processed into a mature transcript, called messenger RNA (mRNA). This is especially true in eukaryotic cells . Processing events include protection of both ends of the transcript and removal of intervening non-protein-coding regions. β-thalassemia, a hemoglobin disease, can be caused by an intron mutation that prevents recognition of a splice site.

On an RNA molecule, the end formed earliest is known as the 5′ (5-prime) end, whereas the trailing end is the 3′ end. The ends of the primary transcript are particularly susceptible to a class of degradative enzymes called exonucleases. During processing, the 5′ end of the primary transcript is protected against the effects of these enzymes by the addition of a CAP. The CAP uses an unusual linkage between nucleotides. Exonucleases do not recognize this unusual structure and therefore cannot remove the CAP. Since exonucleases work only from an end, if the CAP nucleotide cannot be removed, the entire 5′ end of the mRNA is protected. The 5′ CAP also aids in transport out of the nucleus and helps bind the mRNA to the ribosome.

To protect the 3′ end against degradative exonucleases, a poly-A tail is added by poly-A polymerase. Poly-A is a chain of adenine nucleotides, one hundred to two hundred units long. The poly-A tail has typical bonds that are susceptible to degradation by exonucleases, but it does not have any protein coding function so it does not particularly matter if some of the A residues are degraded. It actually takes quite some time for the poly-A tail to be completely lost, and during this time the protein coding portion of the mRNA remains intact. Without the poly-A tail, however, the exonucleases would rapidly degrade into the protein coding portion of the mRNA. An exception to the poly-A strategy is seen in the mRNA for histones, proteins that wrap deoxyribonucleic acid (DNA) into chromosomes. Instead of poly-A, histone mRNA uses a much smaller structure that is regulated by factors present during DNA synthesis.

The most striking event in RNA processing occurs because the protein coding region in eukaryotic genes is not continuous. A typical eukaryotic gene is composed of a number of protein coding regions, called exons, that are separated by noncoding regions called introns. In fact, the number of nucleotides in the introns can be much larger than the number of nucleotides in the combined exons. The DNA gene contains the code for both the exons and the introns, as does the primary RNA transcript, but the noncoding intron sequences must be removed to form the mRNA before protein synthesis.

The process by which introns are removed and exons are joined to one another is called RNA splicing, and it is catalyzed by complexes of proteins and RNA called SNuRPs (small nuclear ribonucleoprotein particles). These complexes locate special RNA sequences that flank the exon/intron junctions, bind to them, and catalyze the splicing reactions. Some primary transcripts can be spliced in a few different ways. Such "alternate splicing" yields a range of related proteins.

After addition of the CAP to the 5′ end, the poly-A tail to the 3′ end, and splicing of the introns, the processing is complete and the mRNA is transported through nuclear pores to the cytoplasm of the eukaryotic cell where translation (protein synthesis) will occur.

mRNA Processing in Prokaryotes and Eukaryotes

In prokaryotes, mRNA is usually short lived resulting in a sequence of events of its synthesis, translation and degradation without much modification. On the contrary, eukaryotic mRNA is generally stable and undergoes sequential elaborate processing as transcription and RNA processing is coupled and localized at discrete places in the nucleus. The precursor, premRNA, in eukaryotes has been indicated as heterogeneous nuclear RNA (hnRNA), as unlike to prokaryotic systems, it has diversity in gene size and presence of partially processed molecules. There is involvement of various mRNA processing factors likewise of spliceosome and polyadenylation enzymes attached to the RNA polymerase C-terminal tail in the elongation complex, that directly links the transcription and processing in eukaryotes. Such an arrangement also ensures export of the RNA molecules which in association with proteins are termed as heterogeneous ribonucleoproteins (hnRNPs). The proteins that are bound to RNA have been characterized to play a role in removing secondary structures and facilitate interactions with components of splicing machinery, so as to act as explicit sites for docking RNA processing factors, such as splicing proteins. Eukaryotic pre-mRNA hence is worked through processing reactions like end-modification by capping and polyadenylation; internal modification by splicing and RNA editing; chemical modification by methylation of internal adenine residues.

The following are the processes coupled to the mRNA processing that relies on specific enzymes and proteins:

- End- modification and methylation of mRNA.
- RNA splicing.

- RNA editing.
- Post-processing regulation.

Table: RNA processing mechanisms.

Processing Reaction	Examples
Cleavage	Release of rRNA and tRNA from polycistronic transcripts; Termination of eukaryotic mRNA transcription.
Exonucleolytic degradation	Processing rRNA and tRNA to generate mature ends.
Nucleotidyl transfer	Transfer of CCA trinucleotide to 3' end of some tRNAs.
Chemical base modification	Occasional methylation in mRNA and rRNA; Extensive base modification in tRNA and snRNA.
Nucleotide excision and replacement	Hypermodification of guanosine residues in tRNA to produce queuosine and wyosine.
Capping	Addition of 7-methylguanosine to 5' end of eukaryotic mRNAPolyadenylation.
Polyadenylation	Addition of polyadenylate tail to 3' end of most eukaryotic mRNAs and few bacterial mRNAs.
Splicing (ligation)	Removal of tRNAs introns.
Editing	Changing information in mRNA by base modification, insertion and deletion of residues from the coding region.

Systemized Modification Processes of mRNA

Capping

Capping is the first level of processing of mRNA, where the commencement of transcription is followed in close conjunction with capping by the addition of an inverted guanosine triphosphate (GTP) residue at the 5' end of the nascent eukaryotic mRNA. The reaction catalyzed by enzyme guanylyl transferase proceeds in a backward alignment relative to the rest of bases in the RNA generating an unusual 5' + 5' phosphodiester bond as seen in figure . The guanine base resulting from the reaction has position G7 being methylated, and the structure is known as "cap0", while higher order eukaryotes have position 02 of the ribose moiety (corresponding to position +1 or original transcript) in the next one or two nucleoside residues also methylated, thus generating the "cap1" and "cap2" structures. If the first base of the original mRNA is adenine this is sometimes methylated at the N6 position. This yields ambiguous cap structure processing of the same mRNA molecule. The capping relays its role in export of RNA molecules through nuclear pores, activity in ribosome binding and prevention in degradation, thereby regulating protein synthesis.

Polyadenylation

The tail of eukaryotic mRNA is a long string of A's at its 3' end. Though the exact mechanism of RNAP II transcriptional termination in eukaryotes is not accurately understood, termination occurs several kilobases downstream of the mature 3' end of the RNA molecules. The growing RNA which is destined as mRNA have a specific tail recognition sequence – AAUAAA- closing at the 3' end. The 3' end is generated by endonucleolytic cleavage of the growing RNA molecule 10 to 30 nucleotides downstream (just after 5' –CA- 3' dinucleotide), through the recognition of the above highly conserved sequence and followed by addition of variable length of the adenylate residues resulting in a polyadenylate or poly A tail. The cleavage site is followed by GU rich tract that is also involved in recognition but is lost after the cleavage. The coupled cleavage and polyadenylation process is achieved by utilizing the platform of a multi-subunit complex composed of trimeric cleavage polyadenylation specifity factor (CPSF) that recognize the poly A site or the AAUAAA sequence, an endonuclease comprising GU tract binding two cleavage stimulation factors (CST) that undertakes cutting step, the enzyme polyadenylate polymerase (PAP) accommodating addition of adenylate residues and some other uncharacterized components. The tail usually consists of 100 to 200 adenine residues. The multi-subunit complex are pitched to be linked to the RNAP II elongation complex, where the poly(A)-binding protein (PABP) stays associated with the mRNA and binds to the C-terminal tail of the enzyme, and in due course enhances the processivity of PAP.

The precise role of polyadenylation is not clear, making the mRNA more stable and helping in translation are the initial functions assigned to the tail. An exception to the theme is that eukaryotic mRNA exemplified by histone mRNAs and genomes of certain plant viruses are not polyadenylated while some bacterial mRNAs are polyadenylated where poly(A) tails enhances their degradation. Polyadenylation aspect can used for in vitro manipulation or purification by the use of synthetic oligo-dT sequences so as to concentrate the minority mRNA (5%) from the total RNA pool and is utilized for making cDNA clone libraries and other technical molecular biology interventions.

RNA Splicing

RNA splicing refers to the further processing of pre-mRNA transcripts directing for the meticulous intron removal, that are the noncoding elements present in the eukaryotic genes. Introns are here graded as per their splicing mechanism and detailed in the table given below. Splicing is considered to be accurate as a single base mistake while splicing would result in frameshift of the coding sequence and henceforth translational changes. The methodology of the entire process is depicted in the figure. The splicing mechanistic processes include sequential trans-esterification reactions characterized by a nucleotide (carrying free hydroxyl) attack on the phosphodiester bond joining the intron and upstream exon, thus liberating the 3' end of the exon. This reaction in sequence if followed by another similar reaction where the free hydroxyl group attacks the phosphodiester linkage of the intron to downstream exon. The intron is ejected and the exons are joined together. The intron types differ in their transesterification and structure, the source of the initial hydroxyl donor, the nature of the intermediate formed and in autoctatalytic versus dependence on splicing apparatus to be supplied in trans.

Distinguished Classes of Introns

Nuclear pre-mRNA introns: These are the highly represented intron types which are unable to splice auto-catalytically are identified as nuclear pre-mRNA introns. Forming highly divergent size and structure, where short conserved elements that are cis-acting sites or the splicing control sites. The 5' left splice site is considered as the donor site and the 3' right splice site is the acceptor site, and a branch site is involved in splicing reaction. In eukaryotes, the process is mostly frequented by GT-AG (or GU-AG in RNA code) group of introns embedded within a weaker consensus sequence, while the AT-AC (or AU-AC) introns are processed in a similar manner albeit with different set of splicing factors.

The cutting and pasting mechanism in the splicing event is coordinated in splicing machinery termed as spliceosome that contains proteins, and some specialized RNA molecules. Small nuclear RNA (snRNA) in conjugation with the protein partners constitute a small nuclear ribonucleoprotein (snRNP) termed as 'snurp'. Five snurps which have been codified as U1 to U6 (U3 being present in nucleolus is called snoRNA), where their snRNAs, often termed as U-RNAs due to their modified uridine content, recognize three sites on pre-mRNA explicitly 5', 3' and the branch site. The proteins part of snurp supervises the cutting and joining reactions. The protein U1 recognizes the 5' splice site while U2 binds the branch site and a protein called U2AF (U2 accessory factor) binds the 3' splice site. U4/U6 complexes with U2 and then U5 arrives, binding first to downstream exon and migrating to the intron/exon boundary. This is depicted as intron DNA forming a lasso as shown in Figure--- and rearrangement of the complex happens.

The splicing is then proceeded as two-step process, wherein the free 2' hydroxyl group on the active adenylate residue at the branch site attacks the phosphodiester bond of the upstream exon to the intron. This is the first transesterification step that leads to bonding of the internal adenylate to the guanidylate residue of the 5' splice site through a 5' + 2' phosphodiester bond. The free 5' end of the intron loops around and generates a lasso shaped structure termed a lariat. The second step is when the 3' hydroxyl group of the upstream exon attacks the phosphodiester linkage of the intron to the downstream exon, which is second transesterification reaction at the 3' splice site. The exons as a result ligate together with the release of the intron as a branched lariat structure which is later degraded. The methodology adopted is most accurate, with usually no exon skipping (due to generic nature of splice signals) and only principle splice sites are utilized.

Group II Self splicing introns: Mostly populated in plants and eukaryotes, these are autocatalytic introns like the group I, though unlike them does the splicing by the 2' hydroxyl group of an internal adenosine residue that attacks the phosphodiester bond linking the intron and the upstream exon, generating a lariat and a free exon with 3' terminal hydroxyl group. This group subsequently attacks the phosphodiester of the intron and downstream exon with the release of a lariat intermediate. The intrinsic capacity of the self-splicing introns is due to the six stem looped secondary structure conformation bringing the exons close together, adopted by these which is similar to the branch site of nuclear introns described above. This is suggestive of the evolutionary drift of class II introns which are homogenous while class I introns shifting the splicing to trans-regulatory complexes have diversified secondary structures.

Group I Self splicing introns: Group I introns are concentrated in mitochondrial genomes mostly, also as rare prokaryotic introns, that are auto-catalytic and use a guanosine containing nucleotide cofactor which provides the free hydroxyl group. The RNA itself provides for the catalytic activity, thus acting as RNA enzyme or ribozyme, where the activated RNA is folded into a series of base-paired stem and loop structures. The splicing reaction is proceeded when the free hydroxyl group of the cofactor attacks the phosphodiester bond joining the intron and upstream exon, extending the intron by adding guanosine (from GMP, GDP or GTP nucleotide) to its 5' end, thereby resulting in a free exon with a terminal 3' hydroxyl group. The next step is the transesterification reaction, where similar to the reactions above, the upstream exon attacks the bond linking the intron to downstream exon. This releases the intron as a linear fragment containing a 5' terminal guanidylate residue. The group I introns have nine hairpin like secondary structure that are directly involved in the recognition of the exon sequences, with hairpin namely P1 considered as internal guide sequence (IGS) juxtaposing the exons by pairing and ensuring splicing specificity.

Intron-encoded proteins: Certain class I and II introns have open reading frames that encode proteins that deliberate the intron splicing. The proteins here are called maturases, which play a structural role rather than a catalytic role. The other group of introns encode for mobility functions associated with them. Introns belonging to class I often endorse for endonuclease expression, functioning in DNA cleavage at intron insertion site, as also passively involved in intron transfer to intron less alleles of the gene in the same cell by repair mediated recombination aka homing. Such introns are termed as homing introns.

Class II introns have their role in expression of proteins involved in reverse transcriptase and endonuclease activity, thus allowing them to get mobilized like a retroposon. This process is called retro-homing.

Twintrons: Genes in higher eukaryotes might have embedded introns of larger genes, either in same or reversed orientation called as overlapping genes or, nested genes. The introns here could be coming from class I, or class II, and are mostly found as archaeal introns in tRNA and rRNA having similarities to eukaryotic pre-tRNA introns. The precursor molecules in tRNA and rRNA fold up into their normal 3D structures with the intron forming a loop, which is sliced out by a ribonuclease and the ends are joined by an RNA ligase. Nuclear tRNA introns – The exquisite class of introns found in nuclear tRNA genes depict splicing by cleavage followed by ligation and not by sequential transesterification reactions. The intron is extracted from pre-mRNA by uncommon endonuclease reaction that generates a 5'-hydroxyl as well as 3' cyclic phosphate group. The conventional 5' phosphate and 3' hydroxyl termini are generated by separate processing reactions. The modified ends juxtaposed by intra-molecular base-pairing in the tRNA, and linked through RNA ligase. tRNAs introns unconventionally have least sequence and structure conservation, hence their recognition is facilitated by exon structure. The introns retain the complementation to the anticodon, so can pair with anticodon loop, so as to form the secondary structure that generates correct splicing.

Phenomenon of Alternate Splicing – Generation of Milieu of RNA types

Eukaryotic RNA regulatory systems have the unique concept of making use of different splice sites within the same gene, by different cell types within the same organism. This invariant processing results in a single original DNA sequence that can give rise to different proteins having distinct but overlapping functions, in spite of precise splice junctions but choice of different exons joining together. Initially thought to provide for ways so as a gene can encode multiple proteins, the variety of choice of alternative splice sites to be utilized in a particular cell is controlled by use of U1 snRNP, splicing factor UUF and SR proteins. The process is dictated to let arrange the splicing components to specific sites on pre-mRNA through stimulation or inhibition of particular splice sites. Alternative splicing is controlled through either constitutive or quantitative controls depending on the quantity of splicing factors often occurring when 5' or 3' sites are used for differential splicing while the other type of control involves dedicated splicing proteins is tissue specific where splice signals are wired by binding splice enhancers or repressors (cis-acting elements) in the primary transcript.

The types of alternative splicing, that form the major regulation step, where primary transcript undergoes differential changes that maybe termed as splice variants or splice isoforms are described further as follows:

- Alternative selection of promoters: Different promoters are selected depending upon the cell type and omission or excision of 5' exon occurs.
- Alternative selection of tail site: The use of alternative polyadenylation sites at the level of transcription results in omission of 3' exons as differential RNA processing occurs here. This is exemplified by antibody production recognizing the same antigen but have different rear ends, thus leading to a secondary aspect of alternative transcriptional control elements.
- Alternative selection of exon cassette: Controls the choice between splice sites at the 5' or 3' exon splice sites, so as to define the exon selection in the processed transcript. The transcripts evolved through alternative exon cassette selection are similar though drawn differently using different splicing plans.
- Alternative selection of exons: Choice between alternative 5' or 3' splice site that results in skipping of exons altogether.
- Aberrant splicing: Occurs due to change in splice site occurring due to mutation and advent of new splice site called cryptic sites having weak consensus to the genuine sites, and as a result aberrant mature transcripts with exon deletion and intron inclusion is seen.
- Hum- splicing and Trans-splicing: Where splicing from two different primary transcripts occur and a rare phenomenon where mutual exclusion of exons is reported, where dispersed exons in different orientations are seen in the transcripts (e.g. capsnatching in Influenza virus).
- Inteins and protein splicing: Protein splicing and the intervening sequences found in proteins is a rare phenomenon. These are considered as protein analogs of introns and exons and names as inteins and exteins. Inteins are those sequences which are presented when the protein is being made, and are later spliced out, while the final protein consists of exteins. These have been reported in yeasts, bacterial, algal species. The intein splicing is an autocatalytic step without involving accessory enzymes, releasing itself as free polypeptide. There is though requirement of specific amino acids at the extein / intein boundaries for the splicing to be achieved. The splicing proceeds here again as a two-step coupled reaction where serine or cysteine must be the first amino acid scored before the downstream extein. The hydroxyl or sulfhydryl group is needed to carry the upstream extein during the branching of the intermediate.

Table: Classification of Intron types and their splicing parameters.

Intron Class	Intron characterization and splicing mechanism
Nuclear pre-mRNA introns	Structurally diverse; Initial donation of free hydroxyl group made by adenosine residue branch site; Lariat intermediate formed; Large trans-acting splicing assembly required.
Self-splicing introns class I	Conserved secondary structure; Initial donation of free hydroxyl group made by guanosine nucleotide cofactor; No lariat formed; Autocatalytic splicing.
Self-splicing introns class II	Conserved secondary structure; Initial donation of free hydroxyl group made by adenosine internal site; Lariat intermediate formed; Autocatalytic splicing.

Self-splicing introns class III	Similar to group II introns but smaller (100-120 bp), containing restricted number of domains.
Twintrons	Multiple embedded self-splicing introns; often group II or mixed group II and group III; Usually spliced in a particular order.
Nuclear tRNA introns	Splicing mechanism similar to tRNA maturation – involves cleavage followed by ligation; No intermediate formed – intron excised as linear fragment; Several processing enzymed required in trans.

RNA Editing

The queerest modification reported for mRNA is the alteration of its base sequence called RNA editing. Such modification within the coding sequence of the bases leads to changed protein product though usually the protein encoding gene at the time of translation remains unaffected, still RNA editing is considered as a co- or posttranscriptional mechanism that shows marked changes in the information encoded in the exon sequences of mRNA. This process is mostly associated with organellar genomes though minor influences are also exerted to mammalian nuclear mRNAs. The genes that undergo extensive RNA editing are called cryptogenes wherein structure of the gene product cannot be deduced from the genomic DNA sequence.

The evolutionary significance of RNA editing is uncategorized, but it is mostly linked with functional consequences and predominance of pyrimidine insertions into purine rich sequences. RNA editing in mammals is restricted to base substitution majorly consisting of Cto-U or A-to-I (inosine) changes while plants have both C to U and U to C editing. The A to I transition occurs by deamination of dsRNA adenosine to inosine. The recognition of such modification is through the formation of double stranded regions by base pairing between alteration sites and sequences from neighboring introns. This implicates the role of intron sequences in the final coding sequence of mature mRNA, making it a co-transcriptional editing step that occurs before splicing of introns. The inosine generated here acts a guanosine during translation process further leading to the change in the sequence of the resultant protein if occurring within a coding frame (e.g. in glutamate and serotonin receptor protein of mammalian nervous system). The C to U and U to c editing of mRNA is mostly grouped in organellar parts of plants (mitochondria and chloroplast), where a range of 4 to around 20 bases are altered in the transcripts, leading to a potential change in amino acid sequence of the encoded proteins that maybe functional changes or silent transitions. The silent editing is conceivably an adjustment to the differential capacities of tRNAs with different anticodons that code for the same amino acids. Primary transcripts are also altered by inserting or removing multiple uridine nucleotides, though in a step-wise manner, before the final mRNA is generated. Such a change is processed as the coding sequences on the DNA might have incorrect reading frames leading to frame-shifted proteins. The stepwise insertions of U in mRNAs occur at position specified by short guide RNAs possessing extra a that show complementarity to short mRNA stretches.

RNA Post Processing Regulation

RNA Transport and Subcellular Localization

The RNA regulation is finally meted with whence it travels or is exported and presented at its requisite sub-cellular localization. Eukaryotic mRNA that is synthesized in the nucleus is complemented to be transported to the cytoplasm for translation. The process requires the restriction of RNA processing complexes to distinct loci in the nucleus playing a substantially direct role in the export of RNA through nuclear pores. The exact mechanism of such export is partially understood though it is known that there is selective transport of mRNPs with its dependence upon ATP hydrolysis. The 5' capping at mRNA plays a major role in nuclear export, while the spliceosome components blocks the export. Some hnRNP proteins are excised from the transcript before export and the rest dissociate following the transport and revert back to the nucleus to be utilized again. The processed mRNA almost immediately lodges itself with ribosomes after its exit from nucleus for further processing during translation, where they can either be freely diffusing or maybe targeted to particular region of the cell. Further processing of such partially translated transcripts that encode secreted protein are mostly transported to rough endoplasmic reticulum membrane using signal polypeptides of the encoded protein, while in other cases signal is carried by the RNA itself (RNA targeting).

RNA Turn-over

The production of transcripts with their abundance is controlled by its synthesis rate and stability that is overwhelmed by the degradation rate also. The mRNA stability dictates the time derived steady state levels of the mRNA change during which transcription rate is altered resulting in the diversified rates of protein synthesis. RNA stability as described above is either constitutive or regulated where mRNA degradation rates can be expressed as mRNA half-life that directly effects the rapid and transitory variations to the abundance of a particular mRNA without substantial transcriptional activity change.

RNA Degradation

The prokaryotic bacterial mRNAs have short half-lives, even for the most stable transcripts that permits rate of protein synthesis to be changed rapidly and regulated by transcriptional rate triggered by the environment. The degradation is triggered when mRNAs are not bound to ribosomes and is mediated through RNA endonucleases (RNA restriction enzymes or ribonucleases) and 3' + 5' exonucleases. The bacterial mRNA is degraded in a two stage process where an endonuclease, cleaves the RNA that is left unprotected by ribosomes followed by exonuclease moving in 3' to 5' direction degrading the fragments, though the overall direction of fragmentation is 5' to 3', as the endonuclease follows the ribosomes.

The stability of mRNA depends upon the secondary structure that is characterized by multiple hairpins and stem-loop structures mostly in the 3' untranslated regions which are responsible for their protection from exonucleases. The stability is compromised in the transcripts that possess endonuclease target sites. Though the documentation of the enzymes degrading RNA in bacteria is not very well demarcated, mutations disrupting ribonuclease proteins are seen to stabilize the transcripts. The coupled associations of transcriptional machinery, protein synthesis and degradation in bacteria is termed as retro-regulation, where gene expression depends upon whether or not a cis-acting element is located downstream to the gene being transcribed. The presence of such elements results in nascent RNA adopting a structure that permits its rapid degradation thus preventing translation, while if transcription terminates preceding to this site, RNA molecule is relatively stable and translation proceeds.

The eukaryotic mRNA stability aka half-life is relatively much stable that depicts the constant environment in multicellular organisms. The stability of the transcripts is further enhanced by the polyadenylation wherein levels of polyadenylation binding protein (PABP) association is regulating the signals for mRNA degradation, though not clearly understood. The exoribonuclease degradation in eukaryotes is the first step in mRNA degradation (e.g. in histone mRNAs lacking polyadenylated tails), while there are suggestions of endoribonuclease activity in eukaryotic cells too. Another mechanism or mRNA degradation is the regulatory proteins appearing in transient bursts, specially related to the genes associated with signal transduction cascades that are entailed to produce short lived responses. The presence of instability elements specifically like the AU-rich elements (AUREs, AREs) wherein multiples copies of the sequence AUUUA, located in the 3' UTR or in some case in the coding region confer instability linked to the stem-loop secondary RNA structures, although the exact mechanism of the instability features are poorly understood. There have been reports of trans-acting factors that regulate mRNA stability by binding to instability elements e.g. in transferrin receptor mRNA, a protein factor associates with an iron response element (IRE) to prevent its degradation by blocking the access to the instability elements.

Eukaryotic cells hold special RNA supervision mechanism that is conveyed by the destruction of mRNAs that contain premature stop codons. Such mRNA molecules are a result of gene expression of sequences with nonsense mutations i.e. codon shift from amino acid coding to coding for stop codon also referred as nonsense codon usually located 50-55 nucleotides upstream of the final exon-exon junction established during splicing. Nonsense mediated decay of mRNA hence gets its name that almost certainly is an evolved process to deal with aberrant mRNAs resulting from errors in expression of normal genes due to the inherent processes (nonsense mutations, frame-shift mutations, RNA splicing errors and RNA editing errors) linked to the processing of RNA molecules. The molecular association of a complex of proteins at 20-24 nucleotides upstream of the exon-exon junction, called exon junction complex (EJC) labels the mRNA that is targeted for nonsense mediated decay. This type of regulation of degradation of mRNA is a protective mechanism that do not allow the production of truncated or altered protein products having deleterious effects.

Table: RNA editing processes.

Editing Processes	Features	Mechanism
Simple editing	Single residue conversions (Posttranscriptional)	Modification by specific cytidine deaminase Modification by dsRNA deaminase, converting adenosine to inosine.
Insertional editing	Insertion of single nucleotides or small areas of nucleotide bases (Co-transcription)	Transcriptional strand slippag.
Pan-editing	Insertion / deletion of multiple uridine / cytidine residues (Posttranscriptional)	Editing sequences by antisense guide RNA (gRNA), usually in complexation with mRNA nucleoprotein termed as editosome.
Polyadenylation editing	Addition of adenosine residues at the transcript ends to arrange for stop codons	Pre-mRNA lacking stop codon is polyadenylated with the first one or two adenylate residues that provide for stop codon generation.

Transfer RNA

Transfer RNAs are indispensable entities for protein synthesis. tRNA biosynthesis recruits specialised RNA polymerase that effectuate preliminary transcripts which are eventually transformed via a host of post-transcriptional refinements prior to tRNAs turning as mature molecules for their commissioned role in protein synthesis. The predominant disposition of transfer RNAs (tRNAs) to act as the salient molecule decoding the genetic instructions held on messenger RNA (mRNA) with an vital capacity of furnishing amino acids, explicit to the mRNA codons, for the cytoplasmic and organellar protein synthesis machineries. This adopter function of tRNA necessitates it to perpetuate the canonical L-profile structure, which is crucial to its decrypting function. The cytoplasmic tRNAs transcribed in the nucleus undergoes immense alterations and these processes contributes to consign processed or partially processed tRNAs to accurate sub cellular location and tRNA trafficking, RNA (re)folding through the enzymes aka RNA chaperones, degradation of misfolded tRNAs and tRNA quality control. The processing events associated with tRNA synthesis and maturation serve myriad paramount errands encompassing creation of tRNA discriminant pool, translation fidelity and sub-cellular dynamics.

The role of tRNAs have also been extended to include functional processes exemplified by roles in targeting proteins through the N-end rule pathway for degradation, role in signalling during amino acid controls, tuning of cellular responses, regulation in apoptosis, translation regulation and as reverse transcription primers as well as strand transfer during retroviral replication. These roles implicates the importance of rate of tRNA transcription, various post-transcriptional tRNA processing, tRNA biosynthesis, turnover and sub-cellular dynamics to the overall molecular control. The genomic, transcriptomic and proteomic analyses in the recent times have identified the genes involved in tRNA processing while the post-genomic structural analyses have demarcated the structural rationale behind the tRNA maturation procedure. The content throws light on the events associated with tRNA processing and modifications to generate the functional tRNA molecules through genetic, genomic, biochemical and cell biology investigations undertaken.

Post-transcriptional Processing, Turnover and Subcellular Dynamics

Transfer RNA Processing

The production of tRNA, in organisms of prokaryotic, eukaryotic and archaeal origin requires post-transcriptional processing, and the resultant tRNA has distinctive nucleotides and a specified folding pattern, still constituting similar final structure and function. The tRNAs can be produced simultaneously with rRNAs or transcribed as part of larger pre-tRNA transcript that is cleaved to generate multiple tRNA molecules. Eukaryotic tRNA genes occur in cluster on specific chromosomes, where individual cytoplasmic tRNAs gets transcribed in the nucleus by dedicated transcription through DNA dependent RNA polymerase III, thus a single pre-tRNA is produced from each gene. The number of different tRNAs produced depends upon the type of organism, where prokaryotes are mostly limited to less than 61 in number, as per the number of codons found in mRNA. The minimal number of at least 22 different tRNAs, as the third base wobble at each codon reduces the number of distinct tRNAs with one for each amino acid have to be maintained though mostly 30 to 40 different tRNAs. Eukaryotes nevertheless produces larger number of tRNAs, in some as much as 61 different tRNA genes, one corresponding to each codon of the genetic code.

Bacterial and eukaryotic tRNAs requiring processing before they assume their functional role of transporting amino acids to the ribosome wherein a lineage of posttranscriptional modulations and alterations, processing into the standard length, and modified so as their canonical L-shaped structure required for its adopter function for mRNA decoding is maintained. The partially transformed and processed tRNAs are relocated to the precise sub-cellular setting, where tRNA trafficking is an elaborate progression. The tRNAs are mostly stable molecules, though there are resident pathways that cater to the degradation of the partially processed or misfiled tRNAs and the quality checkpoints. The content focusses on the modification procedures and post-transcription events that are required for the biogenesis, turnover, and intracellular dynamics contributing towards functional tRNA processing. The information has been assorted from the validated scientific studies through genetic, biochemical, cell biology, genomic and post-genome analyses of the genes and structural factors involved in tRNA processing and modifications. The role of RNA chaperones as modification enzymes is depicted vividly as its contribution towards RNA (re) folding.

The central dogma signifies the reading of the codon (i.e. three bases as single entity of genetic instruction) from mRNA expressing as amino acid via transfer RNA (tRNA). This is considered as the adopter function of tRNA molecule that converts the nucleotide sequence coded genetic information expressed as the amino acid sequence. Structurally demarcating, the tRNA possess four domains of the D or acceptor arm, the anti-codon arm, and the TΨC-arm (Figure 1). The acceptor stem and the TΨC-arm are conjoined proximately to one another to demarcate as continuous A-helix, while the D-arm and anticodon arm stack to comprise an additional continuous helix. This structure is crossed over by two RNA double helices at an angle of 90° to have a conformation of the L-outlined tertiary attribute, depicted with one end of tRNA possessing an anticodon that specifically bonds to a codon in a base specific manner through hydrogen bonding. Structurally, there are no physical forces that contribute towards the pairing of amino acid with the anticodon, thus making the strict selection of the pairing of animoacyl-tRNA synthetases (corresponding to single amino acids) with their cognate tRNA molecules + amino acids in a highly precise combination. The amino acid is coupled to the 3' CCA terminus located on the farther end of the tRNA L-shaped molecule.

Synthesis of Precursors of tRNA by RNA Pol III – Transcription Initiation

The promoter region of tRNA genes possessing two consensus sequences that fall downstream of the transcription start site (namely Box A and Box B). There have been demarcation of two complex DNA binding factors that are precursors to transcription initiation by RNA polymerase III, termed as TFIIIC and TFIIIB. TFIIIC, a large protein complex is the assembly factors for liasoning TFIIIB at the correct location while the latter is the literal initiation factor for Pol III. TFIIIB is characterised as three subunits entity consisting of TBP, B and BRF (TFIIB related factor with homology to TFIIB-RNA Pol II initiation factor), while B is comparable to sigma factor of prokaryotes that operates in initiating transcription bubble. The sequence specificity of TFIIIB is dependent alone upon the spot of the TFIIIC binding to DNA.

The transcriptional initiation process is progressed through various steps:

- TFIIIC binding: TFIIIC binding to transcriptional start sites Box 'A' and Box 'B' of the tRNA promoter.

- TFIIIB binding: TFIIIB tethering to TFIIIC-DNA complex 50 bp upstream from TFIIIC binding site at box A.
- RNA Pol III binding: TFIIIB furthers the RNA Pol III recruitment that initiates transcription, probably repositioning TFIIIC from DNA template while moving on the strand.

Transcriptional termination of mostly achieved without adjunct factors through dA cluster residues with the overall efficiency depending upon adjacent sequences.

tRNA Maturation

The tRNA is transcribed after the initiation step by RNA polymerase III as a precursor molecule (pre-tRNA) with 5' and 3' extension (i.e. with leader and trailer sequences respectively). The two sequences which are extensions at the 5' and 3' end are cleaved by assorted enzymes so as to snip the precursor to desired standard length. This event is followed by modifications through modification enzymes that includes incorporation in the tRNA core landscaped bordering the D-loop and TΨC-loop, and the anticodon loop of the unprocessed transcript, intron removal from the transcript subset transcribed from intron possessing genes, discrete nucleoside modifications (sugar methylations, base deamination, base isomerizations, and addition of exotic moieties to the bases) resulting in a mature tRNA. This process is followed by aminoacyl-tRNA synthetases that are very specifically recognising the mature tRNA, and then ligating the proportional amino acid onto the 5' - CCA terminus.

The events in tRNA maturation are elaborated as below:

Post transcriptional processing of tRNA: Processing of tRNA and rRNA synthesised as precursors include cleavage by nuclease and other modifications of nascent RNA chains. The tRNA molecules are either snipped out from a single primary RNA transcript or from transcripts that embrace arrays of scattering tRNAs or multiple copies of the same tRNA. The eukaryotic tRNA molecules have no caps and as these are non-coding where the nucleotide base chemical alterations affect solely the structural details which in turn would possibly alter the catalytic activities of the molecule.

The E.coli rRNA operons contain the coding sequences for tRNAs as also operons that contain upto seven tRNA genes separated by spacers. The processing of mature tRNA is done from these precursor transcripts by nucleases that cleave and trim them with high precision. In prokaryotes, this process involved deletion of 5' leader sequence, eradication of 3' terminal non-essential residues and chemical modifications of assorted bases and ribose units. Similarly, in eukaryotes pre-tRNA that is trimmed of the extra nucleotides at both 5' and 3' ends as well as in some cases of the introns. The primary transcript shapes up to take characteristic secondary structure revealing stems and loops allowing endonuclease to recognise and bisecting the 5' leader and two 3' nucleotides. The 5'-CCA-3' at the 3' end of the mature tRNAs are appended through distinct enzymatic reactions not encoded by the genes.

The pre-tRNA processing includes the following events:

a) Cleavage of 5' leader sequence: The erasure of 5' leader sequences from initial tRNA transcripts prevails before effacing 3' extensions. Both prokaryotic and eukaryotic contain extra nucleotides at the 5' end that are severed by endonuclease RNase P. This enzyme is a ribozyme comprising a catalytically active RNA molecule and small basic protein subunits, capable of catalysing chemical reactions on its own also termed as ribonucleoprotein (RNP). The RNA subunit embodies tRNAspecificity and catalytic domains, though structurally RNase P holoenzyme or its complexation with tRNA precursor remains to be elucidated.

b) Removal of the 3' trailer residues: The eradication of 3' extensions from pre-tRNA is an intricate process that necessitates inclusion of both exo- and endonucleases that advocate competitive differential access to tRNAs for end maturation. The secondary structure stems and loops possessing extra nucleotides at the 3' end (present in both prokaryotes and eukaryotes) are cleaved by endonuclease, RNase E or F at the middle of the 3' trailer sequence or base of the stem leaving still some extra nucleotides which are snipped through exonucleolytic reduction of the surplus trailer sequence by RNase D, RNase PH from the 3' end in a one-step manner. If the exonuclease crops the trailer sequence surpassing the amino acid attaching CCA sequence site, a template independent RNA polymerase termed CCA-adding enzyme restores the CCA terminus.

c) Template-independent RNA polymerization by CCA-adding enzyme/Attachment of CCA at 3' and 5' termini: All the tRNAs includes a 3' terminal CCA sequence vital for tRNA aminoacylation and interaction with the ribosomes exhibited vividly in protein synthesis. The CCA sequence is synthesised de novo and / or reconstructed by a template independent RNA polymerase, CCA-adding enzyme, utilising CTP and ATP as energy substrates. CCA- adding enzyme is a unique nucleotidyl-transferase that stretch- out a specific primer (tRNA) by annexing a explicate sequence (CCA) as well as supervise tRNA 3' terminus status and reconstructs if needed. Whereas the prokaryotic tRNAs are encoded with a CCA sequence i.e. possess gene encoding CCA adding enzyme, the eukaryotic nuclear encoded tRNAs 3' CCA sequences are formed definitely through post-transcriptional nucleotidyl transferase catalysis.

The 3' terminus tRNA maturation is catalysed by RNase Z, that cleaved precursor tRNAs to yield a primed tRNA for annexing of the CCA motif by tRNA nucleotidyl transferase enzyme. The unusual CCA -adding enzyme encase the tRNA top half helix organised by the acceptor and TΨC-arm, as if monitoring the RNA helix length so as to adhere to only tRNA substrate. The enzyme bind three ribonucleotide triphosphate precursors at distinct active sites and catalyse the phosphodiester bond formation resulting in the CCA (3') sequence. It assumes a Loutlined catalytic pocket, with the the terminal nucleoside of the tRNA primer projected to clamp the succeeding ribonucleotide triphosphate at the bottom of the pocket. The tRNA primer alongside terminal nucleoside and the succeeding NTP are anchored by corresponding pockets structured through the protein amino acid residues ('protein template'), mimicking the DNA template. The nucleotidyl-transfer reaction is followed by the second terminal nucleoside primer site retraction, putting forth again the terminal nucleoside to stack with the next arriving NTP. The templateindependent RNA polymerisation mechanistic conformational change is due to the dynamic CCA adding enzyme and elastic CCA terminus of tRNA primer as well as the protein template that forms the complementary pockets for the primer and the incoming NTPs.

d) tRNA splicing/Removal of introns: This step is mostly limited to some eukaryotic pre-tRNAs and archaeal that possess introns, subsequently removed by endonucleolytic cleavage of both intron ends accompanied by ligation of the half molecules of tRNA. The tRNA introns are consistently presented one base 3' to the anticodon making the extrication of introns by tRNA splicing an requisite process so as to prompt a complete set of tRNAs meant for decoding. The function of tRNA introns is implicated for nucleoside modifications in the anticodon loop. The pretRNA splicing reaction is mediated by an endonuclease that is a heterotetramer, that results in two tRNA half molecules after removal of introns. The 5' half catering to have a 2',3' cyclic phosphate while the 3' half carries a 5' hydroxyl. The tRNA halves are ligated by a 3' - 5' activity, that connects the phosphate from the 2',3' cyclic bond to the 5' hydroxyl on the 3' half molecule, transiently making the 2' phosphotransferase residue while in plants and yeast, 5'-3' ligation and 2' phosphotransferase reaction steps occurs for ligation.

e) Chemical modifications of bases and ribose units: tRNAs are efficiently modified with a plenitude of known tRNA nucleoside refinements. The addition of CCA at the 3' end is followed by the base and ribose unit modification through the enzymatic changes of the standard ribonucleotide in a tRNA precursor resulting in unusual bases. The modifications comprehends methylation, acetylation, deamination, reduction, rearrangement, attachment of isopentenyl or SH group of bases. These alterations occurring approximately in one in ten nucleotides identified in tRNAs conjure diversity permitting significant structural and functional flexibility, illustrated by uridylate residues tailored after transcription to from ribothymidylate and pseudoridylate residues. Such variations are projected to arbitrate identification of individual tRNAs by the amino acid attaching enzymes to these tRNA molecules, thus increasing the interaction radius between tRNAs and codons during translation, in turn facilitating a single tRNA to recognise more than one codon. Except for two modified nucleotides, quenosine and wyosine where an entire nucleotide is cleaved and replaced by their modified version, most of such modifications are borne directly by existing nucleotides within the transcripts. The base modifications are exclusive to individual tRNA type, though the general processing of pre-tRNAs is similar throughout.

The tRNA modifications achieve manifold functions specifically targeting tRNA discrimination, maintaining translational fidelity through selective codon - anticodon interaction, sustenance of mRNA reading frame, and tRNA stability. The tRNA modifications are assigned the role dissected predominantly into two categories - structural alterations served in the tRNA core region (D- and TΨC- loop) concentrate on stabilising the canonical L- shaped tertiary structure by intensive stacking propensities while alterations spanning the anticodon loop disposes the dual functionality of stringent codon pairing and maintaining translational fidelity through high precision recognition by the cognate aminoacyl-tRNA syntheses. The overall impact of tRNA modifications enthuse the accurate decoding process ensuring appropriate ribosome interaction, tRNA discrimination by the cognate aminoacyltRNA synthetases.

Further tRNA modifications are postulated to play a role in discrimination of tRNAs during translation. The best suited example is the presence of separate tRNAMet for either initiation or the elongation steps of translation, where in the initiator and elongated tRNAs have dissimilar primary sequences and structures, although both are amino-acylated by a single methionyl tRNA synthesise. These are discerned at translation vis specific initiator and elongated translation factors.

The codon-anticodon specificity and reading frame sustenance comes as the other tasks achieved through tRNA modification. A well-documented evidence of such a case is the deamination of adenosine (A) to inosine (I) at the wobble position of the triplet codon. As, A specific base pairing is with U, I can base pair with both U, C and A. tRNAs with I at the wobble position have well demarcated codon-anticodon combinatorial capacity and the omission of I results in decoding errors that furthers its affect in translational reprogramming.

The modifications done at the tRNA also affects its stability. tRNA modification activities possessing single subunit responsible on tRNA selection and modification activation reaction. In the present scenario, these are the best documented events and associations of tRNA processing activities, with differential activities that modifies single or multiple substrates termed as 'dual substrate specificity' (e.g. tRNA vs. rRNA), modification enzymes catalysing single or multiple nucleosides modifications on a given tRNA or 'multisite substrate specificity', or only a subset of the same nucleosides at the similar position in varied tRNAs or 'restricted specificity'. The phenomenon of such varied range of tRNA modification processes remains an unexplored arena as intron containing tRNA changes is completely non-aligned to non-intron containing tRNAs.

Transfer RNA Biogenesis

Transfer RNA biogenesis is concluded to occur at multiple sub-cellular sites where mostly the responsible enzymes for the tRNA modifications are substantially reside in nucleus (modifying intron requiring sites in nuclear encoded pre-tRNAs) or in mitochondria as well as in cytoplasm and these enzymes dictate an ordered pathway for tRNA modifications. The investigations of tRNA alterations and their definitive role in tRNA function and stability gives the due worth of their importance in cellular metabolic processes. The continued studies on the processing of tRNAs throw novel information into the protein-RNA specificity and the structured pathways of metabolically linked molecular coordination.

Transcription of pre-tRNAs is succeeded by an intricate and coordinated tRNA maturation process. End-processing (removal of 5'-leaders and 3'-trailers), successive nucleotide modifications, tRNA splicing and aminoacylation are all quality control steps in the production of mature and functional tRNAs. Numerous enzymes are involved in these processes including RNA exo- and endonucleases, RNA ligases and RNA modification enzymes.

tRNAs are Highly Modified for Many Reasons

Up to 20% of all nucleotides in tRNAs can carry modifications. Importantly, chemical tRNA modifications and isomerizations are found in all characterized tRNA species and are conserved within each domain of life indicating their importance for both tRNA maturation and function. Specifically, RNA modifications prevent degradation of pre- and even mature tRNAs , they modulate the efficiency and specificity of amino acid charging and alter the specificity and decoding frame during protein translation as well as affecting translational speed . Generally, ribonucleotide modifications in tRNAs conform to three rules:

- Modifications at different positions specifically affect tRNA identity and ensure proper acylation.
- Modifications in the main body of tRNAs affect folding and stability.
- Modifications in or around the anticodon loop are responsible for correct codon usage and affect translation.

RNA Modification Systems and the 'Epitranscriptome': Reversible or Not?

Specific RNA modifications are catalysed by a set of enzymes ('writers'), which can be interpreted by specific proteins ('readers'). The discovery of enzymatic activities that can remove particular RNA modifications (i.e., 'erasers' for adenosine methylations) has led to the coining of the term 'epitranscriptome' implying the existence of a dynamic network of RNA modification writers, readers and erasers, in close analogy to epigenetic modification systems. However, the defining characteristics of reversibility has only been reported for RNAs (including tRNAs) containing m^1A, m^6A and m^6Am. Of note, reversible modifications display a common chemistry, namely a nitrogen-carbon bond between modifier group and base that can be broken by, for instance, oxidative demethylation. Demethylation reactions are catalysed by Fe^{2+}- and α-ketoglutarate-dependent dioxygenases, such as fat-mass and obesity (FTO)-associated protein and ALKBH5 enzymes. In contrast, modifications of different chemical composition such as carbon-carbon formation in m^5C or pseudo-uridinylation, while being dynamically installed, are thought not to be readily reversible. However, the recently reported in vivo C–C bond cleavage of 5′-formylcytosine resulting in cytosine in DNA suggested the possibility that further modification of the chemical group could allow reversibility under particular conditions and in specific cell types, potentially even in RNA . Consequently, these issues raise the question as to how semantically distinguish RNA modification systems from one another.

RNA Modifications and their Impact on tRNA Stability

The existence of sequence-specific DNA modification/restriction systems regulating endonucleolytic cleavage of genomic DNA are well known from bacterial systems. Because tRNAs are the most highly modified RNAs and since tsRNAs are produced through the activity of various endonucleases attacking single-stranded RNA (ssRNA) in open loop structures, the question arises as to what extent chemical modifications could modulate endonucleolytic cleavage of mature tRNAs thereby influencing tsRNA biogenesis.

Specific tRNA Modifications Attract and Repel Endonuclease Activity

Various bacterial and fungal endonucleases (ribotoxins) depend on specific nucleotide modifications in the anticodon loop of their target tRNAs. For example, a hyper-modification of guanine (queuosine, Q) determined the activity of Colicin E5 RNase at Q34 of four tRNAs in Escherichia coli. Furthermore, a modified adenosine at position 37 (t^6A37) or 5-methylaminomethyl-2-thiou ridine at position 34 (mnm^5s^2U34) in tRNA-Lys enhanced the cleavage activity of bacterial PrrC, while pseudo-uridine at position 39 (Ψ39) was inhibitory. The modified side chain at uridine 34 (mcm^5s^2U34) was also important for targeting fungal zymocin to tRNAs . In contrast, PaT—a ribotoxin in Pichia acacia—specifically cleaved 3' of mcm^5s^2U34 in the anticodon loop of $tRNA\text{-}Gln^{UUG}$ but unlike zymocin, this cleavage occurred independently of mcm^5s^2U34.

Since many tRNA modifications support various 'proofreading' steps during tRNA maturation into a functional cloverleaf structure, loss of particular tRNA modifications might impact efficient folding thereby allowing stress-induced endonucleases to better access ssRNA regions. To what extent single tRNA modifications can impact tRNA folding is exemplified by the findings that the positioning of m^1A at A9 shifts the equilibrium between two major structures of tRNA-Lys. Supporting this notion, cytosine-5 methylation (m^5C), catalysed by Dnmt2/Trdmt1 at position 38, modulated stress-induced tRNA cleavage of tRNA-Asp^{GUC}, tRNA-Gly^{GCC} and tRNA-Val^{AAC} in Drosophila melanogaster and suppressed ANG-mediated cleavage of these tRNAs when extracted from mouse cells. Interestingly, m^{5}C at C38 in tRNA-Asp^{GUC} is modulated by the presence of Q, a guanosine modification at G34, which has been detected in four tRNAs and Q affected ANG-dependent tRNA cleavage in vitro. Furthermore, the activity of another tRNA modification enzyme, NSun2, which targets m^{5}C to most tRNAs, was also required for tRNA stability. These observations suggest that specific RNA modifications likely change the structural context of tRNAs thereby modulating access for endonuclease activities or serving as platforms for specific enzymes.

tRNA Modifications Protect from Exonuclease Activities

The fact that tsRNAs are detectable as specific small RNAs raises the question as to how they evade exonucleolytic degradation after tRNA cleavage. The stability of small RNA species relies on various factors including nucleotide modifications, hybridization with complementary RNAs as well as association with effector proteins. Importantly, tsRNAs in human serum appeared to be devoid of proteins and recent findings introduced the possibility of cross-hybridization between tsRNAs from different parental isoacceptors suggesting that such dsRNA structures confer resistance to nuclease attack. Furthermore, 2′-O-methylation at 3′ ends, mediated by RNA methyltransferase Hen1 homologues, is a conserved mechanism protecting various small RNA species against 3′-5′ exonucleolytic degradation. In contrast, 3′ uridinylation targets small RNAs towards degradation. Interestingly, tRNAs bear 2′-O-methylated nucleotides (Cm, Gm, Um), not at 3′ ends but located in D-, variable and anticodon loops, which is in close proximity to mapped endonucleolytic cleavage sites. These modified nucleotides could limit the extent of exonuclease activity and thus increase the stability of produced tsRNAs. Indeed, tsRNAs^{NonS} originating from tRNA-Gln^{GUG} were ribose 2′-O-methylated at 3′ ends. In this context it is important to mention that particular nucleotide modifications not only improved the stability of synthetic small RNAs but also reduced the antigenicity of tRNA sequences suggesting that a combination of various modifications in tsRNAs might contribute to their stability and functional impact.

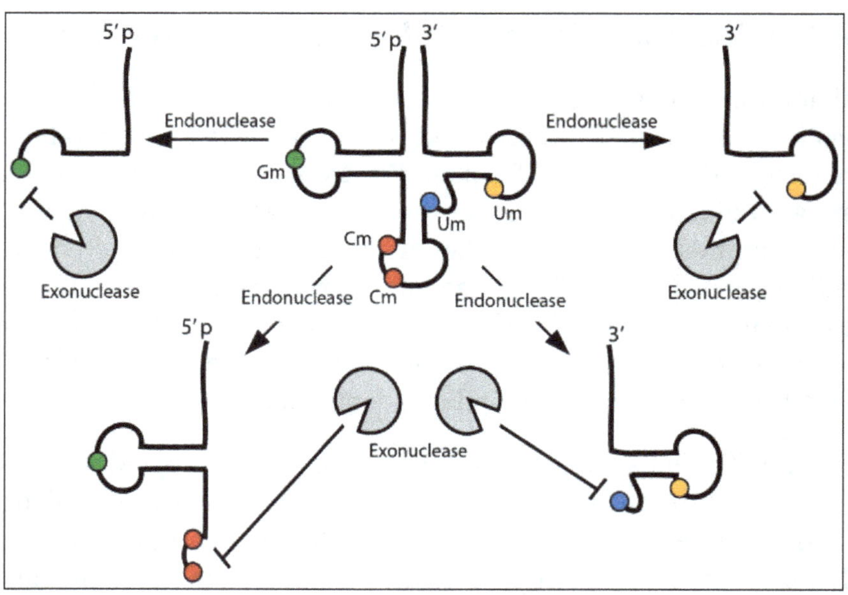

Specific tRNA modifications might influence tsRNA stability. Various tRNAs contain particular

RNA modifications that inhibit the access of 3′-5′ exonucleases. In particular, the positioning of 2′-O-methylated nucleotides (Cm, Gm, Um) in open loop structures suggests that such modifications might stabilize produced tsRNAs against degradation.

tsRNAs: Degradation Leftovers or Functional Small RNAs?

Enzymes that can cleave tRNAs and even tRNA fragments have been described in the 1970s . Around the same time, tRNA-derived degradation products in the form of modified nucleosides were detected in urine from tumour patients and interpreted to be a result of high tRNA turnover in cancer cells . However, only recently it became possible to detect and catalogue tsRNAs as reproducible entities in many cell types and tissues exposed to varying environmental conditions. Here, we will briefly describe the various identities of these tsRNAs, introduce nuclease activities responsible for their production and mention the conditions and biological settings leading to tsRNA production.

tsRNA-Nomenclature

Since the beginning of this century, tsRNAs have been recognized as small RNAs with the potential to be biologically relevant. However, categorization and naming of tsRNAs, both in literature and public databases, is still largely inconsistent.

Starvation-induced tsRNAs^{S} in eukaryotic cells were first reported in Tetrahymena and simply named tRNA fragments. However, researchers chose to label the products of stress-induced tRNA fragmentation in different organisms with different names such as stress-induced tRNA-derived RNAs (sitRNAs,) or tRNA-derived, stress-induced RNAs (tiRNAs,). In addition, tsRNAs existing under steady-state conditions were either called tRNA-derived fragments (tRFs) or tRNA-derived small RNAs (tsRNAs,). Others have called these RNAs, tRNA-derived RNAs (tDRs,) or even tRNA-derived small non-coding RNA (tsncRNAs,).

Although various researchers attempted introducing some order into the nomenclature, the net result was only a distinction between tRNA halves (28–36 nucleotides) that are largely stress-induced (often also called tiRNAs) and tRFs (14–30 nucleotides) that are produced under steady state conditions. We will call the former tsRNAs^{S} and the latter tsRNAs^{NonS} throughout this review. Unfortunately, the term 'tiRNA' remains ambiguous since it has also been used to name 'transcription initiation RNAs' . Therefore, attempts to systematically name tsRNAs and assign recognizable nomenclature for reproducible future use are very welcome.

Processing and Maturation: Nucleases Acting on tRNAs

tRNAs can be extremely stable with half-lives measured in days and this stability is connected to numerous nucleotide modifications as well as to correct tertiary structure. Aberrantly processed or hypo-modified tRNAs become destabilized and are rapidly degraded by various exonucleolytic activities that are part of two molecular pathways: the TRAMP/nuclear exosome and the 'rapid tRNA decay'. The existence of such RNA degradation pathways indicated the need for cells to sense and remove hypo-modified tRNAs . Importantly, dysfunctional RNA surveillance can cause imbalances in small RNA pathways, as reported in a study in Schizosaccharomyces pombe that demonstrated that TRAMP-mediated RNA degradation prevents entry of tRNAs and other RNAs into the siRNA pathway. Whether by-products of these constitutively active tRNA degradation pathways give rise to defined tsRNAs^{NonS} remains to be tested.

On the other hand, correct tRNA maturation is a multistep process involving the removal of pre-tRNA sequences by RNaseP and RNaseZ-like enzymes (ca. 50 nucleotides at the 5′ and 3′ end, respectively). In addition, various tRNAs contain introns that are 'spliced' during their maturation process. tRNA splicing endonuclease activities create nicked tRNA precursors, which are substrates for the tRNA ligase complex creating mature tRNAs. Mutations in the latter complex resulted in the accumulation of tsRNAs^{NonS} representing non-ligated tRNA splice intermediates that were diagnostic for particular human disease syndromes.

In contrast to general tRNA processing, degradation and splicing mechanisms, various endonucleases target correctly formed and functional tRNAs, preferentially at open loop structures (D-, anticodon-, variable- and T-loops) resulting in 'nicked' tRNAs containing a 2′-3′-cyclic phosphate at the 3′-end and a 5′-OH at the 5′ end of the hydrolysed RNAs. Particular killer ribotoxins have first been described in phage-infected bacteria and later also in simple eukaryotes. ACNases cleave various tRNAs in the anticodon loop, causing cytotoxicity and cell death. For instance, VapCs are site-specific endonucleases that cleave tRNA-fMet in the anticodon stem-loop in Shigella flexneri and Salmonella enterica thereby reprogramming translation. Furthermore, E. coli endonuclease PrrC cleaves tRNA-Lys^{UUU} in a suicidal attempt to defend the population against infection by bacteriophage T4 . Furthermore, the actively secreted bacterial plasmid-encoded ACNases colicin D and E5 cleave tRNA-Tyr^{QUA}, tRNA-His^{QUG}, tRNA-Asn^{QUU}, tRNA-Asp^{QUC} and tRNA-Arg$^{ICG/CCG/U*CU/CCU}$ (Q indicates queuosine, U* indicates the modification m^{5}U) in order to defend the strain against competing E. coli. Both nucleases, when expressed in eukaryotic cells, were toxic, which suggested also susceptibility of eukaryotic cells to tRNA fragmentation. Indeed, fungi such as Kluyveromyces lactis and Pichia acaciae produce ACNases (zymocin and PaT, respectively), which cleave specific tRNAs in the anticodon loop as a defence against non-self-species.

Of note, vertebrates also express cytotoxic tRNA endonucleases, namely Angiogenin (ANG), onconase (Ranpirnase) and amphinase (Amph). Recently, RNaseL, an interferon-inducible endonuclease, has been implicated in targeting tRNAs thereby producing tsRNAs . While ANG is clearly an ACNase, Ranpirnase cleaves tRNAs predominantly in the variable loop regardless of the sequence context and Amph and RNaseL appear to have an even wider RNA substrate specificity. ANG is an RNase A-family enzyme that is internalized via receptor-mediated endocytosis by binding to Plexin-B2, while Ranpirnase enters cells in a manner similar to cell-penetrating peptides using an abundance of lysine residues.

The biological role for such secreted cytotoxic endonucleases in multicellular organisms beyond defence against invasive agents is not completely clear. Of note, ANG is highly produced by liver cells and secreted into the bloodstream. ANG promotes blood vessel growth and neuronal cell survival. Particular isoforms of ANG display anti-microbial properties suggesting activities on microbial tRNAs. Interestingly, both Ranpirnase and Amph display antiviral and anti-tumour activities, which suggested additional substrates beyond tRNAs for these enzymes, while ANG has the opposite effect through its angiogenic activities thereby promoting tumour growth.

Environmental Stress: tRNA Fragmentation as Conserved Cellular Response

The exploration of fungal ribotoxins has been a major driver for elucidating the mechanisms, structures and evolution of eukaryal tRNA restriction enzymes, especially in light of the recent discoveries that tRNA cleavage is a general response to stress. Two eukaryotic ACNase protein families (RNase A and T2) specifically cleave mature tRNAs in response to stress. Of note, Rny1 (RNaseT2) is an ACNase in S. cerevisiae while other eukaryotic RNaseT2 family members appear to target tRNAs in every loop structure . In contrast, the RNaseA-family member Angiogenin attacks many tRNAs during the stress response in mammalian cells. Both ACNase family members are normally sequestered (and inhibited by binding to proteins) from cytoplasmic tRNAs and only become activated upon stress exposure. The production of distinct tsRNAs^{S} has been detected after starvation, oxidative stress, nutritional deficiency, hypoxia and hypothermia, heat shock or ultraviolet irradiation. In addition, tsRNAs^{S} appeared to increase during aging. A study in Caenorhabditis elegans reported elevated levels of tsRNAs in an age-dependent manner indicating increased stress responses in older animals. However, especially in whole organisms it is often impossible to clearly distinguish whether tsRNAs^{S} are produced in response to a particular stress or as part of a developmental process exerting stress on certain cell types.

Developmental tRNA Fragmentation

tsRNAs can also be generated in the absence of overt stress exposure. In Streptomyces coelicolor, changes in tRNA fragmentation were observed during vegetative growth, aerial hyphae formation and sporulation. In more detail, increased levels of tsRNAs^{NonS} were detectable at the initiation of aerial hyphae formation followed by a reduction after bacterial development had proceeded through sporulation. tsRNAs^{NonS} were also observed upon conidiation of Aspergillus fumigatus resulting in depletion of full length tRNAs in the conidia, which suggested that tRNA fragmentation might serve to stall protein synthesis of filamentous fungi during their resting state . These observations indicated a role for active tRNA fragmentation during bacterial development and particular life cycle stages.

Stress-independent tsRNAs have also been observed in mammals. For instance, high levels of tsRNAs^{NonS} were detected in both hematopoietic and lymphoid systems as well as in cell-and vesicle-free blood fractions suggesting that tsRNAs can be secreted and exist as stable entities outside of membranous organelles. In contrast, others reported tsRNA^{NonS} content in extracellular vesicles and comparison of cellular and vesicle-borne tsRNAs in the immune system showed that immune cells. contained exclusively 30–35 nt-long tsRNAs while extracellular vesicles carried 40–50 nt-long tsRNAs. These combined observations suggested that specific tsRNAs could serve as signalling molecules in blood and lymphatic circulatory systems. Interestingly, specific tsRNAs^{S} were also detected in human breast milk and in mature mouse sperm indicating tsRNA^{S} production for directed transmission into the next generation.

Biological Functions of tsRNAs

Various studies using hybridization techniques indicated that only 0.1–5% of a given tRNA isoacceptor yield tsRNAs, especially under stress conditions. This suggested that any amount of tsRNAs is likely very low if averaged over the RNA content of single cells or tissues. However, when taking into consideration the possibility that tsRNA production could occur rather locally within cells and thereby would remain confined to, for instance, particular subcellular ribonucleoprotein particles or membrane-less organelles (i.e., ribosomes, stress granules), the amount of locally produced tsRNAs might become extraordinarily high and thereby locally effective. We, therefore, suggest that any potential biological function of tsRNAs should be considered under the assumption that tsRNAs are produced locally in high relative concentrations.

Only very few experiments have been performed which unequivocally (e.g., by direct manipulation of tsRNAs) tested the involvement of tsRNAs in the biological processes that were associated with their occurrence. This experimental conundrum is further complicated by the fact that it is often impossible to separate the function of specific tsRNAs from those of their parental tRNAs (intact or 'nicked') or from phenotypes caused by mutations in tRNA processing and modification enzymes. In this respect, the use of antisense oligomer-mediated knock-down with the aim of targeting tsRNAs but not tRNAs remains questionable even though interference with full length tRNAs were reported to be absent. When considering that only fewer than 5% of all molecules of a given tRNA isoacceptor will be targeted to produce tsRNAs (i.e., during stress conditions), it remains unclear how antisense oligonucleotides seek out those few tsRNAs instead of the majority of parental tRNA molecules. In addition, absolute quantification of tRNAs and tsRNAs using next-generation-sequencing (NGS)-based techniques should be critically assessed because the modification landscape of any tRNA-derived molecule will greatly affect all reverse transcriptase (RT)-based assays thereby shifting readouts in unpredictable ways.

To facilitate orientation, we introduce here an experimental 'categorization', which might allow better judging the existing experimental evidence for tsRNA involvement in a particular process.

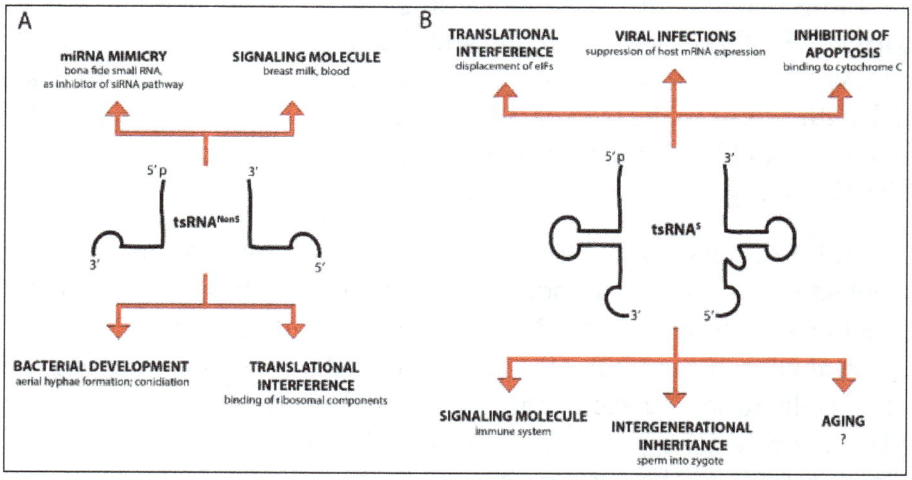

The activity of tsRNAs has been implicated in various biological processes. A selection of reported and suggested functions for tsRNAs^{NonS} (A) and tsRNAs^{S} (B) in different model organisms.

Table: Experimental evidence for suggested/reported function of various tsRNAs. Table depicts experimental approaches to identify the molecular function of tsRNAs, models and model organisms, species of tsRNA and suggested/reported molecular function of individual tsRNAs.

Experimental Category [a]	Model	Methodological Details	tsRNA	Molecular Function [b]
4	Mus musculus, Mammalian cell culture	Injection and transfection of synthetic tsRNAs	S and NonS	C, D
1	Mammalian cell culture	NGS	NonS	B

1, 2	Mammalian cell culture (cancer)	RNA-immunoprecipitation, NGS, Northern blotting	NonS	B
4	Mammalian cell culture (cancer)	tsRNA transfection	S	A
1, 3	Mammalian cell culture (cancer)	Northern Blotting, NGS, qRT-PCR, RNAi knockdown of tsRNAs	NonS	B
3	Cucurbita maxima	RT-PCR, Northern blotting	S	A, B
4	Mammalian cell culture (cancer)	tsRNA transfection	S	A
2, 3	Mammalian cell culture (cancer)	RNA-immunoprecipitation; Northern blotting, mRNA reporters	NonS	B
2, 3	Mammalian cell culture (cancer)	mRNA reporters; RNA affinity chromatography	S	A
2, 3	Haloferax volcanii	In vitro binding studies; mRNA reporters	S	A, D
1, 2	Tetrahymena thermophila	RNA-immunoprecipitation, NGS	NonS	D
1, 2, 3	Mammalian cell culture (cancer)	mRNA reporters; Northern blotting	NonS	A
1, 2, 3	Mammalian cell culture (cancer)	Northern blotting, RNA-immunoprecipitation, NGS, mRNA reporters	NonS	B
1, 4	Drosophila melanogaster	Northern blotting, RNA-immunoprecipitation, NGS	S	D
1, 2	Mammalian cell culture	RNA-Immunoprecipitation, NGS	S	C, D
4	Mammalian cell culture, zebrafish, human patient material	NGS, Northern blotting, tsRNA transfection	NonS	A
1	Mammalian cell culture	NGS, RT-PCR	NonS	B
1, 2	D. melanogaster	RNA-Immunoprecipitation, NGS	NonS	B
1, 2, 3, 4	Mammalian cell culture (cancer)	Co-Immunoprecipitation, tsRNA transfection	S	D
1, 2, 4	Mammalian cell culture (cancer)	NGS, RNA affinity chromatography, CLiP-Seq, transfection with mimetic or antisense oligos	S	D
1, 2, 4	Mammalian cell culture (cancer), human patient material	NGS, Northern blotting, RNAi knockdown of tsRNAs	S	C
1, 3	Mammalian cell culture (cancer)	RNA-Immunoprecipitation, NGS, EMSA	NonS	B
1, 4	M. musculus	NGS, Northern blotting, small RNA microinjection, RNAi knockdown of tsRNAs	S	C
1, 4	M. musculus	small RNA microinjection	S	C
1, 4	Arabidopsis thaliana	synthetic tsRNA transfection	NonS	B
1, 4	Mammalian cell culture (cancer), human patient material	Reporter assays	NonS	C
2, 3	H. volcanii, Saccharomyces cerevisiae	NGS, RNAi knockdown of tsRNAs	S	A, D
4	Mammalian cell culture (cancer), M. musculus, human patient material	in vitro binding studies; toeprinting analysis; cross-linking studies	NonS	A, B

2, 3	Mammalian cell culture (cancer)	In vivo RNA cross-linking, Northern blotting, reporter assays, RNAi knockdown of tsRNAs	S	A
1, 2, 3, 4	Mammalian cell culture, M. musculus, human patient material	tsRNA transfection, RNA affinity chromatography	NonS	A, D
4	Mammalian cell culture, human patient material	iCLIP, NGS	NonS	C
1, 4	M. musculus	Northern blotting, tsRNA transfection, RNAi knockdown of tsRNAs	S	C

[a] 1: Measuring of tsRNA abundance (Hybridisation, PCR amplification, NGS); 2: Enrichment in potential effector complexes (i.e., RNA-immunoprecipitation); 3: Indirect activity test (i.e., synthetic reporter systems); 4: Direct activity test (i.e., tsRNA by transfection or microinjection, interference with antisense oligos). [b] A: Interference with protein translation; B: Small RNA function; C: Signalling function; D: Protein Binder/Aptamer.

tsRNAs Interfere with Protein Translation

Particular 5′-tsRNAs^{S} directly interfered with translation. As one mode of molecular action, specific tsRNAs^{S} displaced various eukaryotic initiation factors from both capped and uncapped mRNA causing translational repression. Two particular isoacceptor-specific tsRNAs^{S} were most potent in causing translational repression, namely tsRNA-Ala^{UGC} and tsRNA-Cys^{GCA} . These tsRNAs^{S} were comprised of 5′ halves sharing two structural features: the D-loop, supposedly binding translational repressor proteins and a terminal oligo-guanine (TOG) motif. TOG motif interactions facilitated RNA G-quadruplex formation (RG4s) involving four individual 5′-tsRNAs^{S}. Interestingly, RG4s were required for the initiation of SGs triggered by tsRNAs^{S} in vivo.

In another mode of action, tsRNAs^{S} indirectly interfered with protein synthesis, in particular through binding to ribosomal components. For instance, specific 5′-tsRNAs^{S} (tsRNA-Val^{GAC}) resulting from tRNA cleavage in or around the D-loop in the archaeon Haloferax volcanii interacted with the small ribosomal subunit. Interestingly, these tsRNAs competed with mRNA for ribosomal binding in response to alkaline stress conditions. Importantly, a similar interaction between tsRNAs and ribosomes without the need for complementarity to mRNAs was reported for eukaryotes and prokaryotes, indicating a conservation of this mechanism during stress conditions.

However, a recent report suggested that tsRNAs do exactly the opposite of interfering with protein translation. In this case, a 3′ tsRNA-Leu^{CAG} enhanced the translation of two ribosomal proteins, which are important for ribosomal biogenesis. These combined observations suggest fragment-specific effects on or condition-dependent usage of tsRNAs during protein translation.

Can tsRNAs Act Like or Mimick Canonical Small RNAs?

Of note, tsRNAs were often wrongly annotated as micro RNAs (miRNAs) in databases. Importantly, particular tsRNAs^{NonS} were detectable almost exclusively in the cytoplasm and showed Dicer-dependent biogenesis. However, various assays using sensor transcripts with complementary regions to tsRNAs^{NonS} did not reveal canonical miRNA or siRNA-like activity indicating that it was unlikely that tsRNAs played direct regulatory roles in post-transcriptional gene silencing

(PTGS). Instead, alternative effects on PTGS were suggested involving competitive association of tsRNAs^{NonS} with Argonaute (AGO) clade proteins. Indeed, various data sets detected tsRNAs in AGO complexes but it remained unclear whether this association could be biologically significant. Only recently, reports emerged on specific tsRNAs acting in a truly microRNA-like fashion. For instance, particular 5′-tsRNAs^{NonS} originating from Dicer-like 1 (DCL1) processing became incorporated into Argonaute 1 (AGO1) and accumulated in pollen of Arabidopsis thaliana. These tsRNAs^{NonS} specifically targeted mRNAs produced from transcriptionally active transposable elements. tsRNAs^{NonS} production and loading into AGO1 and AGO2 in Drosophila melanogaster was age-dependent and bioinformatic prediction of potential target genes suggested regulatory roles in neural processes . In addition, tsRNA mimicry of miRNAs was also reported in human cells. Here, pre-tRNA-Ile^{UAU} was recognized as miRNA precursor by exportin-5 (Xpo5) and transported to the cytoplasm, processed by Dicer and loaded into AGO effector proteins. Furthermore, an abundant and Dicer-dependent tsRNA^{NonS} (CU1276) was also described in germinal centre-derived B cells but not in germinal centre-derived lymphomas. CU1276 was derived from tRNA-Gly^{GCC} and induced repression of the RPA1 gene in a miRNA-like fashion resulting in changes in DNA damage responses.

Lastly, tsRNAs also feature in the tug-of-war between viruses and host systems. For instance, tsRNA^{NonS} (tRF-3019) of host cells bound to the primer-binding sites (PBS) of retroviral RNA from the human T-cell leukaemia virus type 1 (HTLV-1) thereby initiating reverse transcription and promoting virus amplification. In addition, respiratory syncytial virus (RSV) utilized host tsRNA-Glu^{CUC} to suppress mRNA expression of the antiviral apolipoprotein E receptor 2 (APOE-R2) by targeting its 3′ untranslated region in a miRNA-like fashion resulting in the amplification of RSV. These combined observations suggest that the uncontrolled production of tsRNAs could be 'double-sided sword'. tsRNAs could either act as small RNA mimics thereby 'distracting' small RNA pathways or perform functions as actual small RNAs in PTGS.

Could tsRNAs Serve as Sink or Decoy for Specific Proteins?

For processing and maturation, tRNAs must interact with a large number of different proteins. In addition, when tRNAs contribute to decoding of mRNAs not only do various aaRSs associate with different tRNAs before proofreading tRNA identity but specific translation initiation factors and many ribosomal proteins do also interact with tRNAs. Since tsRNAs originate from pre-tRNAs and mature tRNAs, they could bind to tRNA interacting proteins thereby competing with their parental molecules. It follows that tsRNA binding to particular proteins might result in a sink for these molecules resulting in the modulation of specific molecular pathways and thereby cellular physiology. Indeed, evidence for such binding modes exist. For instance, both tRNAs and tsRNAs^{S} can bind to cytochrome C both in vivo and in vitro. The balance of these interactions appears to become relevant during hyperosmotic stress conditions when specific tsRNAs^{S} bound to cytochrome C and interfered with apoptosome formation thereby attenuating cell death signalling.

Furthermore, it was demonstrated that tsRNAs^{S} originating from the 5′ but not 3′-end of mature tRNAs repressed protein synthesis. Mechanistically, particular tsRNAs with a propensity to form RG4s inhibited translation initiation through binding to specific components of the cap complex thereby interfering with eIF4A/G assembly at mRNAs with potential implications for the cellular stress response. Similarly, specific tsRNAs^{S} formed during hypoxic stress conditions suppressed the stability of multiple mRNAs thereby interfering with metastasis and cancer proliferation. Specifically, tsRNAs^S bound to YBX1, an RNA-binding protein, thereby displacing its stabilizing influence from the 3′ UTRs of oncogenic mRNAs. Furthermore, excessive production of small RNAs, for instance by ectopic siRNA production in some experimental RNAi settings or through dysfunctional TRAMP activity, can be lethal partially by disrupting other small RNA pathways resulting, for instance, in increased tsRNA loading into AGO1 complexes. Similarly, if lupus autoantigen (La), which ensures correct processing of pre-tRNA into mature tRNAs, was knocked down, cleavage of pre-tRNAs into tsRNAs increased significantly, resulting in tsRNAs entering the miRNA pathway. Of note, loading of an essential Tetrahymena Piwi protein (Twi12) with tsRNAs^{NonS} was required for nuclear import of Twi12 and the activation of RNA processing by the exoribonuclease Xrn2 indicating aptamer function of tsRNAs. Interestingly, also a human Piwi-like protein (Hiwi2), when overexpressed in human breast adenocarcinoma cells, preferentially interacted with tsRNAs, which the researchers named tRNA-derived piRNAs, further supporting the notion that small RNA pathway components can associate with tsRNAs under specific conditions. Of note, mutations in Dnmt2, a tRNA modification enzyme, caused increased tRNA fragmentation in flies resulting in pleiotropic phenotypes including the deregulation of small RNA pathways. These combined observations suggest the existence of intricate systems controlling tsRNA production, localization and removal thereby avoiding such interactions constitutively but at the same time allowing tsRNA binding to various proteins when required.

Could tsRNAs Act as (Stress) Signals within Organisms?

Analysis of tsRNAs^S in plants indicated functional roles as long-distance signals. Specifically, tsRNAs^S produced in the leaves of the pumpkin Cucurbita maxima entered the phloem, thereby accessing circulatory networks of the plant and suggesting the possibility of function at a distance. In support of this notion, high levels of tsRNAs^S have been detected in A. thaliana roots compared to shoots after controlled phosphate starvation, whereas oxidative stress resulted in high levels of tsRNAs^S in flowers. These observations led to the testable assumption that tsRNAs^S could serve as signalling molecules acting at a distance potentially ‘informing’ other cells and tissues thereby allowing the rest of the plant to adapt its metabolism in response to, for instance, nutrient availability, pathogen infection or other environmental stimuli.

As tsRNAs were also detected in circulatory systems of mammals both in cell-free as well as in vesicle fractions, it was suggested that tsRNAs could serve as signalling molecules [78,121,122,123,167]. For instance, during acute inflammation, increased levels of tsRNAs^S could be detected in the circulatory system. Furthermore, particular tsRNAs^S derived from tRNA-Ala^{UGC} and containing a CCACCA sequence at the 3′ end were sufficient to activate the immune responses of Th1 and toxic T lymphocytes through Toll-like receptors. These combined observations suggest the possibility of directed tsRNA secretion upon stimulation to act in a systemic context as information carriers and signalling molecules.

Can tsRNAs Carry Information between Organisms?

Specific tsRNAs were also found in bodily fluids such as saliva, tears, urine and breast milk. Interestingly, mature mouse sperm contained high levels of tsRNAs. Various intergenerational experiments suggested a role for sperm-borne tsRNAs in relaying extra-chromosomal information in response to diet manipulation. Specifically, in offspring of males that were fed on high-fat or low-protein diet the development of metabolic changes including insulin resistance and glucose intolerance were observed. As to how the few inherited tsRNAs affect the complexity of gene expression patterns remains to be elucidated. Furthermore, metabolic labelling revealed that tsRNAs entered maturing sperm through micro-vesicles (called epididymosomes), which are secreted from somatic cells lining the epididymis. Interestingly, particular tsRNA modifications (i.e., m^5C, m^2G, m^1A) appeared to change upon exposure to high-fat diet indicating that they play regulatory roles during the transfer of small RNA-based extrachromosomal information. These combined observations clearly indicated that modified tsRNAs can act as information carriers between two generations in mammals.

tRNA Modification Systems Impact Human Disease Development

Transfer RNAs are central to any cellular process because of their involvement in protein synthesis. It follows that interference with protein synthesis will likely cause malfunctioning of cells, tissues, organs and organisms. Indeed, tRNA mutations, overexpression of specific tRNAs, incomplete tRNA processing or missing tRNA modifications often result in biological consequences impinging on aberrant protein translation. Of note, genetic mutations in various tRNA modification systems are frequently associated with human disorders, disproportionately affecting proliferation and metabolically active tissues such as muscle and nervous system supporting the notion that chemical modifications have a major impact on small RNA function.

Table: tRNA modifications associated with human disease.

Modification	Yeast Genes S. cerevisiae (Sc) S. pombe (Sp)	Mutant Phenotype in Yeast	Human Genes	Mutant Phenotype/Disease in Humans	Direct Link to tsRNA Function?
Nuclear-encoded tRNAs					
2'-O-methyl	TRM7 (Sc)	Growth defect	FTSJ1	Intellectual disability (i.e., non-syndromic X-linked mental retardation)	No
m^2_2G	TRM1 (Sc)	Non-essential, temperature sensitivity	hTRM1	Recessive cognitive disorders	No
m^5C	TRM4 (Sc)	No effect on growth; higher sensitivity to MMS and H_2O_2	NSUN2	Autosomal-recessive intellectual disability; Dubowitz-like syndrome ; Noonan-like syndrome	Yes
m^7G	TRM82 (Sc)	Growth defects	WDR4	Indirectly linked to Down syndrome	No
A-to-I editing	TAD3 (Sc)	Lethal	ADAT3	Intellectual disabilities, strabismus ; microcephaly and hyperactivity	No
mcm^5U/mcm^5s^2U	ELP1 (Sc)	Delayed adaptation to changes in environment; 'slow-start' phenotype of spores; sensitivity to salt, temperature and 6-aza-uracil	IKBKAP	Familial dysautonomia	No

mcm^{5}U/mcm-^{5}s^{2}U	ELP3 (Sc)	Delayed adaptation to changes in environment; 'slow-start' phenotype of spores; sensitivity to salt, temperature and 6-aza-uracil	ELP3	Amyotrophic Lateral Sclerosis (ALS)	No
mcm^{5}U/mcm-^{5}s^{2}U	ELP4 (Sc)	Delayed adaptation to changes in environment; 'slow-start' phenotype of spores; sensitivity to salt, temperature and 6-aza-uracil	ELP4	Rolandic epilepsy	No
Wybuto-sine	TRM12 (Sc)	Non-essential	TRMT12	Breast cancer	No
m^{5}U	TRM2 (Sc)	Non-essential	TRMT2A	Breast cancer	No
m^{1}G	TRM10 (Sc)	Non-essential	HRG-9MTD2/ TRM10A	Colorectal cancer ; Diabetes type 2, intellectual disability, micro-cephaly	Cancer: NoT2 Diabetes: Yes
m^{1}G/m^{1}A	TRM10 (Sc)	Non-essential	TRMT10C	Multiple respiratory chain deficiencies, severe cardiomyopathy, mental retardation	No
mcm^{5}U/mcm-^{5}s^{2}U	TRM9 (Sc)	Hypersensitive to translational inhibitor at elevated temperatures	hABH8 (hALKBH8)	Urothelial cancer	No
mcm^{5}U/mcm-^{5}s^{2}U	TRM9 (Sc)	Hypersensitive to the translational inhibitor paromomycin at elevated temperatures	HTRM9L	Breast, bladder, colorectal, cervical, testicular cancer ; Ovarian cancer Epigenetic cancer treatment ; Cervical cancer	No
m^{5}C	PMT1 (Sp)	Non-essential	DNMT2	Amyotrophic Lateral Sclerosis (ALS)	No
Mitochondria-encoded tRNAs					
ms^{2}t^{6}A	n.d.	n.d.	CDKAL1	Diabetes type 2	No
t^{5}s^{2}U/nm^{5}s^{2}U	MTO2/ MTU1 (Sc)	Non-essential, reduced respiration	TRMU	Infantile liver failure ; deafness	No
m^{5}C	n.d.	n.d.	NSUN3	Mitochondrial disease: developmental disability, micro-cephaly, failure to thrive, lactic acidosis, muscular weakness	No

2′-O-methyl: 2′-O-methylribose, $m^2{}_2G$: N2, N2-dimethyl guanosine, m^{5}C: 5-methylcytosine, m^{7}G: 7-methylguanosine, A-to-I editing: adenosine-to-inosine edition, mcm^{5}s^{2}U: 5-methoxycarbonylmethyl-2-thiouridine, mcm^{5}U: 5-methoxycarbonylmethyluridine, mcm^{5}s^{2}U: 5-methoxycarbonylmethyl-2-thiouridine, m^{5}U: 5-methyluridine, m^{1}G: 1-methylguanosine, m^{1}A: 1-methyladenosine, m^{2}t^{6}A: N2-methyl-N6-thereonylcarbamoyladenosine, t^{5}s^{2}U: 5-taurinomethyl-2-thiouridine, nm^{5}s^{2}U: 5-aminomethyl-2-thiouridine.

Table depicts the connection between particular tRNA modifications, their knock-out models in yeast and the resulting mutant phenotypes and published associations between human orthologous genes and human disease syndromes. The last column states when a direct link to tsRNA activity has been made in humans.

Are tsRNAs Causative Agents or Just Signs for Particular Human Disease States?

As of now, connections between tsRNA function and human health remain largely phenomenological, descriptive and therefore only tentative. Importantly, the multiplicity of assigned tsRNA functions has often not been dissociated from potential effects caused by the dysfunction of their parental tRNAs. Furthermore, tRNA levels vary widely between different human tissues that are considered healthy and non-diseased. However, when human-derived material was experimentally queried for tsRNA abundance and potential function, only a few studies used patient-derived material (i.e., biopsies, blood samples), while the majority utilized only proxies for human disease states, especially cancer cell culture systems, resulting overwhelmingly in reports that correlated changes in tsRNA abundance and identity with certain disease states. In this respect, it is important to note that the use of long-term cultured primary cancer-derived cells, while preferable when elucidating basic molecular biology, are changing constantly in culture both genetically as well as epigenetically, thereby accommodating to the various stresses they encounter. Such changes certainly include molecular circuitry impinging on protein synthesis including tRNA expression and metabolism. Whether increased tRNA levels in cancer-derived cells consequently results in an increase in tsRNAs remains unknown.

Despite these experimental shortcomings, quantitative and qualitative changes might allow developing particular tsRNAs into useful biomarkers, with the aim to report on and monitor the state and progression of a particular human disease or syndrome. On the other hand, it still remains to be determined whether aberrant production (too much or too little) of tsRNAs can actively contribute to disease pathogenesis. Below, we introduce a number of examples in which the abundance and function of particular tsRNAs has been linked to cancer, infection, neurodegeneration and other pathological conditions.

Signs for Stress and Viral Infection

Persistently activated stress responses and increased inflammation due to infection result in the need of constant controlling and repairing cellular damage and are central to and promote disease pathogenesis. Changes in tsRNA abundance have been observed in many stress-exposed or virus-infected cell types or tissues, therefore making the existence of these small RNAs proxies for sustained stress levels.

For instance, RSV infection, the most common cause of bronchiolitis and pneumonia in infants and elderly people, led to overproduction of ANG-dependent tsRNA^{S}, which repressed particular host mRNAs thereby promoting RSV replication. These findings are supported by a study in cell culture showing that persistent infections with hepatitis B or C viruses (HBV and HCV, respectively) resulted in hepatic cirrhosis and hepatocellular carcinoma (HCC). Profiling of small RNAs in liver from human subjects with advanced hepatitis B or C and HCC revealed that tsRNAs^{S} were significantly increased in humans with chronic viral hepatitis. Importantly, in HBV-associated HCC, tsRNAs^{S} abundance correlated with expression of ANG demonstrating that tsRNAs^{S} are highly abundant in chronically infected livers and in liver cancer tissues.

Links to Cancer

As cancer cells are highly proliferative, their demand for elevated protein synthesis requires adjustment of the translation machinery. Indeed, cancer cells produce increased levels of tRNAs, ribosomal RNAs and ribosomes when compared to non-cancerous cells. Cancer cells also feature high tRNA turnover. Importantly, the angiogenic ACNase ANG is overexpressed in almost all cancer types. Of note, its ribonuclease activity is critical for promoting angiogenesis and inhibiting this activity significantly inhibited tumour formation in xenograft tumour models. Whether increased tRNA levels in human cancer-derived cells consequently results in an increase in tsRNAs remains unknown.

Links to cellular overproliferation can be made through the activity of RNaseZ/ELAC2, a general tRNA biogenesis enzyme, which was responsible for the production of tsRNAs^{NonS} (tRF-1) and that has been identified as a prostate cancer susceptibility gene thereby indirectly connecting tsRNA biogenesis with a particular cancer type.

On the other hand, specific tsRNAs^{S}, termed Sex HOrmone-dependent tRNA-derived RNAs (SHOT-RNAs), were highly expressed in oestrogen receptor (ER)-positive cells from luminal-type breast cancer patient tissues. These SHOT-RNAs were produced by angiogenin-mediated anticodon cleavage in a sex hormone-dependent fashion. These findings indicated tRNA cleavage as a result of particular hormonal changes resulting in functional involvement in cell proliferation. Also, high abundance of tsRNA^{NonS} (tRNA-Leu^{CAG}) was reported in human non-small cell lung cancer (NSCLC) patient-derived tissues. A significant relationship between cancer stage and this tsRNA was reported in NSCLC-derived patient sera. Furthermore, inhibition of this tsRNA suppressed cell proliferation and impeded cell cycle progression indicating a supportive role of this particular tsRNA^{NonS} in proliferation. Interestingly, inhibition of specifically this tsRNA (tRNA-Leu^{CAG}) induced apoptosis in a patient-derived orthotopic hepatocellular carcinoma model in mice. Mechanistically, this tsRNA associated with at least two mRNAs coding for ribosomal proteins (RPS28 and RPS15) thereby positively affecting their translation. Furthermore, tsRNAs targeting translation initiation in embryonic stem cells were dysregulated in aggressive subtypes of human myelodysplastic syndromes indicating a function for tsRNAs in hematopoietic stem cell commitment and germ layer specification. These combined observations suggest a link between tsRNA function and the highly proliferative phenotype observed in stem cells and tumours.

Links to Neurodegeneration

Genetic mutations in tRNA processing and tRNA modification systems are frequently associated with human disorders, many of which affect metabolically active tissues such as the nervous system. Strikingly, mutations in ANG, the stress-induced endonuclease producing tsRNA^{S}, were implicated in the pathogenesis of Amyotrophic Lateral Sclerosis (ALS), a still incurable neurodegenerative disease. Importantly, a subset of ALS-associated ANG mutations was also discovered in patients suffering from Parkinson's Disease (PD). Interestingly, most ALS/PD-associated ANG mutations are clustering in the RNase domain of the enzyme. Furthermore, ANG-dependent tsRNAs, if transfected into cells, increased motoneuron survival upon stress exposure, suggesting that tsRNA^{S} can confer neuroprotection.

In addition, several studies linked cellular stress, tRNA modification networks and ANG-mediated production of tsRNAs^{S} to neurodevelopmental disorders. For instance, mutations in the (cytosine-5) RNA methyltransferase NSun2 have been genetically linked to human forms of intellectual disability (ID) and to Dubowitz-like syndrome. On the molecular level, NSun2 null mutations in mice resulted in the ectopic production of tsRNA^{S}, increased stress responses and neuronal cell death. While removing a pivotal tRNA methyltransferase, which also targets other RNAs, linked tRNA modifications and tRNA metabolism with aberrant nervous system development, the particular contribution of tsRNAs to both growth retardation and impaired intellectual development in patients carrying NSun2 gene mutations remains unclear.

Furthermore, tsRNAs^{NonS} derived from intron-containing tRNAs accumulated in animal models of pontocerebellar hypoplasia (PCH), a heterogeneous group of inherited human neurodegenerative disorders characterized by developmental and neuromuscular defects. One of the causative mutations for PCH affected the function of the RNA kinase CLP1, which is important for tRNA splicing. On the cellular level, interference with CLP1 function led to the accumulation of unspliced pre-tRNAs and ultimately to depletion of mature tRNAs in patient-derived neurons. Interestingly, accumulating tsRNAs^{NonS} in CLP1 mutants were represented by linear introns and 5′-tsRNAs and transfection of particular 5′-tsRNAs into patient-derived neurons resulted in reduced survival, especially under oxidative stress. While the exact mechanistic details on the interplay between tRNA biogenesis factors, the resulting depletion of mature tRNAs and the accumulation of particular tsRNAs are yet to be elucidated, these studies clearly link aberrant tRNA metabolism with the aetiology of neurodegenerative syndromes.

Links to Metabolic Syndromes

By utilizing historical records, various correlative findings have been made indicating that environmental factors affected human paternal transmission of obesity and associated metabolic disorders in the past. While the mechanisms of the transmission of such complex phenotypes in humans remain unclear, various studies using animal models supported the notion that mammalian metabolism can be influenced paternally through nutrition-induced signals, which impinge on epigenetic mechanisms in the fertilized zygote.

Recently, two studies indicated that tsRNA transmission into fertilized mouse zygotes contributed to establishing gene expression programs that recapitulated metabolic syndrome-like phenotypes seen in the paternal organism. Specifically, particular tsRNAs exhibited changes in abundance in the sperm from mice exposed to high-fat diet (HFD) or low-protein diet . Interestingly, also changes in RNA modification patterns were observed. Of note, injection of small RNAs (in the size range of tsRNAs extracted from HFD sperm) into fertilized zygotes resulted in gene expression changes of metabolic pathways and islets of the offspring and RNAi-mediated knockdown of particular tsRNAs resulted in the upregulation of specific genes that contained sequences derived from endogenous retro-elements (MERVL). Importantly, synthetic tsRNA injection into fertilized zygotes did not produce such changes indicating that RNA modifications in tsRNAs are pivotal to the function of these small RNAs. Furthermore, deletion of DNMT2/TRDMT1, a (cytosine-5) tRNA methyltransferase, abolished sperm-mediated transmission of HFD-induced metabolic disorders into offspring indicating that m^5C contributed to biological properties of tsRNAs. In addition, paternal exercise appeared to influence the abundance of small RNAs in sperm and their transmission of 'metabolic memory' into the next generation. These findings support the notion of direct effects of modified tsRNAs on intergenerational inheritance of acquired traits. Whether human sperm also contain tsRNAs with similar potency or if tsRNAs contained in breast milk carry some form of information from mother to infant remains to be tested.

Loss-of-function mutations in TRMT10A, a (1-methylguanosine) tRNA methyltransferase, are a monogenic cause of early onset diabetes. Using induced pluripotent stem cell-derived pancreatic β-like cells from TRMT10A-deficient patients it was reported that TRMT10A deficiency induced oxidative stress resulting in increased apoptosis of pancreatic β-cells. Importantly, tRNA hypomethylation at position G9 led to tRNA fragmentation of specific tRNAs and the resulting tsRNAs mediated the β-cell death by unknown mechanisms supporting an active role for tsRNAs in changing the susceptibility of particular cell types to environmental stress.

Links to Microbiome Dysregulation?

Recent studies revealed the potential role of small RNAs in interspecies and cross-domain interactions. These 'social RNAs' could serve as gene silencing and molecules serving other functions in recipient cells across domains, a phenomenon named 'cross-domain RNAi'. For instance, association between colorectal cancer, miRNA expression and the gut microbiota has been reported in humans.

Of note, mammals maintain intricate symbiotic relationships with other eukaryotes (i.e., fungi) and with a plethora of prokaryotic organisms (representing the microbiota), all of which also maintain diverse molecular machineries for the production of tsRNAs. Interestingly, specific tsRNAs secreted into human saliva affected the growth of a key oral commensal and opportunistic pathogen Fusobacterium nucleatum. Such observations pose exciting questions as to how tsRNAs produced in microbes could affect health and disease states of the mammalian host. Since one molecular consequence of tRNA fragmentation is interference with protein translation, microbiome-derived tsRNAs detectable in mammalian systems might not only be the result of the ongoing battle between microorganisms expressing killer ribotoxins but might also influence host cell physiology.

Transfer RNA Turnover

Transfer RNA have stable half-lives and follow separate pathways for tRNA turnover that are bound to function in tRNA quality control, eliminating inappropriately processed, modified or folded tRNAs from the generated pool.

Control by 3'-5' exonucleoytic degradation - The role of 3' to 5' exonucleolytic degradation through nuclear exosome and Trf4 (mediating Poly (A) tailing) in tRNA turnover has been delineated through precursor hypo-modified tRNAi Met which after polyadenylation is degraded as it is a type of aberrant transcript. The process of tRNA turnover requires RNA-dependent helicase, TRAMP complex proteins - poly(A) polymerases (Trf4 and Trf5), RNA binding proteins (Air 1/ Air2) and this activated complex interacts with nuclear exosome. The TRAMP/nuclear exosome machinery is a quality control pathway that dictates tRNA nuclear modification as well as 3' end maturation.

Transfer RNA Cleavage

Endonucleolytic cleavage - The transfer RNA endonucleolytic cleavage is a process that generates 5' and 3' half size molecules of tRNAs - a process induced through oxidative stress, nutrient deprivation, molecular level tRNA anticodon loop modifications as has been reported in various organisms. The propose relative stability of tRNAs have been compromised with the recent genomic level investigations that the multiple means of destroying tRNAs serve as quality control pathway that in turn ensures appropriately modified, mature and functional tRNAs to undertake protein synthesis and molecular singling control machineries.

Transfer RNA Sub-cellular Trafficking

Nuclear encoded tRNAs are the key to protein synthesis in the cytoplasm which in the process of generation and maturation undergo minutiae amounts of changes. The aminoacylation of tRNAs in the nucleus as found through numerous studies have pointed to their trafficking from cytoplasm to the nucleus via the tRNA retrograde process, and these exported molecules being reverted back to the cytoplasm. The intricate pathways followed for the to and fro tRNA transport is explored as follows:

1) tRNA nuclear export: RNA movement is an energy driven mechanism occurring through nucleopores, aqueous channels connecting the two compartments of nucleus and cytoplasm and proceeds via Ran pathway. Ran, a small GTPase in association with importin-b subfamily of proteins regulates the nuclear export process as exporting. The RNA or protein cargo binding dedicated to the export is Ran-GTP dependent where the cargo-exportin-Ran-GTP complex reaching cytoplasm is acted upon by hydrolysis of Ran-GTP to Ran-GDP, thereby releasing the cargo.

2) tRNA retrograde nuclear import: The unidirectional nucleus to cytoplasm movement was challenged when yeast tRNAs were discovered to get spliced on mitochondrial surface or in the cytoplasm raising a new conundrum, where it was suggested that tRNAs might travel in retrograde way from cytoplasm to the nucleus. The evidences for such a movement came from accumulation of foreign tRNA encoded by one nucleus in the nuclear area that does not encode tRNA; nutrient deprived assortment of tRNA where normal tRNA transcription in inhibited as also the absence of spliced tRNAs in nuclear import impaired haploid cells. The above mechanisms have been poorly manoeuvred through scientific studies, so need more explorations to delineate the mechanism(s) of tRNA nuclear import.

3) tRNA re-export to cytoplasm: tRNAs that are imported into the nuclei from the cytoplasm upon nutrient deprivation can be reverted to the cytoplasm through refeeding by a process termed tRNA reexport. The phenomenon is mostly dependent upon molecular analyses of conditional yeast mutants where the nuclear export and re-export to the cytoplasm of tRNAs have been studied using flow and accumulation of nucleus tRNAs derived from cytoplasm or amino-acylated tRNAs stimulating the report steps, though it is poorly understood till date and needs further investigations.

Transfer RNA Processing, Regulation of Biogenesis

Transfer RNA biology is a challenging field of investigations that circumscribe their importance in processes associated with translation, decoding the mRNA language into the amino acid sequence, reading frame maintenance, tRNA stability and in coordinating metabolism alongside. tRNA biology demarcates the essential roles of tRNA processing and regulation at its biosynthesis, with emphasis on the universal presence of tRNA aminoacylation processes not just in nucleus, but also in cytoplasm and their processing, trafficking and regulation also feathered across the nucleicytoplasmic boundaries. This cross-talk of molecular events associated with tRNA biogenesis, their regulation and control as well as degradation is likely to dwell into newer discoveries and surprises to come.

References

- What-is-rna: rnasociety.org, Retrieved 05, August 2020
- Glucose-storage: inflibnet.ac.in, Retrieved 29, March 2020

4

Techniques used in Molecular Biology

Some of the diverse techniques which are used within the domain of molecular biology are blotting, northern blotting, cationic electrophoresis and eastern blotting, and DNA sequencing. This chapter closely examines these key techniques of molecular biology to provide an extensive understanding of the subject.

Blotting

Blotting means to transfer DNA or RNA to nitrocellulose membrane and probe binds to complementary strand in the sample to give result in the form of signal using radioactive or biotin labeled probe.

Principle of Blotting

At high temperature two strands of DNA get separated they and re-anneal when temperature is brought back to normal. There are 2 important features of hybridization:

- The reaction is very specific that means the probes will only bind to targets with a complementary sequence.
- The probe can identify one molecule of target in a mixture of millions of related and non-complementary molecules.

Types of Blotting

- Southern Blot: It is a method in molecular biology which is used for the identification of a specific DNA sequence in a mixture of DNA samples. In this DNA is first run on agarose gel and than transferred to membrane and further detection using specific probe. This method is known by its inventor's name who was biologist by profession, Edwin Southern. After this other methods of blotting are named Northern and western further in continuation to Southern blot.
- Northern Blot: The northern blot is technique used in molecular biology research to study expression of by detecting RNA (or isolated mRNA) in a sample. Northern blotting involves the use of agarose gel electrophoresis to separate RNA samples by size, and detects with a hybridization probe which is complementary target sequence. Blot means to transfer RNA from the electrophoresis gel to the blotting membrane. This technique was given in 1977 by James Alwine, David Kemp, and George Stark at Stanford University.
- Western Blot: The western blot is a widely used analytical technique used to detect specific proteins in a sample. It uses SDS to separate proteins. The proteins from the SDS gel are

then transferred to a nitrocellulose or PVDF membrane, and incubated with antibodies specific to the target protein.

- Eastern Blot: To study post translational modification this technique is (lipids, phospho-moieties and glycoconjugates). It is mostly used to detect carbohydrate epitopes. Eastern blotting is considered an extension of the Western blotting. Transferred proteins are analyzed for posttranslational modifications using specific probes that detect lipids, carbohydrate, phosphorylation or any other protein modification.

Probe

Is defined sequence which is used to search mixtures of nucleic acids for molecules containing a complementary sequence. Any positive sample can be used as control (e.g. In diagnosis of viruses).

Blotting Membranes

Membrane or support onto which the separated proteins are transferred which bind proteins with high affinity:

- Nitrocellulose membrane.
- Has very good protein binding capacity and retention ability.
- Polyvinylidene fluoride (PVDF) membrane.
- Its strength is more than nitrocellose membrane so it can be used if reprobing and stripping is required.

Steps

- Run samples on a gel/make a slot/dot blot.
- Transfer to a membrane.
- Link sample to membrane.
- Make a probe.
- Hybridize probe to membrane.
- Detect probe.

Labeling of DNA or RNA Probes

- Radioactive labeling: Use of radioactive material for labelling.
- Non-radioactive labeling: Enzymatic labeling using biotin labeled probes.
- End labeling: Labels at the ends.
- Uniform labeling: Labels internally.

Hexanucleotide Primered Labeling

After denaturing DNA add random hexamer primers and DNA polymerase Known viral gene sequence (or any other DNA/RNA) run in 1% agarose gel. Viral DNA fragment is eluted from the gel and incubated in boiling water bath for 10 min, for denaturation of DNA.

Table: Probe preparation mix.

Denatured DNA	200-500 ng
Random Primer	100 ng
10x Klenow buffer	3 μl
dNTP mix (-CTP)(3.3nM each)	4.5 μl
α-^{32}P dCTP (10μCi/μl, specific activity 3x103 Ci/mmole)	10 μCurie
Klenow enzyme	5 units
Final volume with ddH_2O	30 μl

- The reaction is incubated at 37°C for one hour.
- To this, equal volume of Buffer-A is added.
- The mixture is then denatured by incubating the tube in boiling water bath for 7 min.
- The contents are immediately transferred to ice before adding to the hybridization bottle.

dNTP mix: (for α^{32}-P dCTP as the radioactive molecule, 100mM stock of dATP, dTTP and dGTP): 1+1+1+27μl water. 4.5μl of this mix is used for one reaction.

Buffer-A: 500mM Tris HCL (pH 7.5), 500mM NaCl, 5mM EDTA and 0.5% SDS.

Blotting in Virus Diagnostics

Procedure for northern and Southern blot is same except that in Southern DNA is used and in Northern RNA is used.

- Probes prepared from cloned viral cDNAs are used in Southern blots to detect the viral DNA in infected plants.
- A modification in this technique called tissue printing employs blotting of leaf squashes rather than the viral DNA on to nylon membranes.
- Use of radioactive material for preparing the probes has been the major limitation of this technique.
- To avoid this hazardous material, non-radioactive labelling techniques have been developed.

Diagnosis of RNA and DNA Viruses

Slot-blot Hybridization

- Samples are crushed in DNA/RNA isolation buffer and centrifuged to remove debris (CTAB Buffer or TRI reagent etc).

- Approx 200µl liquid sample is denatured in a boiling water bath for 5 min before loading.
- The slot-blot manifold is assembled with positively charged nylon membrane, previously wetted with sterile water.

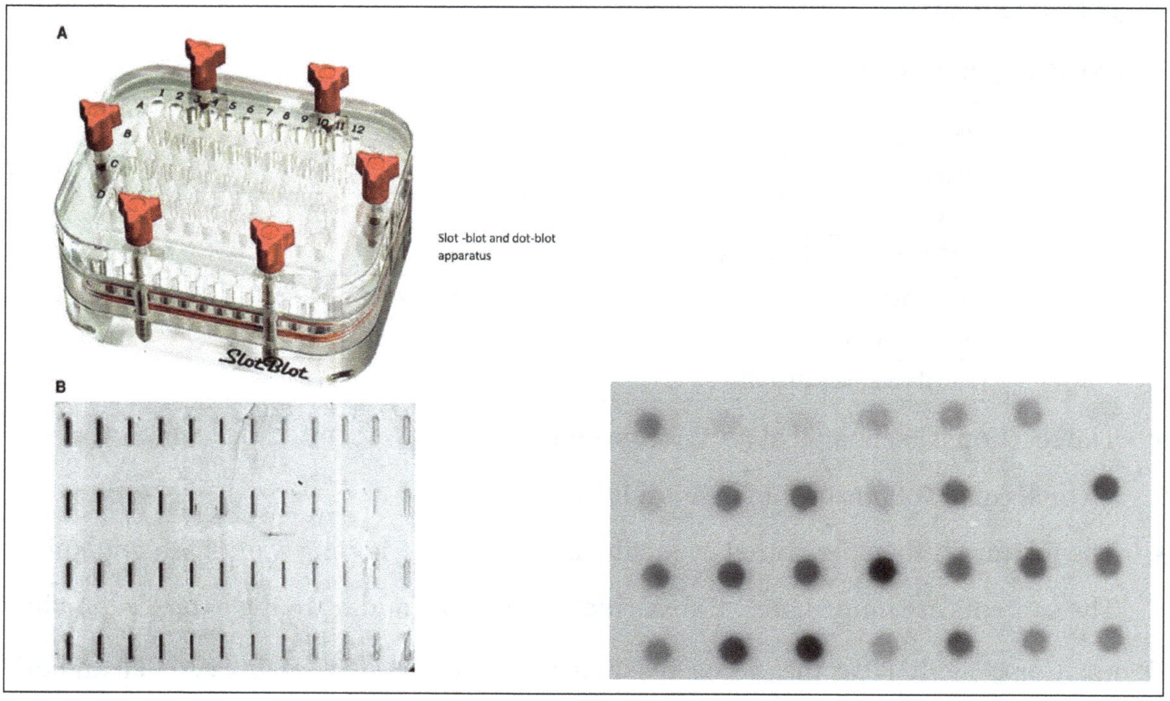

Membrane after dot-blot transfer.

Procedure

- The manifold is connected to vacuum pump.
- To each well, 10x SSC buffer (200µl) is added and vacuum was applied till the buffer is completely absorbed but not dried.
- After that, 200µl each of the denatured samples is loaded to the wells.
- Vacuum is applied till the samples are absorbed completely.
- Afterwards, 200µl of 10x SSC is added and allowed to completely pass through.

20x SSC Buffer: 3M NaCl and 0.3 M Trisodium citrate:

- Vacuum is released, apparatus disassembled and the membrane is rinsed in 2x SSC.
- The membrane is air dried and exposed to UV light for 2 min in a UV crosslinker for binding of the transferred nucleic acids.
- The membrane is wrapped in saran film and stored at 4°C until hybridization.
- After completion of electrophoresis, the DNA in the gel is denatured by placing the gel in a plastic tray containing the denaturation solution with slow shaking on a gel rocker for 30 min.

- After denaturation, the gel is soaked in neutralizing solution for 30 min with slow shaking.
- The two chambers of the electrophoresis tank are filled with transfer solution (10x SSC).
- A wick of Whatman 3MM paper is soaked in 10x SSC and placed over the gel platform of the tank with its ends submerged in the solution.

Denaturing Solution: 1M NaCl and 0.5M NaOH.

Neutralizing Solution: 1.5M Tris HCl (pH 8) and 3M NaCl:

- Three more sheets of Whatman 3MM paper (slightly bigger than the gel) are soaked in 10x SSC and placed over the wick.
- The gel is carefully inverted and placed over the sheets avoiding any air bubbles between the gel and Whatman papers.
- Nylon membrane is cut to the gel size, wetted in sterile water and placed over the gel avoiding any air bubbles.
- Three more sheets of Whatman 3MM paper are cut to gel size, soaked in 10x SSC and placed over the membrane.
- A stack of gel-sized dry blotting papers (8-10 cm height) are placed over the 3MM papers and approximately 500 g weight is placed on top of the stack.
- The transfer is allowed to take place for 14-16 h.

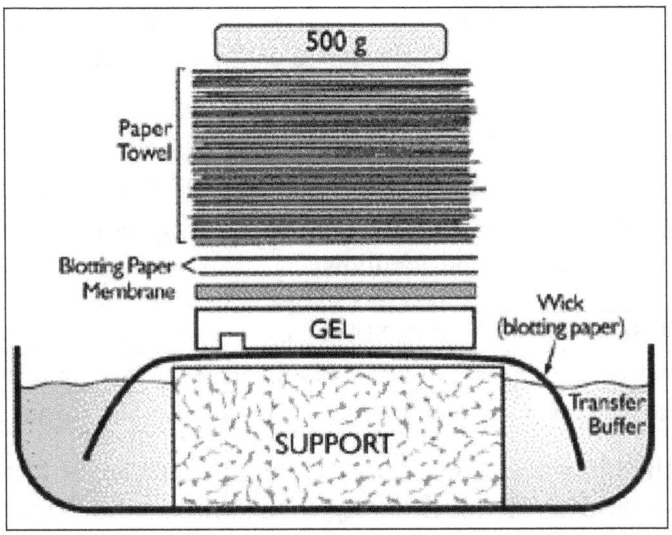

Capillary Transfer of DNA from gel to the membrane.

- After completion of the transfer, the blotting paper stack is removed, the membrane is marked so as to indicate the orientation and rinsed in 2x SSC briefly.
- The buffer is removed by placing the membrane between the folds of a dry 3MM sheet but the membrane is not completely dried.
- The membrane is wrapped in a cling film and placed under UV light in a UV cross-linker for 2 min for cross-linking the DNA to the membrane.

The blot is then stored at 4°C until hybridization.

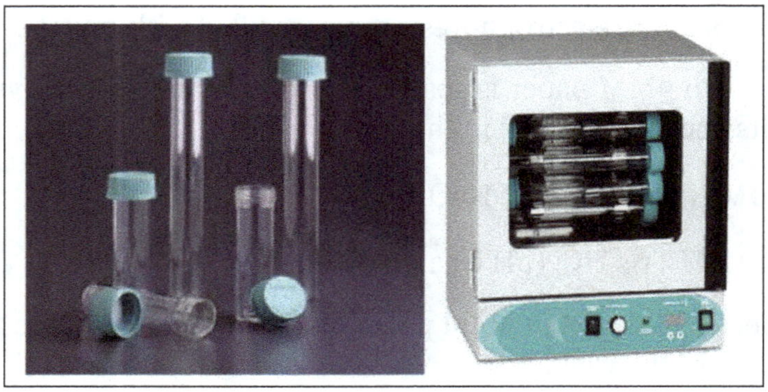

Pre-hybridization

- Pre-hybridization is carried out for 1 hr at 42°C in the pre-hybridization buffer.

  Pre-hybridization buffer consists of:

Formamide	50% (v/v)
Na_2HPO_4 pH 7.2	120 mM
NaCl	250 mM
EDTA, pH 8.0	1 mM
SDS (w/v)	7%

- After pre-hybridization, probe is added to fresh pre-hybridization solution and incubated at 42°C overnight (10-18 hrs) and this step is called hybridization.

Washing of Blots

- After hybridization is complete the blots are removed from the hybridization solution, rinsed briefly in 2x SSC and placed in the first washing buffer followed by 2nd and third wash with progressive stringency as follows:

First Wash Buffer 2.0x SSC	0.1% SDS (25°C)	Two washings of 15 min each.
Second wash buffer 0.5x SSC	0.1% SDS (25°C)	Two washings of 15 min each.
Third Wash buffer 0.1x SSC	0.1% SDS (65°C)	Two washings of 30 min each.

- After washing, the blots are placed on a Whatman filter paper to remove excess of liquid and wrapped in a saran wrap immediately, to prevent it from drying and for the purpose of autoradiography.

Autoradiography

- The blots are exposed to X-ray films by putting them between two intensifying screens for the autoradiography in a cassette at -80°C.

The films are developed after sufficient exposure, depending on the intensity of the signal prior to autoradiography as determined by a portable radioactivity monitor.

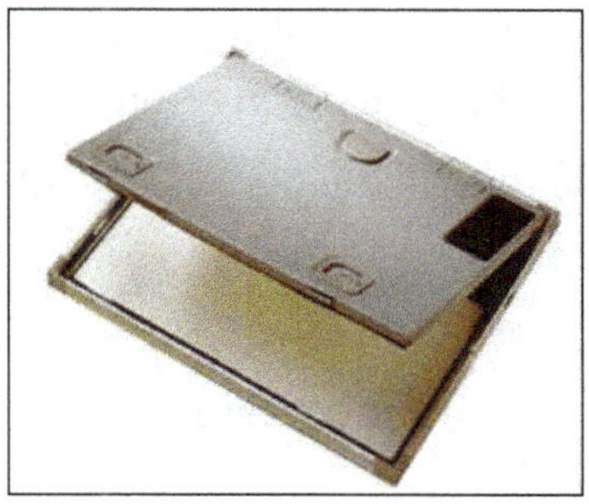

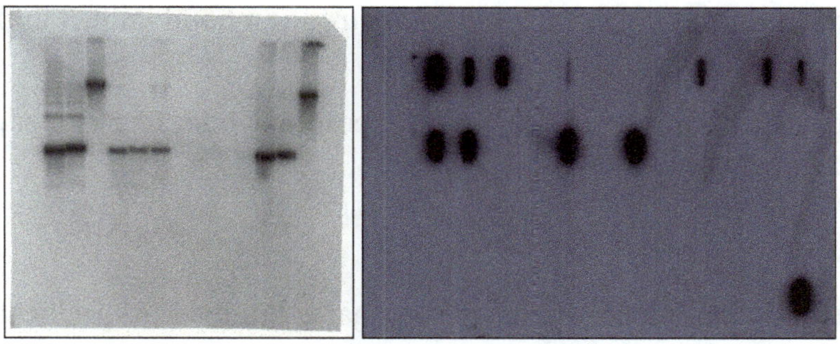

Results of Southern Blotting.

Biotin Labeling Kits

- The Biotin labelled detection Kits are more convenient method for the chromogenic detection of biotinylated nucleic acid probes.
- These kits are optimized to provide high sensitivity with lesser background in in Southern, Northern, dot and slot blotting.
- Sensitivity is 30-100 fg.

Principle of Biotin Labeling Kits

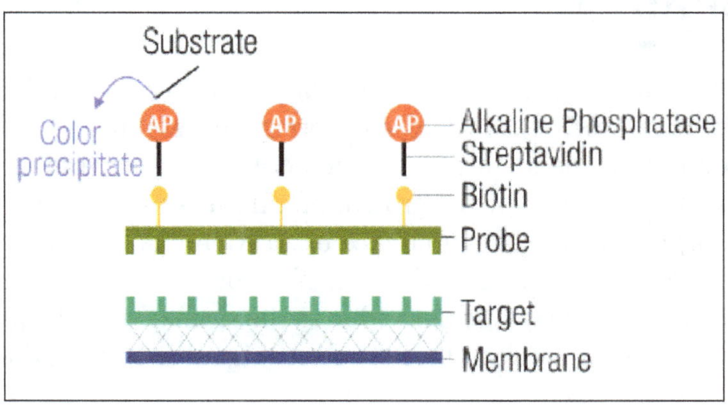

- Probes which are Biotin-labeled are detected with alkaline phosphatase-conjugated streptavidin.
- Streptavidin-AP conjugates bind irreversibly and specifically to probes which are the biotin-labelled.

- The probes then are visualized using a chromogenic substrate for alkaline phosphatase BCIP/NBT, which produces a blue-purple precipitate.
- In enzymatic method visualization does not any specific equipment or X-ray film.

Streptavidin

- Streptavidin biotin binding tetrameric protein isolated from Streptomyces avidinii and its molecular weight is about 60kDa.
- The streptavidin-biotin noncovalent bond is strongest bond known in the biological interaction.
- Formation of such a complex is a very fast process and, once it is formed, the complex is not affected by any external factors.

Streptavidin-AP

- Streptavidin and alkaline phosphatase conjugates binds very specifically to the biotin-labelled probes and it is irreversible.
- To visualize probes chromogenic substrate for alkaline phosphatase BCIP/NBT is used.
- Alkaline phosphatase cleaves the substrate BCIP-T (5-bromo-4-chloro-3-indolyl phosphate, p-toluidine salt) which is supplemented with the enhancer chromogen NBT (nitro blue tetrazolium).
- A blue precipitate is produced thus visualization does not require any X-ray film or any other equipment.

Northern Blotting

Northern Blotting is a technique used for the study of gene expression. It is done by detection of particular RNA (or isolated mRNA). mRNA is generally represented as 5% of the overall RNA sequence. This method reveals the identity, number, activity, and size of the particular gene. This blotting technique can also be used for the growth of a tissue or organism. In different stages of differentiation and morphogenesis the abundance of an RNA changes and this can be identified using this technique. It also aids in the identification of abnormal, diseased or infected condition at the molecular level. The technique got its name due to the similarity of the process with Southern blotting. The primary difference between these two techniques is that northern blotting concerns only about RNA.

Principle of Northern Blotting

As all normal blotting technique, northern blotting starts with the electrophoresis to separate RNA samples by size. Electrophoresis separates the RNA molecules based on the charge of the nucleic acids. The charge in the nucleic acids is proportional to the size of the nucleic acid sequence.

Thus the electrophoresis membrane separates the Nucleic acid sequence according to the size of the RNA sequence. In cases where our target sequence is an mRNA, the sample can be isolated through oligo cellulose chromatographic techniques, as mRNA are characterized by the poly(A)-tail. Since gel molecules are fragile in nature, the separated sequences are transferred to the nylon membranes. The selection of nylon membrane is contributed to the factor that nucleic acids are negatively charged in nature. Once the RNA molecules are transferred it is immobilized by covalent linkage. The probe is then added, the probe can be complementary an ss DNA sequence. Formamide is generally used as a blotting buffer as it reduces the annealing temperature.

Procedure

- The tissue or culture sample collected is first homogenized. The samples may be representative of different types of culture for comparison or it can be for the study of different stages of growth inside the culture.
- The RNA sequence is separated in the electrophoresis unit an agarose gel is used for the purpose of the nucleic acid separation.
- Now the separated RNA sequence is transferred to the nylon membrane. This is done by two mechanisms capillary action and the ionic interaction.
- The transfer operation is done by keeping the gel in the following order. First, the agarose gel is placed on the bottom of the stack, followed by the blotting membrane. On top of these paper towels a mild weight (glass plate) is placed. The entire setup is kept in a beaker containing transfer buffer.
- RNA transferred to the nylon membrane is then fixed using UV radiation.
- The fixed nylon membrane is then mixed with probes. The probes are specifically designed for the gene of interest, so that they will hybridize with RNA sequences on the blot corresponding to the sequence of interest.
- The blot membrane is washed to remove unwanted probe.
- Labeled probe is detected by chemiluminescence or autoradiography. The result will be dark bands in x ray film.

GEl and Probes

The RNA samples are separated using agarose gels using formaldehyde as denaturing agents but in small RNA or micro RNA sequences, polyacrylamide sequences with urea as a denaturing agent also can be used. Ethidium bromide can be used as a staining agent. Two types of markers are for size marking. An RNA ladder and ribosomal subunit are used for the identification of the size of the RNA sequences.

Probes can be complementary to the whole or part of the RNA of interest. They can be RNA, DNA or oligonucleotides of 25 complementary basepairs to the target RNA. In case of RNA probes, in-vitro produced probes are used as invivo probes can denature due to the rigorous washing. In case of cDNA, the probes are labeled with radioactive isotopes, alkaline phosphatase or horseradish peroxidase in case of chemiluminescence.

Western Blotting

First step in any blotting mechanism is to separate the molecules by electrophoresis which is agarose electrophoresis in DNA/RNA macromolecules and SDS PAGE in case of proteins. After electrophoresis is complete the macromolecules which are separated are blotted to PVD or NC membranes. After this, to prevent any non-specific binding of antibodies to membrane blocking is done using suitable blocking solution. The transferred protein is complexed with an enzyme-labeled antibody as a probe. Suitable substrate is then added to the enzyme and which produces a detectable product such as a chromogenic precipitate on the membrane for detection.

Chromogenic Western Blotting

Precipitating substrates have been used mostly for many years these are the simplest and most economic method of detection. When these substrates come in contact with the appropriate enzyme, they gets converted to colored products which are insoluble, and precipitate onto the membrane. As a result a colored band is visible to naked eyes. Chromogenic blotting substrates are available in a various specifications and different formats. The use of suitable substrate depends on the enzyme label and sensitivity. Like other fluorescent blotting applications, chromogenic substrates do not require any equipment to visualize results. Like developing film, the blot is incubated in substate till the color is developed.

Procedure of Western Blotting

- Preparation of sample,
- Electrophoresis (Agarose/SDS),
- Blotting,
- Blocking,
- Probing of Antibody,
- Detection using substrates.

Sample Preparation

- Every type of protein whether single cells or whole tissues can be analyzed using western blotting.
- The cells are harvested, washed, and lysed to release the target protein.
- For optimum results all steps should be perform in the cold room or on ice.
- This will reduce denaturation, proteolysis, dephosphorylation of proteins.
- The use of extraction technique will depend primarily on the sample.
- Once the cells are disrupted the endogenous proteases gets liberated and it can degrade the molecules. Which we want to study. So suitable protease inhibitor should be used to prevent protein loss.

SDSPAGE

Proteins are run on SDS gels.

Blotting

After gel electrophoresis, the separated proteins are transferred to a membrane to further analyse them Proteins can be transferred in using suitable western blot apparatus For transfer, the membrane is placed just next to the gel Both gel and membrane is sandwiched between absorbent materials, and the to maintain tight contact between the gel and membrane sandwich is clamped between solid supports.

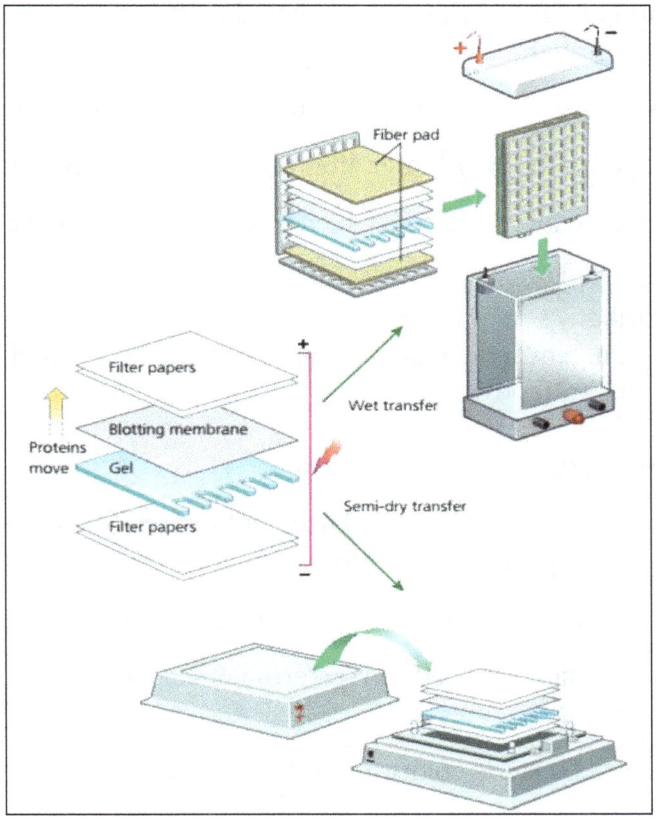

The apparatus is connected to a power supply is run for 1-2 hour at 70 mA.

The Buffers for Transfer Contain

- Tris
- Methanol (20%)
- Glycine
- 0.1% SDS

Blocking

- Blocking is major step in the Western blotting to prevent non-specific binding of antibody to the blotting membrane.

- BSA 3-5%/dried milk are commonly used blocking solutions made in PBS (phosphate buffered saline).
- Usually , Tween 20 which is a detergent, in a small amount is added to blocking and washing solutions as it reduces background staining.

Antibody Probin

- After separating protein samples and transferring onto a membrane, the protein of interest is detected using a specific antibody.
- In a dilute solution of antibody blot is incubated usually for a 1-2 hours at room temperature or at 4°C for overnight.
- The manufacturer's instructions should be used for dilutions of antibodies.
- Membrane is washed with PBST buffer after each step to remove residual primary antibody.
- Specific, labeled secondary antibody which is against the constant region of the primary antibody is used.
- The secondary antibody serves two purposes first as a carrier of the label second as a mechanism to amplify the emitted signals, because secondary antibodies can bind simultaneously to the primary antibody.

Detection with Substrates

- HRP is the most common antibody label used in Western blots, it is a small, stable enzyme with very high specificity and fast turnover.
- When HRP is exposed to a substrate solution the signal is detected.
- Substrate solutions which are used in Western blotting are reagents which gives appropriate signal with enzymes.
- HRP label is detected by using colorimetric or chemiluminescent substrates.

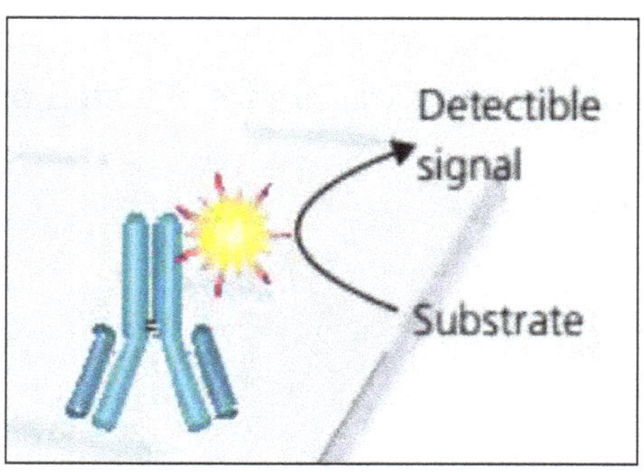

- Colorimetric substrates used for HRP such as Tetramethylbenzidine(TMB) produce purple or black bands on the surface of the blot.
- These substrates are very easy to use and take very less time to give visible bands.
- These substrates can detect proteins up to Nano-gram range.

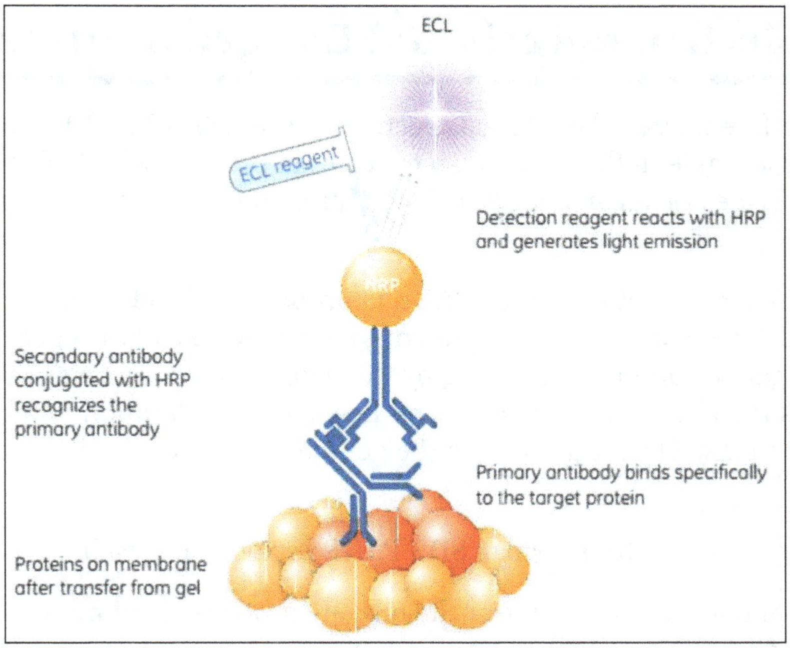

- Commonly HRP is used with ECL (enhanced chemiluminescence) detection.
- The substrate used is luminol For ECL detection, which is oxidized by HRP in the presence of H_2O_2 to produce light.
- The emitted light is detected by exposing the Western blot to X-ray film.
- The emitted light forms a band on the film gives indication of positive result.
- ECL detection of HRP is very sensitive, which can visualize proteins in picogram to femtogram amounts.

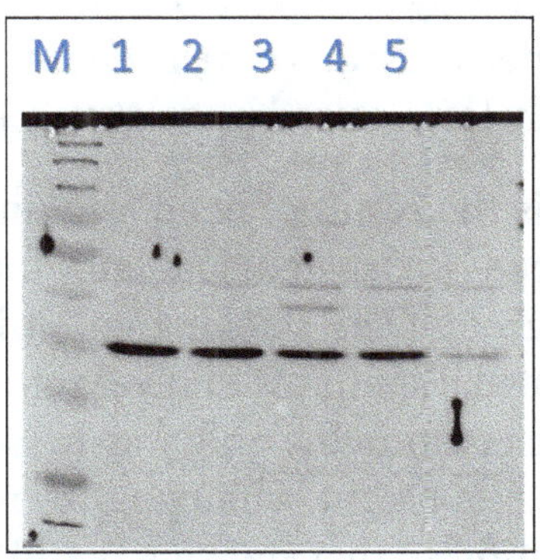

Western blotting result showing dark bands.

Table: Comparison of Southern, Northern and Western blot/ hybridization and their applications.

Blot type	Target	Probe	Applications
Southern	DNA	DNA	In the mapping of genomic clones Estimating gene numbers, Pathogen Detection
Northern	RNA	DNA/RNA	Gene expression profiling
Western	Protein	Antibodies	Molecular weight of the protein

Cationic Electrophoresis and Eastern Blotting

Sodium dodecylsulfate-polyacrylamide gel electrophoresis (SDS-PAGE) and western blotting have become indispensable in the protein lab to separate proteins with high resolution. Separation is usually performed on an analytical scale, but preparative equipment is available commercially.

With most proteins the R_f -value in SDS-PAGE is proportional to to the size because proteins bind about 1:4 g SDS per g of protein, equivalent to one molecule of SDS per three amino acids. The number of negative charges introduced by the detergent far outweighs the charges on the protein itself, thus the charge/weight ratio, and hence the acceleration in an electrical field, is identical for all proteins. The restriction by the gel matrix, however, increases with the size of the protein.

There are, however, some situations where SDS-PAGE performs less well:

- Very hydrophobic proteins (i.e. transmembrane proteins) bind more than the usual 1:4 g SDS per g of protein, increasing the charge/mass ratio of the protein. Thus these proteins run faster in SDS-PAGE than expected for their molecular mass.
- Proteins with a large number of positively charged amino acids (e.g. histones) also run faster than expected for their molecular mass.
- Glycoproteins contain a highly variable number of negative charges in their sugar side chains. Thus they do not run as crisp bands in SDSPAGE, but as broad smears, reducing the achievable resolution of the method.

In the latter two cases, it should be possible to circumvent the problems by using a positively charged detergent in lieu of the negatively charged SDS. Of course, the proteins then run from the positive to the negative electrode. Such attempts have been reported several times in the literature, but resolution of the gels was usually low, caused by relatively broad protein bands.

The high resolution of SDS-PAGE is the result of band stacking in a discontinuous (multiphasic) buffer system, an effect first described by Ornstein and later theoretically elaborated by Jovin and others. While the protein moves through the stacking gel, it is electrophoretically concentrated from below 1 mg=ml (as present in usual samples) to several 100 mg=ml. Since the amount of protein cannot change due to mass conservation the only way in which this can happen is by reducing the volume of the protein band, i.e., its height. Changing from an anionic to a cationic detergent requires a change in buffer composition so that stacking is still possible. With such selected buffer systems high resolution electrophoresis of proteins in cationic detergents is possible. As with SDS-PAGE separation in cetyl trimethyl ammonium bromide (CTAB)-PAGE is based on protein size, as noted also by others.

An additional advantage of CTAB compared to SDS is that it efficiently solubilises membrane proteins, often without damaging their structure. Thus one can use the same detergent for electrophoresis that is also used for solubilisation and purification, an advantage since extraneous detergent can interfere in PAGE by competing with SDS for binding to the protein.

Casting of CTAB gels

- 40 % Acrylamide/Bis (37:1): 1.08 g bisacrylamide and 38.9 g acrylamide made to 100 mL with water. Stable for months at 4°C, especially when stored over an anion exchanger.
- 40 % Acrylamide/Bis (19:1): 2.11 g bisacrylamide and 37.9 g acrylamide made to 100 mL with water. Stable for months at 4°C, especially when stored over an anion exchanger.
- Potassium hydroxide (KOH; 1 M) 5.611 g KOH made to 100 mL with water. Stable at room temperature (RT) if protected from air.
- 16.6 M Acetic acid (commercial 99.5 % glacial acetic acid): Stable at RT. Exact molarity is determined once by titration and noted onto the bottle.
- 10 % CTAB: 10 g CTAB made to 100 mL with water. Store at 37°C to increase solubility.
- Malachite green (1 %): 10 mg/mL malachite green in water, stable at 4°C. water saturated butanol n-butanol shaken with some water, after phase separation the upper, organic phase is used. Stable at RT.

For Photo-polymerisation

- 100 mM Methylene blue: 780 mg methylene blue made to 20 mL with water. Stable for months at 4°C.
- 100 mM sodium toluene 4-sulfinate (T4S): 356 mg T4S (anhydrous) made to 20 mL with water. Stable for months at 4°C.
- 1 mM diphenyliodonium chloride (DPIC): 6.3 mg DPIC made to 20 mL with water. Stable for 1 week at 4 °C.

For Fenton-system

- 10 mM ferrous sulfate ($FeSO_4$): 27.8mg $FeSO_4 \times 7H_2O$ made to 10 mL with water make fresh daily.
- 40 mM Ascorbic acid 70.5mg ascorbic acid made to 10 mL with water, make fresh daily.
- 30 % hydrogen peroxide commercially available, store at 4 °C.

The recipes for both photo-polymerisation (5) and a Fenton-system (8) are given, obviously only one needs to be prepared.

Running of CTAB Gels

- Upper tank buffer: 40 mM (3.56 g/L) β-alanine, 70 mM (2.29 mL/L) acetic acid, 0.1 % CTAB. Make fresh each time.
- Lower tank buffer: 50 mM KOH, 187 mM (3.18 ml/L) acetic acid, 0.1 % CTAB. Make fresh each time.

- Sample buffer (2×): 1.27 mL 1 M KOH (127 mM final), 107 µL acetic acid (187 mM final), 2 mL 10 % CTAB (2 % final), 100 µL β-ME (1 % final), 1 mL glycerol (10 % final), 7.21 g urea (12 M final) and 50 µL 1 % basic fuchsin (0.005 % final), make to 10 mL with water. Stable at RT for a week.

Staining of CTAB gels with Ponceau S

- Fixative: 100 mL glacial acetic acid and 400 mL methanol made to 1 L with water. Stable at RT.
- Ponceau S solution: 0.1 g Ponceau S and 10 mL glacial acetic acid made to 1!L with water. The solution is stable at RT and may be reused several times.

Eastern Blotting

- Blotting buffer: 10 mL 1 M KOH (10 mM final) and 636 µL acetic acid (11 mM final), water to 1 L. Make fresh.
- Washing solution: 50 mg SDS (0.05 % final) in 100 mL methanol. Stable at RT.
- Phosphate buffered saline (PBS; 10 × stock): 2.0 g KH_2PO_4 (14.7 mM final), 2.0 g KCl (26.9 mM final), 80.0 g NaCl (1369 mM final) and 21.6 g $Na_2HPO_4 \times 7H_2O$ (or 11.44 g of the anhydride, 80.6 mM final) made to 1 L with water. Store at RT.
- Blotto: 1 g low fat milk powder (0.1 % final), 0.5 mL Tween-20 (0.05 % final) and 0.1 g thimerosal (0.01 % final) made to 1 L with PBS. Make fresh each day.
- Blocking solution: 0.5 g low fat milk powder in 10 mL Blotto. Make fresh each time.
- horseradish peroxidase (HRP)-substrate (20 ×): 6.01 g Tris (1 M final), 14.6 g NaCl (3.2 M final), 950 mg $NiCl_2 \times 6H_2O$ (80 mM final) dissolved in 40 mL water. Adjust to pH 7.6 with HCl, and add 250 mg DAB (13 mM final). Dilute to 50 mL with water, filter and freeze in 1 mL aliquots. Stable for years at −20 °C.
- Chemoluminescence reagent: 400 mg luminol (2.2 mM final) and 9.5 mg piodophenol (432 µm final) in 19 mL dimethylsulfoxide diluted to 100 mL with water. Immediately before use take 10 mL and add 4.8 µL 30 % hydrogen peroxide (420 nM final).
- Primary antibody: 1 µg monoclonal antibody in 10 mL Blotto. Stable for many months at 4 °C if the milk powder in Blotto is replaced by BSA (fraction V). For antisera the dilution has to be worked out on a caseby-case basis.

2D-electrophoresis

- IEF sample buffer (2 ×): 100 µL β-ME (1 % final), 100 µL IPG-buffer (same pH-rage as IEF-strip, 1 % final), 7.21 g urea (12 M final), 0.4 g CHAPS (4 % final) and a trace of bromophenol blue, made to 10 mL with water. Store in aliquots at −20 °C.
- Equilibration buffer: 3.6 g urea (6 M final), 3 mL glycerol (3 % final), 0.2 g CTAB (2 % final), 640 µL 1 M KOH (64 mM final), 54 µL glacial acetic acid (94 mM final) and 25 µL 1 % basic fuchsin (0.0025 % final), make to 10 mL with water.

- Mounting medium: 50 mg agarose in 10 mL stacking gel buffer (0.5 % final), heated to boiling in a microwave oven. Best prepared fresh for each run.

Methods

Casting of CTAB Gels

Pour the mixture, containing either the reagents for photo-crosslinking or the Fenton-system, into an Erlenmeyer flask with magnetic stirrer. The stirrer is adjusted for vigorous movement without foam production. Then a vacuum is applied for at least 10 min to remove dissolved oxygen. This is essential as oxygen inhibits the polymerisation reaction. Do not apply the vacuum before starting the stirrer!

Finally the catalyst is added: 18 µL 10 mM methylene blue (9 µL for the stacking gel) or 2 µL hydrogen peroxide depending on whether the photopolymerisation or the Fenton-system is used. Mix by gentle inversion; do not shake oxygen into the solution again. At this stage, do not add the catalyst for the stacking gel.

Table: Mixing table for CTAB-PAGE gels (all volumes in mL). Note that either the reagents for photopolymerisation or those for the Fenton-system should be used.

Solution	5 %	7.5 %	10%	12.5%	15%	17.5%	20%	Stack
Acrylamid 19:1	—	—	—	—	—	—	—	3
Acrylamid 37:1	7.5	11.3	15.0	18.8	22.5	26.3	30.0	—
1 M KOH	2.6	2.6	2.6	2.6	2.6	2.6	2.6	1.91
Glacial acetic acid	0.962	0.962	0.962	0.962	0.962	0.962	0.962	0.161
Urea	10.8 g	10.8 g	10.8 g	10.8 g	10.8 g	10.8 g	10.8 g	5.4 g
10 % CTAB	0.6	0.6	0.6	0.6	0.6	0.6	0.6	0.3
100 mM T4S	0.30	0.30	0.30	0.30	0.30	0.30	0.30	0.15
1 mM DPIC	1.5	1.5	1.5	1.5	1.5	1.5	1.5	0.75
10 mM $FeSO_4$	0.048	0.048	0.048	0.048	0.048	0.048	0.048	0.024
40 mM Ascorbic acid	6.0	6.0	6.0	6.0	6.0	6.0	6.0	3.0
Water to	60	60	60	60	60	60	60	30

Pour the separating gel into the casting sandwich and overlayed with watersaturated n-butanol. The Fenton-system will polymerise by itself, the photo-polymerisation system requires exposure to a strong source of white light. A sun-exposed window or a halogen lamp may be used. Polymerisation is finished when the interface between gel and butanol becomes prominent. Methylene blue will become colourless during polymerisation. Normally polymerisation should be complete within 30 min.

Pour the butanol from the top of the gel, add catalyst to the stacking gel mixture, and cast the stacking gel. Insert combs immediately and the gel is allowed to polymerise.

Running of CTAB Gels

Mount the gel in the running chamber according to manufacturer's instruction, add upper and lower tank buffer. Sometimes air bubbles get trapped in the wells; these can be rinsed out using a tuberculin syringe with 27G needle. This process and the loading of samples is aided by malachite green in the stacking gel, which makes the wells easier to see (for SDS-PAGE one can use phenol red for the same purpose).

Sample loading is easiest with a 25 μL Hamilton syringe, the needle should have a flat point. Between samples the syringe is rinsed with upper tank buffer, finally with water. For a standard minigel, electrophoresis is performed at 20 mA per gel (10 mA during stacking), with a maximum voltage of 200 V. Do not forget to reverse the electrode polarity compared to SDS-PAGE.

Staining of CTAB Gels with Ponceau S

Fix the gel twice for 10–15 min on an orbital shaker, then replace the fixative with dye solution for 5 min. The gel can be differentiated by incubation with several changes of fixative (in the same way as is commonly done with CBB-R250), but this procedure is time consuming.

Eastern Blotting

- Soak the PVDF blotting membrane first in methanol, then in water and finally in blotting buffer (nitrocellulose membranes are soaked directly in buffer).
- Place one of the sponges and 3 sheets of filter paper (e.g. Whatman No. 3), all soaked in blotting buffer, onto the red part of the blotting cassette. Place the gel on top and flood it with a small amount of blotting buffer (plastic Pasteur-pipettes are ideal for this). Add the blotting membrane carefully on top without trapping air bubbles between gel and membrane, then place three sheets of filter paper and the second sponge, all soaked in blotting buffer, on top. Close the cassette and place it (in the correct orientation, red to red and black to black) into the blotting tank already filled with blotting buffer. In this way the gel is on the positive and the membrane on the negative side. This orientation is opposite to the one used in western blotting, hence the name "eastern blotting". The entire procedure is best performed on a plastic tea-tray or similar, so that the unavoidable spillage of buffer is contained.
- Electrotransfer is achieved at 40 V constant voltage (resulting in a current of ≈200 mA), no reversal of current is required if the sandwich is assembled as described above. The BioRad blotting tank comes with a plastic vessel for ice to keep the buffer cold during blotting. In my experience 2 h blotting time is sufficient to transfer even high molecular mass proteins at the interface between stacking and separating gel. It is good practice to stain the gel afterwards to check for residual protein, at least occasionally.
- Remove bound CTAB by incubating the membrane twice for 1 h in washing solution.
- To check whether proteins have been transferred to the membrane, incubate it with staining solution for a few minutes and then rinse with running tap water. It is convenient to mark the position of the molecular weight standards at this time with a soft pencil.

- Block non-specific protein binding sites on the membrane with 5 % low fat milk powder in Blotto for about 15 min. Longer times are often suggested in the literature, but are not required and may reduce sensitivity by covering the bands to be detected with milk proteins. This step also removes any Ponceau that may still be on the membrane.
- Wash the membrane briefly with Blotto and incubated with the primary antibody. Incubation can be performed for 1 h at 37°C if one is in a hurry. However, if the incubation is performed overnight at 4°C in the cold room on an orbital shaker, sensitivity is higher and the antibody solution can be reused many times. In fact it is quite possible to do ones entire PhD work with a single 10 ml batch of primary antibody solution.
- Wash the membrane three times for 15 min with blotto, incubate with the secondary antibody (HRP-conjugated, diluted as by manufacturers recommendation) for 1 h and wash again three times with blotto.
- For chemical staining of the blot (24) defrost a 1 mL aliquot of substrate solution, dilute to 20 mL and add 1 µL of 30 % H_2O_2. Incubate the membrane with that solution on an orbital shaker until you see well-developed bands. Rinse briefly with water and air-dry. The precipitate of elementary Ni photographs well and is stable for years in a note book.

2D-electrophoresis

- To use an IPGphor horizontal electrophoresis unit mix the sample 1:1 with IEF-sample buffer and placethis solution into the ceramic strip holder.
- Remove the protective plastic cover from the gel and place it gel side down into the sample. The sample volume depends on the length of the strips, for the 7 cm strips compatible with minigels use 125 µl. Add covering solution and place the strip holder onto the IPGphor unit.
- Allow the sample to soak into the strip at a temperature of 18 °C and an applied voltage of 50 V for 8 h. Then increase the voltage to 500, 1000 and 4000 V for 30 min each, finally 8000 V for a total of 25 000 V h. The maximum voltage available on the IPGphor unit is 8000 V; depending on the conductivity of the sample this value may not be reached. This is no problem, however, since runs can be normalised by the applied V h. It is convenient to run the IEF overnight, since it does not need supervision.
- Remove the strip from the sample holder and place it into a test tube with equilibration buffer. Seal the tube with Parafilm and place it on an end-over-end mixer for 10 min at 37 °C. This allows sufficient time for CTAB binding to the protein bands without diffusional band spreading.
- Mount the equilibrated strip carefully onto a CTAB-gel without introduction of air bubbles. A thin layer of stacking gel buffer on top of the gel can aid this process and is wicked away once the strip is in place. Use gels with one small lane for the molecular mass standards and a wide lane for the IEF-strip. Strip and gel should be hand-warm to allow mounting the strip with agarose solution (300 µL). Once the agarose has set the gel is ready to be run.

Safety Issues

(a) Acrylamide is neurotoxic and needs to be handled with great care. Readymade acrylamide/bisacrylamide solutions are commercially available and avoid the development of dust during weighing. Store the solution over an anion exchanger to remove any acrylic acid which may form to prevent electroendosmosis during runs.

(b) Store hydrogen peroxide solutions in the bottles supplied by the manufacturer, which have release valves to prevent the build-up of pressure from decomposition.

(c) DAB is a suspected carcinogen. It is safest to add the buffer directly to the bottle and assume that the manufacturer has sold the correct weight. When opening a bottle containing DAB, do not wear gloves as the electrostatic charge on the gloves would blow the fine powder directly into your face. Eye protection is recommended. DAB is also available as tablets for safer handling.

(d) Agarose: Some experimentation is required with the settings of the microwave oven so that the agarose dissolves in the mounting medium without boiling over.

Electrophoresis

(a) Fenton-polymerisation results in gels which are somewhat more brittle than those produced by photo-polymerisation. However, if the Pharmacia multi-casting stand is used the non-transparent aluminia back-plates prevent the use of a photo-polymerisation system, in that case the Fenton-system must be used. The separation achievable does not depend on the polymerisation reaction used, however.

(b) Both systems are even more sensitive to the presence of oxygen than the TEMED/APS-system used for Laemmli-gels. Proper degassing of the gel mixture is essential. In addition, it is important that the combs prevent access of air to the polymerising stacking gel. Neither those produced by Pharmacia nor those from BioRad do that however, resulting in ill-formed wells, sometimes so badly that the gel has to be discarded.

(c) Gel concentration depends on the molecular mass range of the proteins of interest. If the 10-gel multicaster is used, the gradient can be easily formed with a gradient maker, but for single gels the volume required is too small. In that case step gradients can be cast by mixing table. These solutions are carefully layered on top of each other, this is made easier when the heavy solution contains 10 % glycerol (which does not otherwise interfere with electrophoresis). The total volume of that gradient is 2.35 mL, enough for 1 mini-gel. With the photo-polymerisation chemistry you can start polymerisation to be delayed until the gel is cast.

Table: Volumes of 5 and 15 % gel solution for a step gradient gel (1 minigel).

% Desired	5 % (µl)	15 % (µl)
15.0	0	392
14.0	36	356
13.1	71	320
12.3	107	285
11.3	142	249

10.5	178	214
9.5	214	178
8.6	249	142
7.7	285	107
6.8	320	71
5.9	356	35
5.0	392	0

(e) Proteins with very high molecular mass (> 200 kDa) require low acrylamide concentration results in very soft, difficult to handle gels. Add 0.5 % agarose to stabilise them without any effect on separation.

(f) The buffer system used here was originally described in as suitable for cationic electrophoresis. For example, the Na/K-ATPase acylphosphate is most stable at acidic pH and destroyed in the presence of potassium.

(g) CTAB has a relatively high Krafft-point and precipitates if the temperature drops below 18°C. If lower temperatures are desired during electrophoresis, consider replacing CTAB with a detergent with lower cmc and/or Krafft point like 16-BAC.

(h) Prestained molecular weight markers, available for SDS-PAGE from several manufacturers, are unusable for CTAB-PAGE even after detergent exchange.

The bound dye influences the Rf -value and separation is no longer according to molecular mass. It can only be hoped that such standards will also become available for CTAB-PAGE.

Gel Staining

(a) The staining of gels with Ponceau S is sensitive and fast, but staining with CBB-R250 is also possible. Phenol red can be used in the same way as Ponceau S, the same is probably true for other acidic dyes. Phenol red is fluorescent under acidic conditions, making very sensitive detection of proteins possible.

(b) Silver staining of CTAB-gels is possible, the method of Heukeshoven & Dernick achieves a higher sensitivity than that of Merril. You can execute the former method at a constant temperature of 37 °C for all steps; replace the glutardialdehyde with formaldehyde for maximum sensitivity.

(c) PAS-staining of CTAB-gels works, but considerable savings in time and chemicals are possible if the staining is performed on eastern-blots rather than on gels.

(d) CTAB is a very mild detergent, which can retain the enzymatic activity of enzymes solubilised with it. Try zymograms, proteins in gels specifically stained by their enzymatic activity, at least with monomeric enzymes.

(e) Gels can be dried after incubation with 1 % glycerol between two sheets of cellophane. Span the assembly between two plastic frames and leave on the bench to dry overnight. This is a more convenient approach than drying gels onto filter paper, rehydration is easier should that become necessary. For autoradiography replace with ^{14}C or ^{35}S replace one of the cellophane sheets by Saran wrap and remove after drying.

(f) Alternatively destain by putting the gel between filter paper (3 sheets of Whatman No. 3 on both sides of the gel) and place it in a tank blotter with blotting buffer. 30 min at 40 V (≈ 200 mA) removes the background stain, while the protein/stain complex is immobile. If need be, proteins become mobile again after incubating the gel in 1 mM KOH for 30 min.

Blotting

(a) Large membrane proteins like Na/K-ATPase and Mdr1 blot well in a tank blotter but there is no obvious reason why you cannot use other blotting methods if desired.

(b) Before eastern blots can be immuno-stained, the CTAB needs to be washed off, as the protein/CTAB complexes would unspecifically bind antibodies. Wash PVDF membranes with SDS in methanol. For nitrocellulose 50 % methanol, 10 % acetic acid is suitable, but results in weaker bands.

(c) After removal of CTAB most staining procedures used for western blots can also be used for eastern, for example India ink (permanent,) or Ponceau S. The latter dye is washed off during the blocking step and does not interfere with immune detection.

(d) When you use horseradish peroxidase conjugated antibodies for detection, do not use sodium azide as preservative, since it inactivates the enzyme. Thimerosal® (sodium ethylmercurithiosalicylate) is a suitable alternative.

(e) In some cases, phosphate or milk powder present in the Blotto may interfere (e.g. anti-phosphotyrosine antibodies). In those cases use Trisbuffered saline (20 mM Tris-HCl, 137 mM NaCl, 4 mM KCl, pH 7.4). Replace milk powder with fish skin gelatine (mammalian gelatine or bovine serum albumin work less well).

(f) Detection of HRP-conjugated secondary antibodies can be done either by forming a coloured precipitate directly on the blot or by chemiluminescence. The direct method is easier to perform and the stained blot can be glued into the notebook for a permanent record. Chemiluminescence is more sensitive by at least 1 order of magnitude. Note that for chemiluminescence detection the secondary antibody must be more dilute to reduce background. Refer to manufacturers instruction. Very pure luminol and p-iodophenol (should be colourless, not brown) are required, commercially available reagent mixes may also be used.

(g) Phosphatase-labelled secondary antibodies with the appropriate detection reagents should also be usable in Eastern blotting.

(h) Alternatively incubate the membrane with the chemiluminescence reagent for several minutes, then wrap in cellophane and placed against X-ray film. The light signal is stable for some time, during which you can try different exposure times.

DNA Sequencing

With the invention of DNA sequencing, research and discovery of new genes is increased many folds. Using this technique whole genomes of many animal (human), plant, and microbial genomes have been sequenced. The first attempts to sequence DNA mirrored techniques developed in 1960s to sequence RNA. These involved:

- Specific cleavage to the DNA in to smaller fragments by enzymatic digestion or chemical digestion.
- Nearest neighbour analysis.
- Wandering spot method.

Indeed in some studies the DNA was transcribed in to RNA with E.coli RNA polymerase and then sequenced as RNA.

- How many base pairs (bp) are there in a human genome?(~3 billion (haploid)).
- How much did it cost to sequence the first human genome? ~$2.7 billion.
- How long did it take to sequence the first human genome?(~13 years).
- When was the first human genome sequence complete? (2000-2003):
    - Goal: Figuring the order of nucleotides across a genome.
    - Problem: Current DNA sequencing methods can handle only short stretches of DNA at once (<1-2Kbp).
    - Solution: Sequence and then use computers to assemble the small pieces.

Maxam-Gilbert Sequencing

Also known as chemical sequencing, here there is use radioactive labeling at one 5' end of the DNA fragment which is to be sequenced.

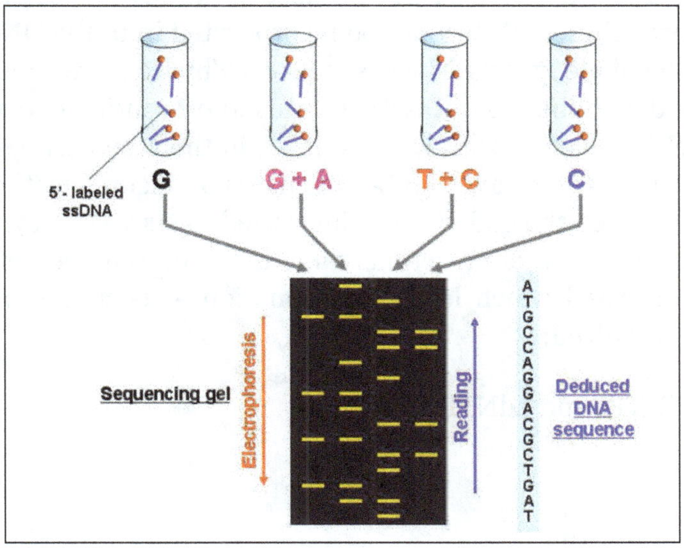

Maxam Gilbert Sequencing Method.

This method grew out of studies of the interaction between Lac repressor and lac operator in vitro. In this procedure DNA is radiolabelled at one end is partially cleaved in five different chemical reactions, and these are specific for each base. This generates five polulations of radiolabelled molecules that extend from a common point to the site of terminal cleavage. Each population consists of mixture of different molecules and their length is determined by how the bases are located in DNA. These populations are then resolved by electrophoresis through polyacrlamide gels and the end labelled molecules are detected by autoradiography. This method has not changed in any form or technique in later years, its same as it was devloped. The chemical degradation approach has one advantage over the chain termination sequencing: that the DNA sequence is obtained from the original DNA molecule and it is not an enzymatic copy. Therefore with this method one can sequence synthetic nucleotides, analyze DNA modifications such as DNA methylation and study both DNA secondary structure and interaction of proteins wth DNA be chemical protection or modification interference experiments.

Sanger's Sequencing

Frederick Sanger and his colleagues developed the method of sequencing known as chain termination method. This method soon became popular and common method of sequencing used by molecular biologists because of its ease to use and reliability. In this there is less use of toxic chemicals and also less amount of radioactivity is used in comparison to Maxam method. Sanger's method was automated and it became the commonly used method of DNA sequencing. Next generation sequencing methods are also based on Sanger sequencing method. This method first described the use of: the use of specific primer for extension by DNA polymerase , base specific chain termination, use of polyacrylamide gels to discriminate between single stranded DNA chains differing in length by a single nucleotide. Despite these advances the method was too inaccurate to gain general acceptance and it was not until the introduction of chain terminating dideoxynucleoside triphosphates ddNTPs that enzymatic methods of DNA sequencing were used extensively.

2', 3'ddNTPs differs from conventional DNTPs because of the absence of a hydroxyl residue at the 3'position of deoxyribose. These ddNTPs can be incorporated in to the DNA chain by DNA polymerases by their 5'triphosphate groups. The absence of a 3'hydroxyl residue will not allow to form the phosphodiester bond with the use of DNPT. It leads to prevention of extension of the growing DNA chain. When ddNTP is added with DNTPs, it set in the race between chain extension and termination of chain. The resultant are number of chains of oligonucleotide whose lengths is inferred by the distance between the end of the primer used to start DNA synthesis and the sites of premature termination. In a separate reactions there is use of four separate ddNTPs, number of oligonucleotides are generated which lead to termination at positions of A, C, G or T in the tem-plate strand. Steps Involved:

- Mix DNA with dNTPs and ddNTPs,
- Amplify,
- Run in Gel,
- Fragments migrate distance that is proportional to their size.

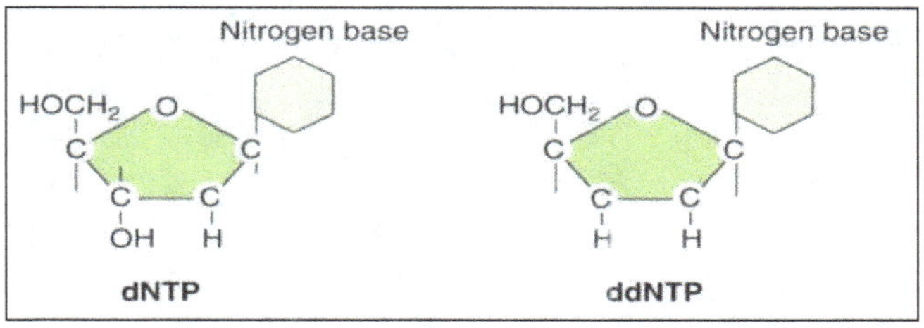

The 3′-OH group is missing in ddNTPs.which is important for the phosphodiester bond formation.

- Labelled primer and template are used in sequencing reaction mixture.
- Dideoxynucleotides are added to all tubes.

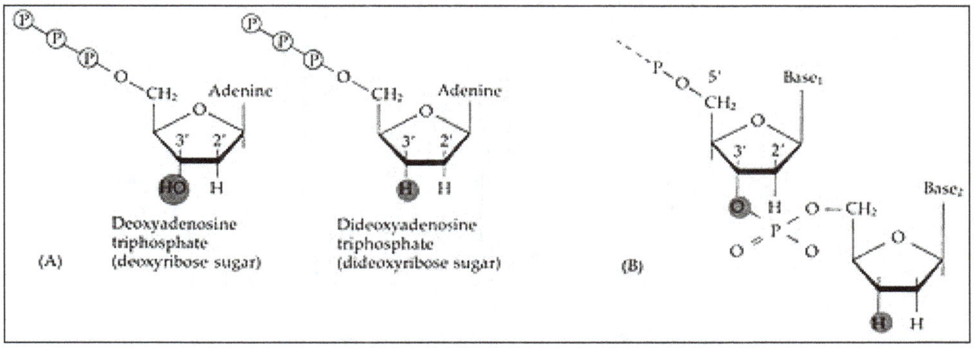

Fluorescent Dyes

- Multicyclic molecules which at different wavelengths absorb and emit light.
- Commonly sued are derivatives of fluorescein and rhodamine.
- In sequencing, these fluorescent molecules attached to the nucleotides covalently.
- Separate color is used for four ddNTP.
- Because we are using a color dye so no need for separate tubes.

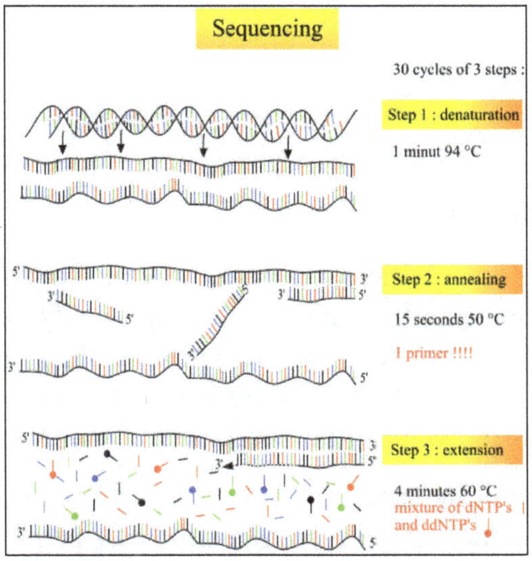

Diagrammatic representation of method.

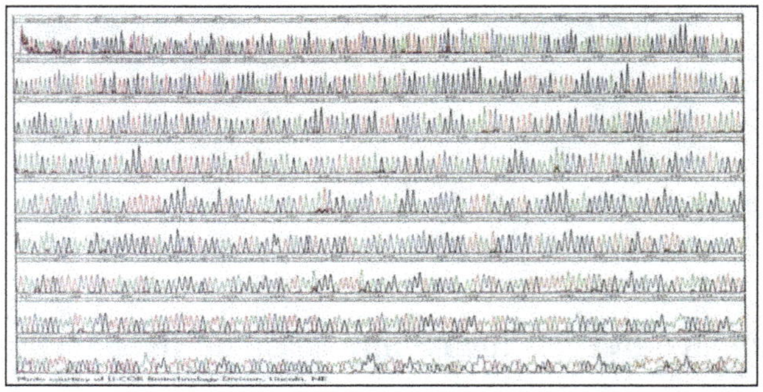

Computerized visualization from a single lane of an automated sequencer.

Method uses non-radioactive fluorescent labelling.

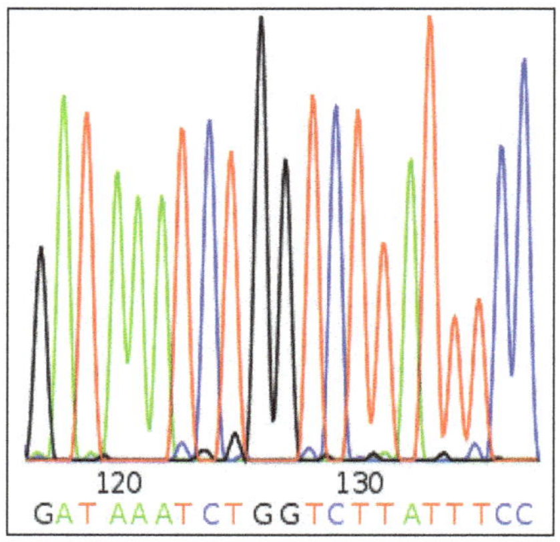

Chromatogram.

Sanger Sequencing

Advantages

- Long reads (~900bps).
- Suitable for small projects.

Disadvantages

- Low throughput.
- Expensive.

Next Generation Sequencing

Next-generation sequencing is term which applies to genome sequencing, genome resequencing, transcriptome profiling, protein-DNA interactions. Resequencing is important to study a genome variation which is not possible by studying single genome. There is increasing demand for low cost sequencing and this has led to invention of next-generation sequencing or high-throughput sequencing technologies that can produce thousands or millions of sequences simultaneously.

NGS is divided further into polony-sequencing based technologies in which there is requirement to amplify DNA before sequencing, and single molecule sequencing which does not require amplification. By this method full E. coli genome was studied in 2005.

Applications of NGS include gene expression analysis and re-sequencing. And of course – advancements in chemistry, microscopy and other related technologies enabled the new sequencing technologies.

NGS Platforms

- Roche 454.
- Illumina Solexa.
- Applied Biosystems SOLiD.

Pyrosequencing

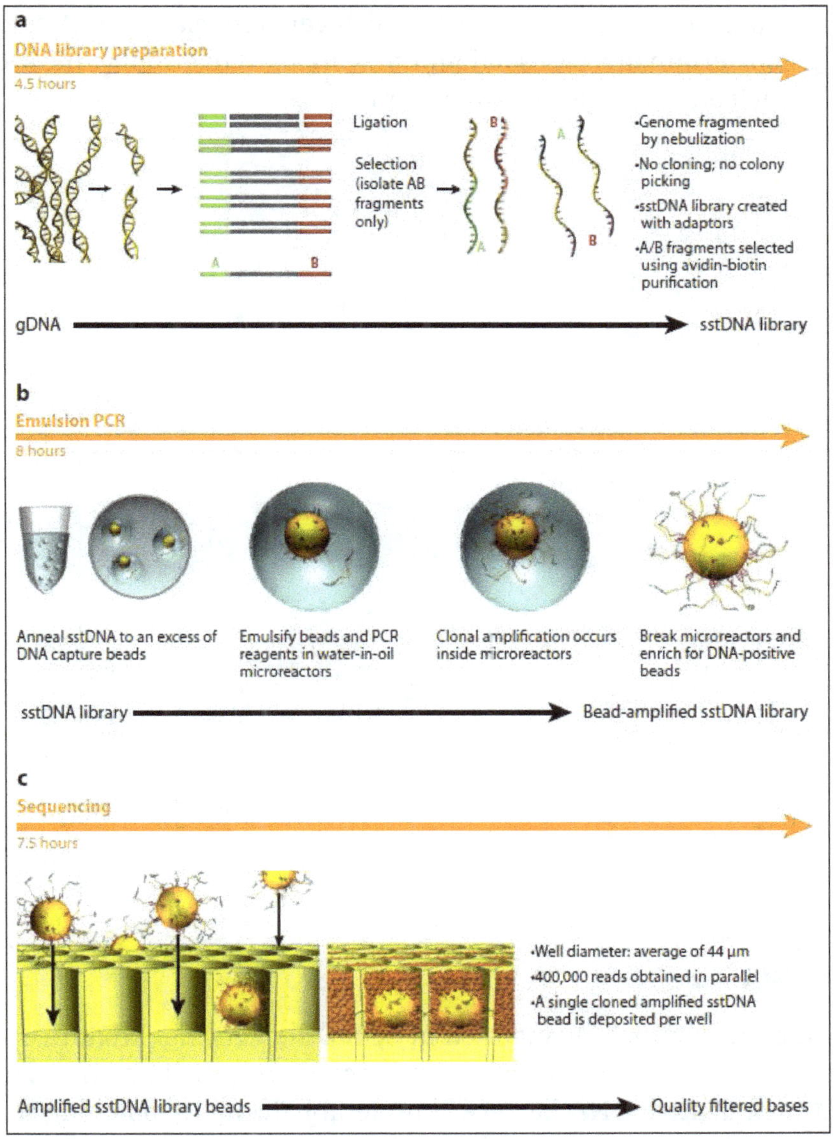

In pyrosequencing sequence-based detection technology is used which enables fast and error free quantification of variation in a sequence. It is based on the principle of "sequencing by synthesis." Here chain termination method is not used rather in this technique when there is incorporation of nucleotide there is detection of pyrophosphate release. This technique employs emulsion polymerase chain reaction. DNA is amplified in water droplets in oil emulsion and each droplet contains DNA attached to beads which are coated with primers. The sequencing equipment contains many wells of picoliter-volume with each well containing a single bead and sequencing enzymes. Luciferase enzyme is used in this technique for the generation of light to detect single nucleotides added to nascent DNA. Pooled data is used to generate read outs of sequence.

Illumina (Solexa) Sequencing

Shankar Balasubramanian and David Klenerman co-founded Solexa in 1988 (it is part of Illumina). Their method of sequencing was based on reversible dye terminator technology and engineered DNA polymerase enzymes. Here primers with DNA template are fixed to slide or flow cell and then amplification is done with polymerase which generate clonal DNA colonies or clusters. Four types of reversible terminator bases of four types are used to evaluate the sequence. Image of fluorescent labeled nucleotide is captured by camera.

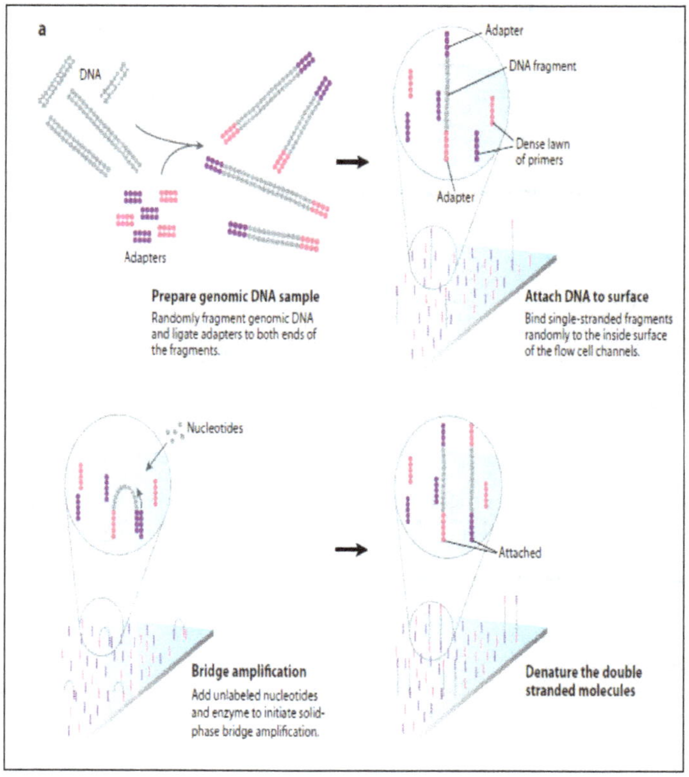

SOLiD Sequencing

Sequencing by ligation technique is being used in SOLiD sequencing. Mixture of oligonucleotides of known and fixed length is labeled according to the sequenced position. The oligonucleotides are first annealed and then ligated. DNA ligase will preferably ligate to matching sequences and will lead to signal corresponding to that nucleotide at that position. Emulsion PCR is used for amplification of DNA prior to sequencing. The resulting beads, each of which contain single copies of the similar DNA molecule, are being deposited on a glass slide. This results in sequences of quantities and lengths which are comparable to Illumina sequencing.

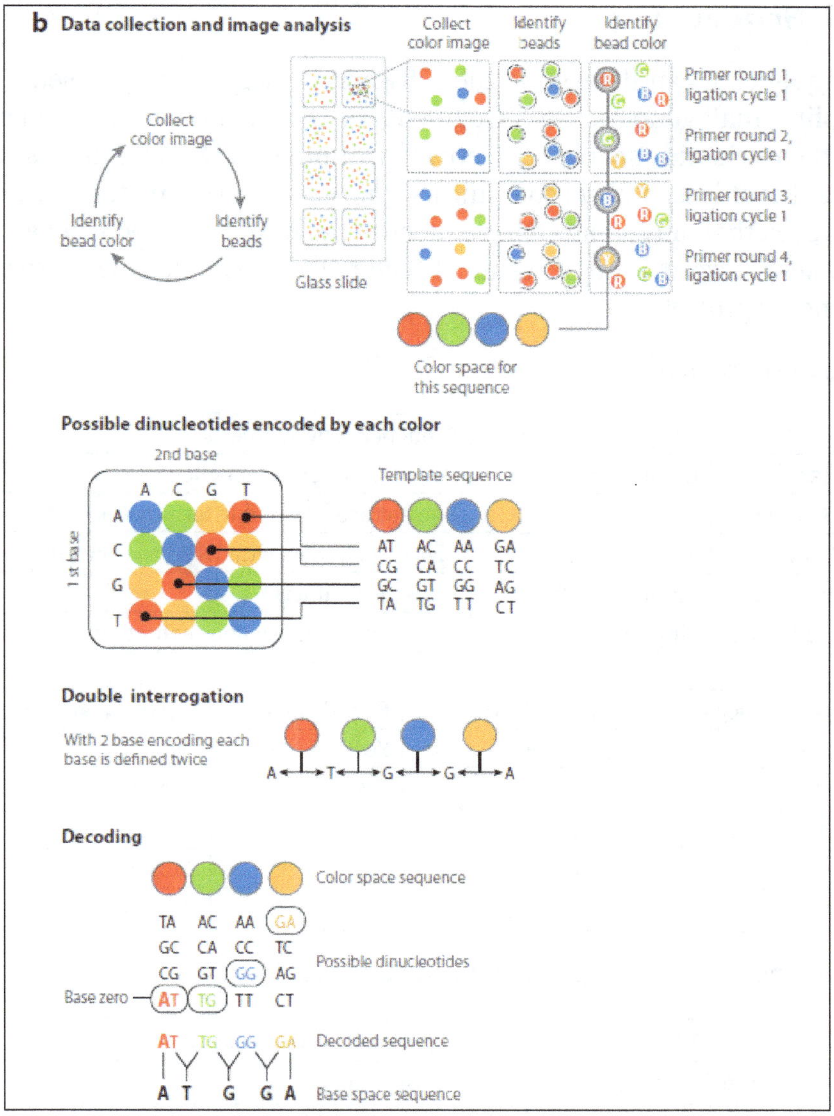

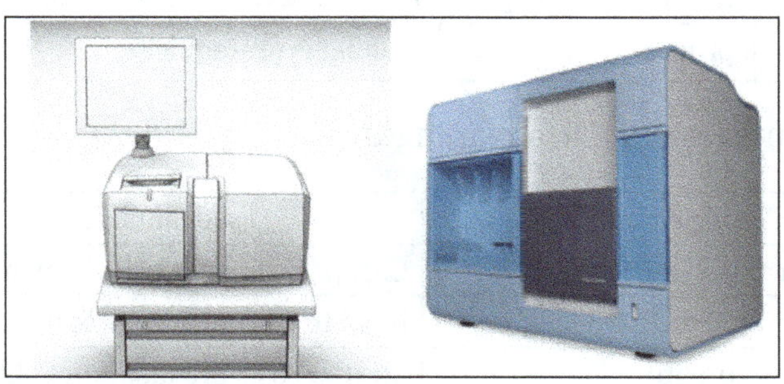

Genome Sequencer 20 (454). Genome analyzer Solexa.

Ion Torrent Semiconductor Sequencing

Life Technologies have invented a system which is based on general technique of sequencing but they have used a novel system based on semiconductor detection. This sequencing method is based on the detection of hydrogen ions which are released during the polymerisation of DNA, not like optical methods which are used in most of sequencing systems.

DNA Nanoball Sequencing

DNA nanoball sequencing can determine the full genome sequence. This sequencing technique uses RCR to amplify small genomic DNA fragments to DNA nanoballs. Sequencing by ligation is then used to see the nucleotide sequence. Here lafge number of DNA nanoballs can be sequenced per run and reagent cost is also very economic as compared to NGS methods which are available. There is one drawback that only small DNA sequences can be determined from each DNA nanoball which further makes interpretation of results difficult. This technology is more successfully used in genome sequencing projects.

Table: Comparative analysis of different methods of NGS.

	Feature generation	Sequencing by synthesis
454	Emulsion PCR	Polymerase (pyro sequencing)
Solexa	Bridge PCR	Polymerase (reversible terminator)
SOLiD	Emulsion PCR	Ligase (octamers with two-base encoding)
Polonator	Emulsion PCR	Ligase (nonamers)
Heliscope	Single molecule	Polymers (asynchronous extention)

	Cost per megabase	Cost per instrument	Paired ends	1°error modality	Read - length
454	-$60	$500,000	Yes	Indel	250bp
Solexa	-$2	$430,000	Yes	Subset.	36bp
SOLiD	-$2	$591,000	Yes	Subset.	35bp
Polonator	-$1	$155,000	Yes	Subset.	13bp
HeliScope	-$1	$1,350,000	Yes	Del	30bp

Table: Sanger Sequencing: Advantages and disadvantages.

Advantages	Disadvantages
Error rate is very low	Cost is high per base
Can read long length (~750 bp)	Takes long time for generation of data
Can target a primer	Cloning is prerequisite

Table: Sequencing: Advantages and Disadvantages.

Advantages	Disadvantages
Error rate is low	Cost per base is relatively high
It can read medium length	Need to be run at large scale
(~400-600 bp)	Startup costs are medium to high

Table: Ion Torrent Sequencing: Advantages and disadvantages.

Advantages	Disadvantages
Startup cost is low	It is New still developing technology
Scalable (10 – 1000 Mb of data per run)	Cost is high in comparisonto Illumina
Cost per base is medium to low)	Read lengths only ~100-200 bp

Error rate is less	
Runs very fast (<3 hours)	

Table: Illumina Sequencing: Advantages and disadvantages.

Advantages	Disadvantages
Error rate is low	Must run at very large scale
Its cost is lowest per base	Short read length
Can generate tons of data	(50-75 bp)
	Multiple days for runs Startup cost is very high De Novo assembly difficult

Table: PacBio Sequencing: Advantages and disadvantages.

Advantages	Disadvantages
Single molecules can be used as template	Error rate is very high (~10-15%)
It is potential for very long reads	Cost per base is high
(several kb+)	Startup costs are high

NGS Platforms Overview

- Differ in design and chemistries.
- Fundamentally related-sequencing of thousands to millions of clonally amplified molecules in a massively parallel manner.
- Orders of magnitude more information-will continue to evolve.
- Attractive for clinical applications – individual sequencing assays costly and laborious- serial "gene by gene" analysis.

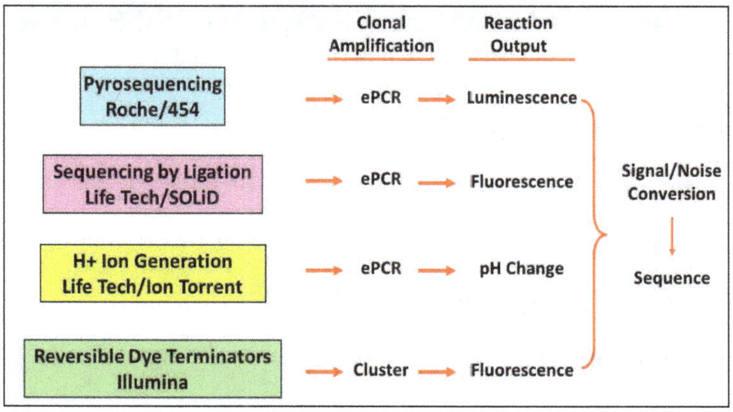

Sequencing

- Sanger:
  - Small projects (less than 1Mbp).

- 454:
  - De-novo sequencing, metagenomics.
- Solexa, SOLiD, Heliscope:
  - Gene expression, protein-DNA interactions.
  - Resequencing.
- Significant improvements were made in Sanger Method but it could not bring down the cost.
- Among many other technologies developed at that time, the one which became first became commercial in 2005 was the 454 Life Sciences sequencer.
- Sequencing costs have come down tremendously.
- Large amount of data can be generated per run.

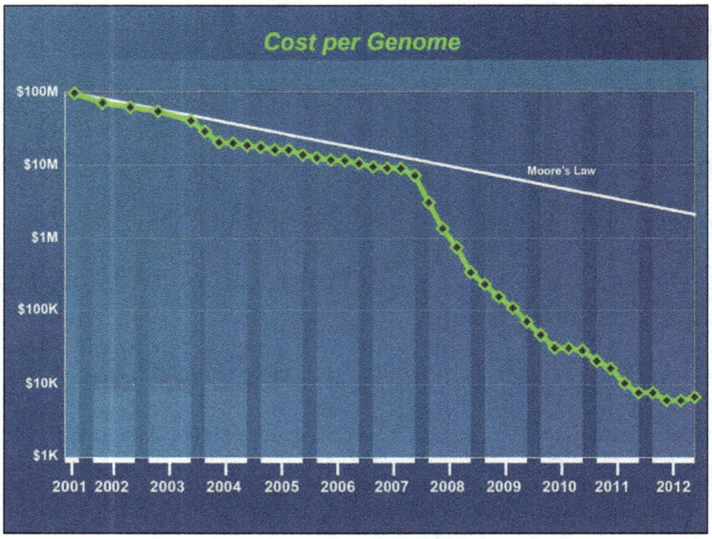

Graph showing fall in cost of sequencing.

Applications of Next Generation Sequencing

- Detection of mutation.
- Detection of foreign DNA.
- Non-invasive diagnosis aneuploidy.
- Characterization of population.
- Cancer genetics.
- Analysis of expression.
- Transcription binding.
- Chromosomal interaction.

Applications of DNA Sequencing

- Forensic: To know the identity of person killed/or any unknown person, or to identify criminals by leftover items on the crime scene.
- Medicine: Genes responsible for up-regulating or down-regulating due to some diseases can be identified by comparing healthy and diseased.
- Agriculture: Development of transgenic plants for increased yield or virus resistant plant development.

References

- Northern-blotting, testing-procedures: mybiosource.com, Retrieved 18, January 2020
- Cationic-Electrophoresis-and-Eastern-Blotting-24308958: researchgate.net, Retrieved 30, May 2020

5

Macromolecule Blotting

The process of transferring macromolecules like proteins, DNA and RNA from gel onto a carrier such as nitrocellulose or nylon membrane is known as blotting. In order to completely understand macromolecule blotting and probing, it is necessary to understand the processes related to it. The following chapter elucidates the varied processes and mechanisms associated with this area of study.

Southern Blot

A Southern blot is a method used in molecular biology for detection of a specific DNA sequence in DNA samples. Southern blotting combines transfer of electrophoresis-separated DNA fragments to a filter membrane and subsequent fragment detection by probe hybridization.

The method is named after its inventor, the British biologist Edwin Southern. Other blotting methods (i.e., western blot, northern blot, eastern blot, southwestern blot) that employ similar principles, but using RNA or protein, have later been named in reference to Edwin Southern's name. As the label is eponymous, Southern is capitalised, as is conventional for proper nouns. The names for other blotting methods may follow this convention, by analogy.

Method

1. Restriction endonucleases are used to cut high-molecular-weight DNA strands into smaller fragments.
2. The DNA fragments are then electrophoresed on an agarose gel to separate them by size.
3. If some of the DNA fragments are larger than 15 kb, then prior to blotting, the gel may be treated with an acid, such as dilute HCl. This depurinates the DNA fragments, breaking the DNA into smaller pieces, thereby allowing more efficient transfer from the gel to membrane.
4. If alkaline transfer methods are used, the DNA gel is placed into an alkaline solution (typically containing sodium hydroxide) to denature the double-stranded DNA. The denaturation in an alkaline environment may improve binding of the negatively charged thymine residues of DNA to a positively charged amino groups of membrane, separating it into single DNA strands for later hybridization to the probe, and destroys any residual RNA that may still be present in the DNA. The choice of alkaline over neutral transfer methods, however, is often empirical and may result in equivalent results.

5. A sheet of nitrocellulose (or, alternatively, nylon) membrane is placed on top of (or below, depending on the direction of the transfer) the gel. Pressure is applied evenly to the gel (either using suction, or by placing a stack of paper towels and a weight on top of the membrane and gel), to ensure good and even contact between gel and membrane. If transferring by suction, 20X SSC buffer is used to ensure a seal and prevent drying of the gel. Buffer transfer by capillary action from a region of high water potential to a region of low water potential (usually filter paper and paper tissues) is then used to move the DNA from the gel onto the membrane; ion exchange interactions bind the DNA to the membrane due to the negative charge of the DNA and positive charge of the membrane.

6. The membrane is then baked in a vacuum or regular oven at 80 °C for 2 hours (standard conditions; nitrocellulose or nylon membrane) or exposed to ultraviolet radiation (nylon membrane) to permanently attach the transferred DNA to the membrane.

7. The membrane is then exposed to a hybridization probe—a single DNA fragment with a specific sequence whose presence in the target DNA is to be determined. The probe DNA is labelled so that it can be detected, usually by incorporating radioactivity or tagging the molecule with a fluorescent or chromogenic dye. In some cases, the hybridization probe may be made from RNA, rather than DNA. To ensure the specificity of the binding of the probe to the sample DNA, most common hybridization methods use salmon or herring sperm DNA for blocking of the membrane surface and target DNA, deionized formamide, and detergents such as SDS to reduce non-specific binding of the probe.

8. After hybridization, excess probe is washed from the membrane (typically using SSC buffer), and the pattern of hybridization is visualized on X-ray film by autoradiography in the case of a radioactive or fluorescent probe, or by development of colour on the membrane if a chromogenic detection method is used.

Result

Hybridization of the probe to a specific DNA fragment on the filter membrane indicates that this fragment contains DNA sequence that is complementary to the probe. The transfer step of the DNA from the electrophoresis gel to a membrane permits easy binding of the labeled hybridization probe to the size-fractionated DNA. It also allows for the fixation of the target-probe hybrids, required for analysis by autoradiography or other detection methods. Southern blots performed with restriction enzyme-digested genomic DNA may be used to determine the number of sequences (e.g., gene copies) in a genome. A probe that hybridizes only to a single DNA segment that has not been cut by the restriction enzyme will produce a single band on a Southern blot, whereas multiple bands will likely be observed when the probe hybridizes to several highly similar sequences (e.g., those that may be the result of sequence duplication). Modification of the hybridization conditions (for example, increasing the hybridization temperature or decreasing salt concentration) may be used to increase specificity and decrease hybridization of the probe to sequences that are less than 100% similar.

Applications

Southern transfer may be used for homology-based cloning on the basis of amino acid sequence of the protein product of the target gene. Oligonucleotides are designed that are similar to the target sequence. The oligonucleotides are chemically synthesised, radiolabeled, and used to screen a DNA library, or other collections of cloned DNA fragments. Sequences that hybridize with the hybridization probe are further analysed, for example, to obtain the full length sequence of the targeted gene.

Southern blotting can also be used to identify methylated sites in particular genes. Particularly useful are the restriction nucleases *MspI* and *HpaII*, both of which recognize and cleave within the same sequence. However, *HpaII* requires that a C within that site be methylated, whereas *MspI* cleaves only DNA unmethylated at that site. Therefore, any methylated sites within a sequence analyzed with a particular probe will be cleaved by the former, but not the latter, enzyme.

Northern Blot

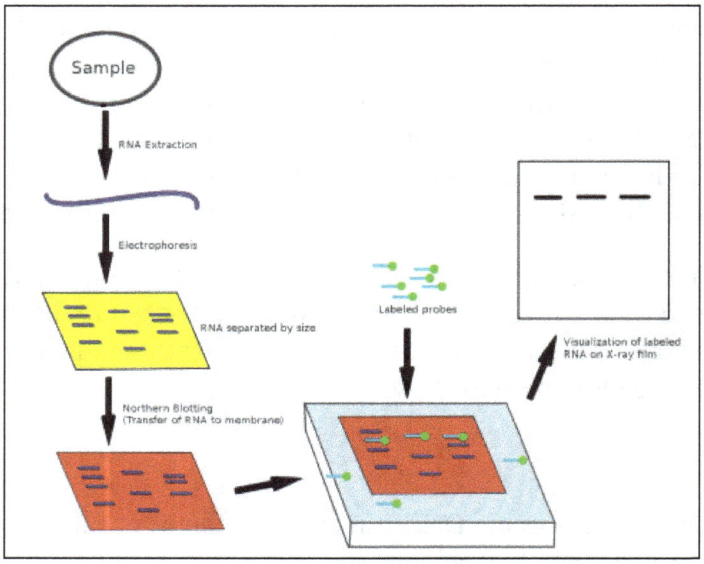

Flow diagram outlining the general procedure for RNA detection by northern blotting.

The northern blot is a technique used in molecular biology research to study gene expression by detection of RNA (or isolated mRNA) in a sample.

With northern blotting it is possible to observe cellular control over structure and function by determining the particular gene expression rates during differentiation and morphogenesis, as well as in abnormal or diseased conditions. Northern blotting involves the use of electrophoresis to separate RNA samples by size, and detection with a hybridization probe complementary to part of or the entire target sequence. The term 'northern blot' actually refers specifically to the capillary transfer of RNA from the electrophoresis gel to the blotting membrane. However, the entire process is commonly referred to as northern blotting. The northern blot technique was developed in 1977 by James Alwine, David Kemp, and George Stark at Stanford University. Northern blotting takes its name from its similarity to the first blotting technique, the Southern blot, named for biologist Edwin Southern. The major difference is that RNA, rather than DNA, is analyzed in the northern blot.

Procedure

A general blotting procedure starts with extraction of total RNA from a homogenized tissue sample or from cells. Eukaryotic mRNA can then be isolated through the use of oligo (dT) cellulose chromatography to isolate only those RNAs with a poly(A) tail. RNA samples are then separated by gel electrophoresis. Since the gels are fragile and the probes are unable to enter the matrix, the RNA samples, now separated by size, are transferred to a nylon membrane through a capillary or vacuum blotting system.

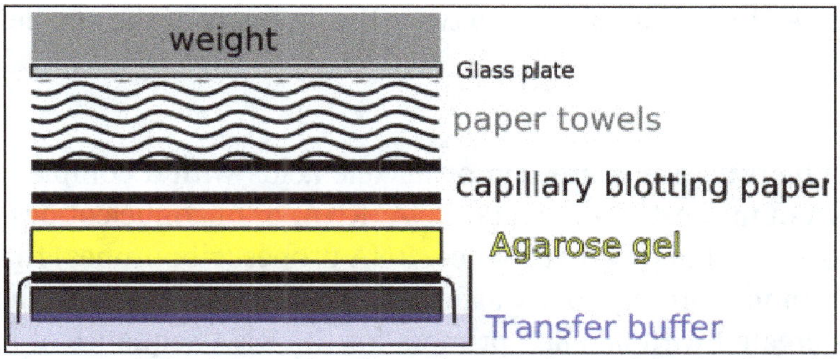

Capillary blotting system setup for the transfer of RNA from an electrophoresis gel to a blotting membrane.

A nylon membrane with a positive charge is the most effective for use in northern blotting since the negatively charged nucleic acids have a high affinity for them. The transfer buffer used for the blotting usually contains formamide because it lowers the annealing temperature of the probe-RNA interaction, thus eliminating the need for high temperatures, which could cause RNA degradation. Once the RNA has been transferred to the membrane, it is immobilized through covalent linkage to the membrane by UV light or heat. After a probe has been labeled, it is hybridized to the RNA on the membrane. Experimental conditions that can affect the efficiency and specificity of hybridization include ionic strength, viscosity, duplex length, mismatched base pairs, and base composition. The membrane is washed to ensure that the probe has bound specifically and to prevent background signals from arising. The hybrid signals are then detected by X-ray film and can be quantified by densitometry. To create controls for comparison in a northern blot, samples not displaying the gene product of interest can be used after determination by microarrays or RT-PCR.

Gels

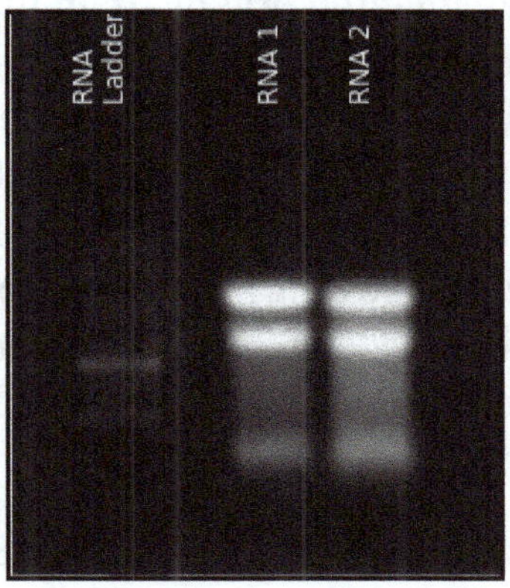

Formaldehyde gel (1%) with RNA samples run at 100V for 1 hour in 1x MOPS buffer.

RNA run on a formaldehyde agarose gel to highlight the 28S (top band) and 18S (lower band) ribosomal subunits.

The RNA samples are most commonly separated on agarose gels containing formaldehyde as a denaturing agent for the RNA to limit secondary structure. The gels can be stained with ethidium bromide (EtBr) and viewed under UV light to observe the quality and quantity of RNA before blotting. Polyacrylamide gel electrophoeresis with urea can also be used in RNA separation but it is most commonly used for fragmented RNA or microRNAs. An RNA ladder is often run alongside the samples on an electrophoresis gel to observe the size of fragments obtained but in total RNA samples the ribosomal subunits can act as size markers. Since the large ribosomal subunit is 28S (approximately 5kb) and the small ribosomal subunit is 18S (approximately 2kb) two prominent bands appear on the gel, the larger at close to twice the intensity of the smaller.

Probes

Probes for northern blotting are composed of nucleic acids with a complementary sequence to all or part of the RNA of interest, they can be DNA, RNA, or oligonucleotides with a minimum of 25 complementary bases to the target sequence. RNA probes (riboprobes) that are transcribed in vitro are able to withstand more rigorous washing steps preventing some of the background noise. Commonly cDNA is created with labelled primers for the RNA sequence of interest to act as the probe in the northern blot. The probes must be labelled either with radioactive isotopes (^{32}P) or with chemiluminescence in which alkaline phosphatase or horseradish peroxidase (HRP) break down chemiluminescent substrates producing a detectable emission of light. The chemiluminescent labelling can occur in two ways: either the probe is attached to the enzyme, or the probe is labelled with a ligand (e.g. biotin) for which the ligand (e.g., avidin or streptavidin) is attached to the enzyme (e.g. HRP). X-ray film can detect both the radioactive and chemiluminescent signals and many researchers prefer the chemiluminescent signals because they are faster, more sensitive, and reduce the health hazards that go along with radioactive labels. The same membrane can be probed up to five times without a significant loss of the target RNA.

Applications

Northern blotting allows one to observe a particular gene's expression pattern between tissues, organs, developmental stages, environmental stress levels, pathogen infection, and over the course of treatment. The technique has been used to show overexpression of oncogenes and downregulation of tumor-suppressor genes in cancerous cells when compared to 'normal' tissue, as well as the gene expression in the rejection of transplanted organs. If an upregulated gene is observed by an abundance of mRNA on the northern blot the sample can then be sequenced to determine if the gene is known to researchers or if it is a novel finding. The expression patterns obtained under given conditions can provide insight into the function of that gene. Since the RNA is first separated by size, if only one probe type is used variance in the level of each band on the membrane can provide insight into the size of the product, suggesting alternative splice products of the same gene or repetitive sequence motifs. The variance in size of a gene product can also indicate deletions or errors in transcript processing. By altering the probe target used along the known sequence it is possible to determine which region of the RNA is missing.

BlotBase is an online database publishing northern blots. BlotBase has over 700 published northern blots of human and mouse samples, in over 650 genes across more than 25 different tissue types. Northern blots can be searched by a blot ID, paper reference, gene identifier, or by tissue. The results of a search provide the blot ID, species, tissue, gene, expression level, blot image (if available), and links to the publication that the work originated from. This new database provides sharing of information between members of the science community that was not previously seen in northern blotting as it was in sequence analysis, genome determination, protein structure, etc.

Advantages and Disadvantages

Analysis of gene expression can be done by several different methods including RT-PCR, RNase protection assays, microarrays, RNA-Seq, serial analysis of gene expression (SAGE), as well as northern blotting. Microarrays are quite commonly used and are usually consistent with data obtained from northern blots; however, at times northern blotting is able to detect small changes in gene expression that microarrays cannot. The advantage that microarrays have over northern blots is that thousands of genes can be visualized at a time, while northern blotting is usually looking at one or a small number of genes.

A problem in northern blotting is often sample degradation by RNases (both endogenous to the sample and through environmental contamination), which can be avoided by proper sterilization of glassware and the use of RNase inhibitors such as DEPC (diethylpyrocarbonate). The chemicals used in most northern blots can be a risk to the researcher, since formaldehyde, radioactive material, ethidium bromide, DEPC, and UV light are all harmful under certain exposures. Compared to RT-PCR, northern blotting has a low sensitivity, but it also has a high specificity, which is important to reduce false positive results.

The advantages of using northern blotting include the detection of RNA size, the observation of alternate splice products, the use of probes with partial homology, the quality and quantity of RNA can be measured on the gel prior to blotting, and the membranes can be stored and reprobed for years after blotting.

For northern blotting for the detection of acetylcholinesterase mRNA the nonradioactive technique was compared to a radioactive technique and found as sensitive as the radioactive one, but requires no protection against radiation and is less time consuming.

Reverse Northern Blot

Researchers occasionally use a variant of the procedure known as the reverse northern blot. In this procedure, the substrate nucleic acid (that is affixed to the membrane) is a collection of isolated DNA fragments, and the probe is RNA extracted from a tissue and radioactively labelled.

The use of DNA microarrays that have come into widespread use in the late 1990s and early 2000s is more akin to the reverse procedure, in that they involve the use of isolated DNA fragments affixed to a substrate, and hybridization with a probe made from cellular RNA. Thus the reverse procedure, though originally uncommon, enabled northern analysis to evolve into gene expression profiling, in which many (possibly all) of the genes in an organism may have their expression monitored.

Western Blot

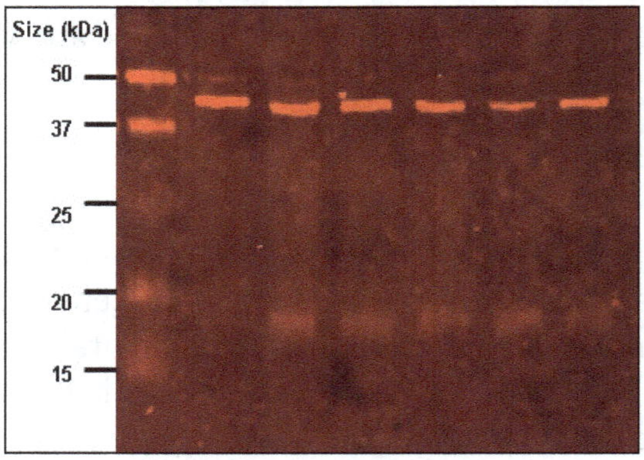

Western blot using an antibody that recognizes proteins modified with lipoic acid.

The western blot (sometimes called the protein immunoblot) is a widely used analytical technique used in molecular biology to detect specific proteins in a sample of tissue homogenate or extract. It uses gel electrophoresis to separate native proteins by 3-D structure or denatured proteins by the length of the polypeptide. The proteins are then transferred to a membrane (typically nitrocellulose or PVDF), where they are stained with antibodies specific to the target protein. The gel electrophoresis step is included in western blot analysis to resolve the issue of the cross-reactivity of antibodies.

There are many reagent companies that specialize in providing antibodies (both monoclonal and polyclonal antibodies) against tens of thousands of different proteins. Commercial antibodies can be expensive, although the unbound antibody can be reused between experiments. This method is used in the fields of molecular biology, immunogenetics and other molecular biology disciplines. A number of search engines, such as CiteAb, Antibodypedia, and SeekProducts, are available that can help researchers find suitable antibodies for use in western blotting.

Other related techniques include dot blot analysis, immunohistochemistry and immunocytochemistry where antibodies are used to detect proteins in tissues and cells by immunostaining, and enzyme-linked immunosorbent assay (ELISA).

The method originated in the laboratory of Harry Towbin at the Friedrich Miescher Institute. The name *western blot* was given to the technique by W. Neal Burnette and is a play on the name Southern blot, a technique for DNA detection developed earlier by Edwin Southern. Detection of RNA is termed northern blot and was developed by George Stark at Stanford.

Steps

Tissue Preparation

Samples can be taken from whole tissue or from cell culture. Solid tissues are first broken down mechanically using a blender (for larger sample volumes), using a homogenizer (smaller volumes), or by sonication. Cells may also be broken open by one of the above mechanical methods. However, virus or environmental samples can be the source of protein and thus western blotting is not restricted to cellular studies only.

Assorted detergents, salts, and buffers may be employed to encourage lysis of cells and to solubilize proteins. Protease and phosphatase inhibitors are often added to prevent the digestion of the sample by its own enzymes. Tissue preparation is often done at cold temperatures to avoid protein denaturing and degradation.

A combination of biochemical and mechanical techniques – comprising various types of filtration and centrifugation – can be used to separate different cell compartments and organelles.

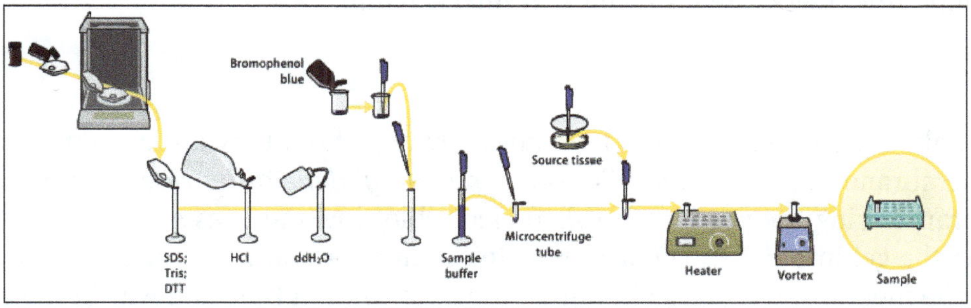

Gel Electrophoresis

The proteins of the sample are separated using gel electrophoresis. Separation of proteins may be by isoelectric point (pI), molecular weight, electric charge, or a combination of these factors. The nature of the separation depends on the treatment of the sample and the nature of the gel. This is a very useful way to identify a protein.

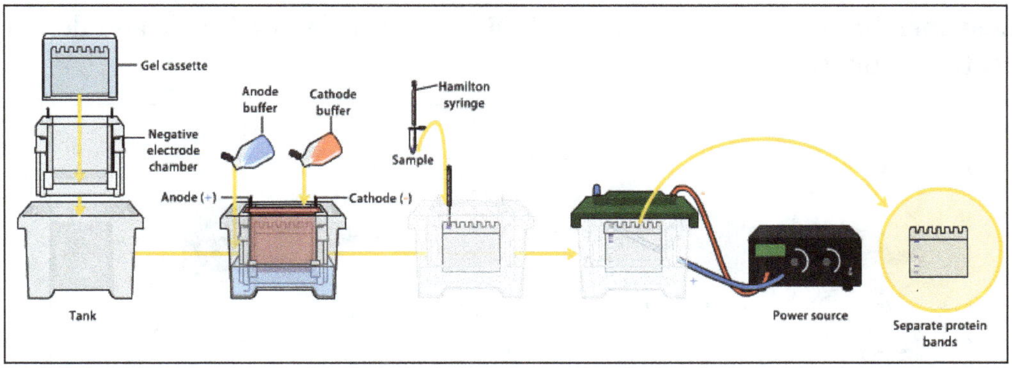

It is also possible to use a two-dimensional (2-D) gel which spreads the proteins from a single sample out in two dimensions. Proteins are separated according to isoelectric point (pH at which they have neutral net charge) in the first dimension, and according to their molecular weight in the second dimension.

By far the most common type of gel electrophoresis employs polyacrylamide gels and buffers loaded with sodium dodecyl sulfate (SDS). SDS-PAGE (SDS polyacrylamide gel electrophoresis) maintains polypeptides in a denatured state once they have been treated with strong reducing agents to remove secondary and tertiary structure (e.g. disulfide bonds [S-S] to sulfhydryl groups [SH and SH]) and thus allows separation of proteins by their molecular weight. Sampled proteins become covered in the negatively charged SDS and move to the positively charged electrode through the acrylamide mesh of the gel. Smaller proteins migrate faster through this mesh and the proteins are thus separated according to size (usually measured in kilodaltons, kDa). The concentration of acrylamide determines the resolution of the gel - the greater the acrylamide concentration, the better the resolution of lower molecular weight proteins. The lower the acrylamide concentration, the better the resolution of higher molecular weight proteins. Proteins travel only in one dimension along the gel for most blots.

Samples are loaded into *wells* in the gel. One lane is usually reserved for a *marker* or *ladder*, a commercially available mixture of proteins having defined molecular weights, typically stained so as to form visible, coloured bands. When voltage is applied along the gel, proteins migrate through it at different speeds dependent on their size. These different rates of advancement (different *electrophoretic mobilities*) separate into *bands* within each *lane*.

Transfer

In order to make the proteins accessible to antibody detection, they are moved from within the gel onto a membrane made of *nitrocellulose or polyvinylidene difluoride (PVDF)*. The primary method for transferring the proteins is called electroblotting and uses an electric current to pull proteins from the gel into the PVDF or nitrocellulose membrane. The proteins move from within the gel onto the membrane while maintaining the organization they had within the gel. An older method of transfer involves placing a membrane on top of the gel, and a stack of filter papers on top of that. The entire stack is placed in a buffer solution which moves up the paper by capillary action, bringing the proteins with it. In practice this method is not used as it takes too much time; electroblotting is preferred. As a result of either "blotting" process, the proteins are exposed on a thin surface layer for detection. Both varieties of membrane are chosen for their non-specific protein binding properties (i.e. binds all proteins equally well). Protein binding is based upon hydrophobic interactions, as well as charged interactions between the membrane and protein. Nitrocellulose membranes are cheaper than PVDF, but are far more fragile and do not stand up well to repeated probings.

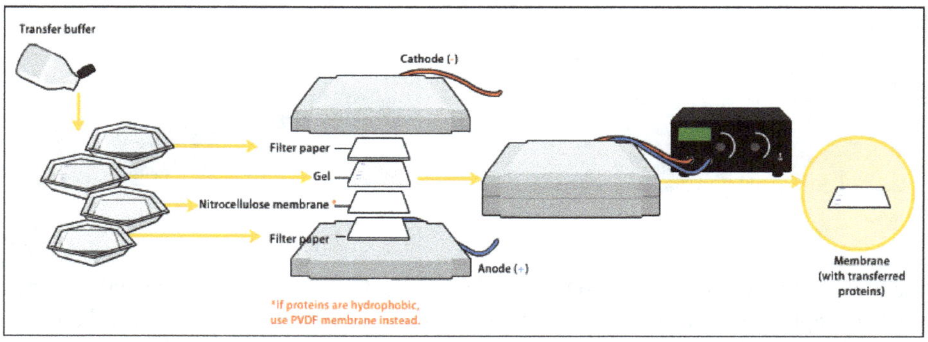

The uniformity and overall effectiveness of transfer of protein from the gel to the membrane can be checked by staining the membrane with Coomassie Brilliant Blue or Ponceau S dyes. Ponceau S is the more common of the two, due to its higher sensitivity and water solubility, the latter making it easier to subsequently destain and probe the membrane, as described below.

Blocking

Since the membrane has been chosen for its ability to bind protein and as both antibodies and the target are proteins, steps must be taken to prevent the interactions between the membrane and the antibody used for detection of the target protein. Blocking of non-specific binding is achieved by placing the membrane in a dilute solution of protein - typically 3-5% Bovine serum albumin (BSA) or non-fat dry milk (both are inexpensive) in Tris-Buffered Saline (TBS) or I-Block, with a minute percentage (0.1%) of detergent such as Tween 20 or Triton X-100. The protein in the dilute solution attaches to the membrane in all places where the target proteins have not attached. Thus, when the antibody is added, there is no room on the membrane for it to attach other than on the binding sites of the specific target protein. This reduces background in the final product of the western blot, leading to clearer results, and eliminates false positives.

Incubation

During the detection process the membrane is "probed" for the protein of interest with a modified antibody which is linked to a reporter enzyme; when exposed to an appropriate substrate, this enzyme drives a colourimetric reaction and produces a color. For a variety of reasons, this traditionally takes place in a two-step process, although there are now one-step detection methods available for certain applications.

Two Steps

- Primary antibody.

The primary antibodies are generated when a host species or immune cell culture is exposed to protein of interest (or a part thereof). Normally, this is part of the immune response, whereas here they are harvested and used as sensitive and specific detection tools that bind the protein directly.

After blocking, a dilute solution of primary antibody (generally between 0.5 and 5 micrograms/mL) is incubated with the membrane under gentle agitation. Typically, the solution is composed of buffered saline solution with a small percentage of detergent, and sometimes with powdered milk or BSA. The antibody solution and the membrane can be sealed and incubated together for anywhere from 30 minutes to overnight. It can also be incubated at different temperatures, with lesser temperatures being associated with more binding, both specific (to the target protein, the "signal") and non-specific ("noise").

- Secondary antibody.

After rinsing the membrane to remove unbound primary antibody, the membrane is exposed to another antibody, directed at a species-specific portion of the primary antibody. Antibodies come from animal sources (or animal sourced hybridoma cultures); an anti-mouse secondary will bind to almost any mouse-sourced primary antibody, which allows some cost savings by allowing an entire lab to share a single source of mass-produced antibody, and provides far more consistent results. This is known as a secondary antibody, and due to its targeting properties, tends to be referred to as "anti-mouse," "anti-goat," etc. The secondary antibody is usually linked to biotin or to a reporter enzyme such as alkaline phosphatase or horseradish peroxidase. This means that several secondary antibodies will bind to one primary antibody and enhance the signal.

Most commonly, a horseradish peroxidase-linked secondary is used to cleave a chemiluminescent agent, and the reaction product produces luminescence in proportion to the amount of protein. A sensitive sheet of photographic film is placed against the membrane, and exposure to the light from the reaction creates an image of the antibodies bound to the blot. A cheaper but less sensitive approach utilizes a 4-chloronaphthol stain with 1% hydrogen peroxide; reaction of peroxide radicals with 4-chloronaphthol produces a dark purple stain that can be photographed without using specialized photographic film.

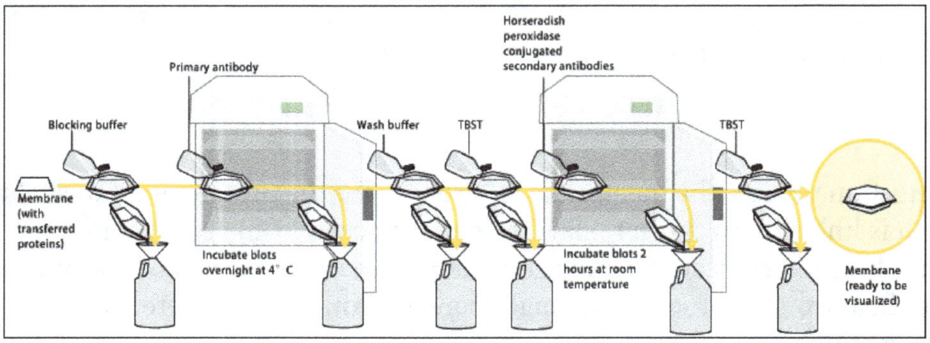

As with the ELISPOT and ELISA procedures, the enzyme can be provided with a substrate molecule that will be converted by the enzyme to a colored reaction product that will be visible on the membrane (see the figure below with blue bands).

Another method of secondary antibody detection utilizes a near-infrared (NIR) fluorophore-linked antibody. Light produced from the excitation of a fluorescent dye is static, making fluorescent detection a more precise and accurate measure of the difference in signal produced by labeled antibodies bound to proteins on a western blot. Proteins can be accurately quantified because the signal generated by the different amounts of proteins on the membranes is measured in a static state, as compared to chemiluminescence, in which light is measured in a dynamic state.

A third alternative is to use a radioactive label rather than an enzyme coupled to the secondary antibody, such as labeling an antibody-binding protein like *Staphylococcus* Protein A or Streptavidin with a radioactive isotope of iodine. Since other methods are safer, quicker, and cheaper, this method is now rarely used; however, an advantage of this approach is the sensitivity of auto-radiography-based imaging, which enables highly accurate protein quantification when combined with optical software (e.g. Optiquant).

One Step

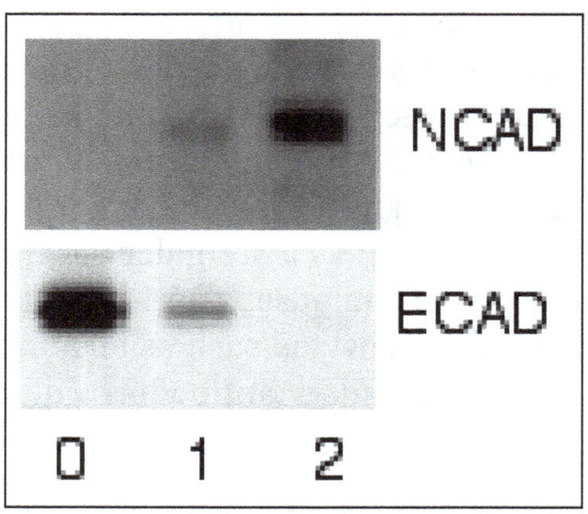

Western blot using radioactive detection system

Historically, the probing process was performed in two steps because of the relative ease of producing primary and secondary antibodies in separate processes. This gives researchers and corporations huge advantages in terms of flexibility, and adds an amplification step to the detection process. Given the advent of high-throughput protein analysis and lower limits of detection, however, there has been interest in developing one-step probing systems that would allow the process to occur faster and with fewer consumables. This requires a probe antibody which both recognizes the protein of interest and contains a detectable label, probes which are often available for known protein tags. The primary probe is incubated with the membrane in a manner similar to that for the primary antibody in a two-step process, and then is ready for direct detection after a series of wash steps.

Detection / Visualization

After the unbound probes are washed away, the western blot is ready for detection of the probes that are labeled and bound to the protein of interest. In practical terms, not all westerns reveal protein only at one band in a membrane. Size approximations are taken by comparing the stained bands to that of the marker or ladder loaded during electrophoresis. The process is commonly repeated for a structural protein, such as actin or tubulin, that should not change between samples. The amount of target protein is normalized to the structural protein to control between groups. A superior strategy is the normalization to the total protein visualized with trichloroethanol or epicocconone. This practice ensures correction for the amount of total protein on the membrane in case of errors or incomplete transfers.

Colorimetric Detection

The colorimetric detection method depends on incubation of the western blot with a substrate that reacts with the reporter enzyme (such as peroxidase) that is bound to the secondary antibody. This converts the soluble dye into an insoluble form of a different color that precipitates next to the enzyme and thereby stains the membrane. Development of the blot is then stopped by washing away the soluble dye. Protein levels are evaluated through densitometry (how intense the stain is) or spectrophotometry.

Chemiluminescent Detection

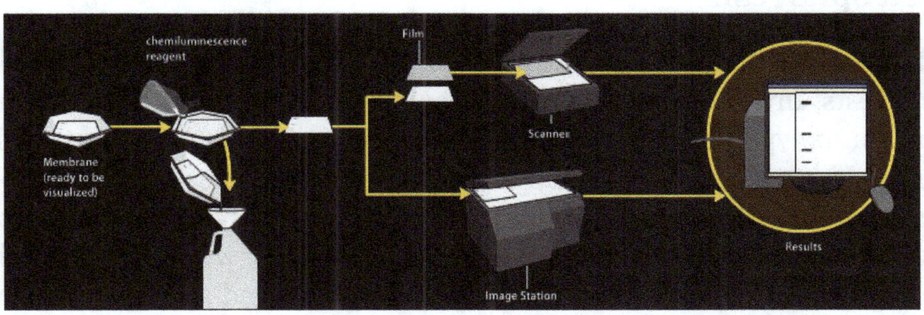

The main companies offering Chemiluminescence imaging systems are Analytik Jena, Syngene, UVP, Azure Biosystems, UVITEC, GE, Biorad and Vilber Lourmat.

Chemiluminescent detection methods depend on incubation of the western blot with a substrate that will luminesce when exposed to the reporter on the secondary antibody. The light is then detected by CCD cameras which capture a digital image of the western blot or photographic film. The use of film for western blot detection is slowly disappearing because of non linearity of the image (non accurate quantification). The image is analysed by densitometry, which evaluates the relative amount of protein staining and quantifies the results in terms of optical density. Newer software allows further data analysis such as molecular weight analysis if appropriate standards are used.

Radioactive Detection

Radioactive labels do not require enzyme substrates, but rather, allow the placement of medical X-ray film directly against the western blot, which develops as it is exposed to the label and creates dark regions which correspond to the protein bands of interest. The importance of radioactive detections methods is declining due to its hazardous radiation, because it is very expensive, health and safety risks are high, and ECL (enhanced chemiluminescence) provides a useful alternative.

Fluorescent Detection

The fluorescently labeled probe is excited by light and the emission of the excitation is then detected by a photosensor such as a CCD camera equipped with appropriate emission filters which captures a digital image of the western blot and allows further data analysis such as molecular weight analysis and a quantitative western blot analysis. Fluorescence is considered to be one of the best methods for quantification but is less sensitive than chemiluminescence.

Secondary Probing

One major difference between nitrocellulose and PVDF membranes relates to the ability of each to support "stripping" antibodies off and reusing the membrane for subsequent antibody probes. While there are well-established protocols available for stripping nitrocellulose membranes, the sturdier PVDF allows for easier stripping, and for more reuse before background noise limits experiments. Another difference is that, unlike nitrocellulose, PVDF must be soaked in 95% ethanol, isopropanol or methanol before use. PVDF membranes also tend to be thicker and more resistant to damage during use.

2-D Gel Electrophoresis

2-dimensional SDS-PAGE uses the principles and techniques outlined above. 2-D SDS-PAGE, as the name suggests, involves the migration of polypeptides in 2 dimensions. For example, in the first dimension, polypeptides are separated according to isoelectric point, while in the second dimension, polypeptides are separated according to their molecular weight. The isoelectric point of a given protein is determined by the relative number of positively (e.g. lysine, arginine) and negatively (e.g. glutamate, aspartate) charged amino acids, with negatively charged amino acids contributing to a low isoelectric point and positively charged amino acids contributing to a high isoelectric point. Samples could also be separated first under nonreducing conditions using SDS-PAGE, and under reducing conditions in the second dimension, which breaks apart disulfide bonds that hold subunits together. SDS-PAGE might also be coupled with urea-PAGE for a 2-dimensional gel.

In principle, this method allows for the separation of all cellular proteins on a single large gel. A major advantage of this method is that it often distinguishes between different isoforms of a particular protein - e.g. a protein that has been phosphorylated (by addition of a negatively charged group). Proteins that have been separated can be cut out of the gel and then analysed by mass spectrometry, which identifies the protein.

Medical Diagnostic Applications

- The confirmatory HIV test employs a western blot to detect anti-HIV antibody in a human serum sample. Proteins from known HIV-infected cells are separated and blotted on a membrane as above. Then, the serum to be tested is applied in the primary antibody incubation step; free antibody is washed away, and a secondary anti-human antibody linked to an enzyme signal is added. The stained bands then indicate the proteins to which the patient's serum contains antibody.
- A western blot is also used as the definitive test for bovine spongiform encephalopathy (BSE, commonly referred to as 'mad cow disease').
- Some forms of Lyme disease testing employ western blotting.
- A western blot can also be used as a confirmatory test for Hepatitis B infection and HSV-2 (Herpes Type 2) infection.
- In veterinary medicine, a western blot is sometimes used to confirm FIV+ status in cats.

Eastern Blot

The eastern blot is a biochemical technique used to analyze protein post translational modifications (PTM) such as lipids, phosphomoieties and glycoconjugates. It is most often used to detect carbohydrate epitopes. Thus, eastern blotting can be considered an extension of the biochemical technique of western blotting. Multiple techniques have been described by the term eastern blotting, most use proteins blotted from SDS-PAGE gel on to a PVDF or nitrocellulose membrane. Transferred proteins are analyzed for post-translational modifications using probes that may detect lipids, carbohydrate, phosphorylation or any other protein modification. Eastern blotting should be used to refer to methods that detect their targets through specific interaction of the PTM and the probe, distinguishing them from a standard Far-western blot. In principle, eastern blotting is similar to lectin blotting (i.e. detection of carbohydrate epitopes on proteins or lipids).

History and Multiple Definitions

Definition of the term *eastern blotting* is somewhat confused due to multiple sets of authors dubbing a new method as *eastern blotting*, or a derivative thereof. All of the definitions are a derivative of the technique of western blotting developed by Towbin in 1979. The current definitions are summarized below in order of the first use of the name; however, all are based on some earlier works. In some cases, the technique had been in practice for some time before the introduction of the term.

- (1982) The term *eastern blotting* was specifically rejected by two separate groups: Reinhart and Malamud referred to a protein blot of a native gel as a *native blot*; Peferoen et al., opted to refer to their method of drawing SDS-gel separated proteins onto nitrocellulose using a vacuum as *Vacuum blotting*.

- (1984) *Middle eastern blotting* has been described as a blot of polyA RNA (resolved by agarose) which is then immobilized. The immobilized RNA is then probed using DNA.
- (1996) *Eastern-western blot* was first used by Bogdanov et al. The method involved blotting of phospholipids on PVDF or nitrocellulose membrane prior to transfer of proteins onto the same nitrocellulose membrane by conventional Western blotting and probing with conformation specific antibodies. This method is based on earlier work by Taki et al. in 1994, which they originally dubbed *TLC blotting*, and was based on a similar method introduced by Towbin in 1984.
- (2000) *Far-eastern blotting* seems to have been first named in 2000 by Ishikawa & Taki. The method is described more fully in the article on Far-eastern blotting, but is based on antibody or lectin staining of lipids transferred to PVDF membranes.
- (2001) *Eastern blotting* was described as a technique for detecting glycoconjugates generated by blotting BSA onto PVDF membranes, followed by periodate treatment. The oxidized protein is then treated with a complex mixture, generating a new conjugate on the membrane. The membrane is then probed with antibodies for epitopes of interest. This method has also been discussed in later work by the same group. The method is essentially Far-Eastern blotting.
- (2002) *Eastern blot* has also been used to describe an immunoblot performed on proteins blotted to a PVDF membrane from a PAGE gel run with opposite polarity. Since this is essentially a western blot, the charge reversal was used to dub this method an *eastern blot*.
- (2005) *Eastern blot* has been used to describe a blot of proteins on PVDF membrane where the probe is an aptamer rather than an antibody. This could be seen as similar to a Southern blot, however the interaction is between a DNA molecule(the aptamer) and a protein, rather than two DNA molecules. The method is similar to South-western blotting.
- (2006) *Eastern blotting* has been used to refer to the detection of fusion proteins through complementation. The name is based on the use of an enzyme activator (EA) as part of the detection.
- (2009) *Eastern blotting* has most recently been re-dubbed by Thomas et al. as a technique which probes proteins blotted to PVDF membrane with lectins, cholera toxin and chemical stains to detect glycosylated, lipoylated or phosphorylated proteins. These authors distinguish the method from the *Far-eastern blot* named by Taki et al. in that they use lectin probes and other staining reagents.
- (2009) *Eastern blot* has been used to describe a blot of proteins on nitrocellulose membrane where the probe is an aptamer rather than an antibody. The method is similar to South-western blotting.
- (2011) A recent study used the term eastern blotting to describe detection of glycoproteins with lectins such as concanavalin A.

There is clearly no single accepted definition of the term. A recent highlight article has interviewed Ed Southern, originator of the Southern blot, regarding a re-christening of *eastern blotting* from Tanaka et al. The article likens the *eastern blot* to "fairies, unicorns, and a free lunch" and states that eastern blots "don't exist." The *eastern blot* is mentioned in an immunology textbook which compares the common blotting methods (Southern, northern, and western), and states that "the eastern blot, however, exists only in test questions."

The principles used for eastern blotting to detect glycans can be traced back to the use of lectins to detect protein glycosylation. The earliest example for this mode of detection is Tanner and Anstee in 1976, where lectins were used to detect glycosylated proteins isolated from human erythrocytes. The specific detection of glycosylation through blotting is usually referred to as *lectin blotting*. A summary of more recent improvements of the protocol has been provided by H. Freeze.

Applications

One application of the technique includes detection of protein modifications in two bacterial species *Ehrlichia- E. muris* and IOE. Cholera toxin B subunit (which binds to gangliosides), Concanavalin A (which detects mannose-containing glycans) and nitrophospho molybdate-methyl green (which detects phosphoproteins) were used to detect protein modifications. The technique showed that the antigenic proteins of the non-virulent *E.muris* is more post-translationally modified than the highly virulent IOE.

Significance

Most proteins that are translated from mRNA undergo modifications before becoming functional in cells. These modifications are collectively known as post-translational modifications (PTMs). The nascent or folded proteins, which are stable under physiological conditions, are then subjected to a battery of specific enzyme-catalyzed modifications on the side chains or backbones.

Post-translational protein modifications can include: acetylation, acylation (myristoylation, palmitoylation), alkylation, arginylation, ADP-Ribosylation, biotinylation, formylation, geranylgeranylation, glutamylation, glycosylation, glycylation, hydroxylation, isoprenylation, lipoylation, methylation, nitroalkylation, phosphopantetheinylation, phosphorylation, prenylation, selenation, S-nitrosylation, succinylation, sulfation, transglutamination and ubiquitination (sumoylation, neddylation).

Post-translational modifications occurring at the N-terminus of the amino acid chain play an important role in translocation across biological membranes. These include secretory proteins in prokaryotes and eukaryotes and also proteins that are intended to be incorporated in various cellular and organelle membranes such as lysosomes, chloroplast, mitochondria and plasma membrane. Expression of posttranslated proteins is important in several diseases.

References

- "Imaging Systems for Westerns: Chemiluminescence vs. Infrared Detection, 2009 - , Methods in Molecular Biology, Protein Blotting and Detection, vol. 536" (PDF). Humana Press. Retrieved 26, February 2020
- Gilda, J. E.; Gomes, A. V. (2013). "Stain-Free total protein staining is a superior loading control to β-actin for Western blots". Analytical Biochemistry 440 (2): 186–8. doi:10.1016/j.ab.2013.05.027. PMC 3809032 PMID 23747530.

- Mariappa D, Sauert K, Mariño K, Turnock D, Webster R, van Aalten DM, Ferguson MA, Müller HA. Protein O-GlcNAcylation is required for fibroblast growth factor signaling in Drosophila.Sci Signal. 2011 Dec 20;4(204) ra89.
- Moritz, C. P.; Marz, S. X.; Reiss, R; Schulenborg, T; Friauf, E (2014). "Epicocconone staining: A powerful loading control for Western blots". Proteomics 14 (2–3): 162–8. doi:10.1002/pmic.201300089. PMID 24339236

6

Diverse Aspects of Gene Expression

The process by which the information encoded in the DNA is translated into a functional product is known as gene expression. This chapter elucidates the concepts of GeneCalling, genetic code, antitermination, artificial gene synthesis, gene silencing, RNA splicing, alternative splicing and epitranscriptome. The topics elaborated in this chapter will help in gaining a better perspective about gene expression.

Gene expression refers to a complex series of processes in which the information encoded in a gene is used to produce a functional product such as a protein that dictates cell function. It involves several different steps through which DNA is converted to an RNA which in turn is converted into a protein or in some cases RNA, for example, genes encoding the necessary information for transfer RNAs and ribosomal RNAs (tRNAs and rRNAs).

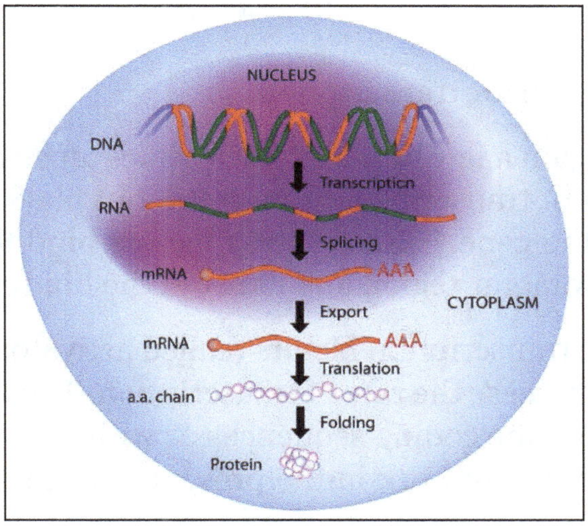

Steps of Gene Expression.

The information flow from DNA to RNA to protein can be controlled at several points helping the cell to adjust the quality and quantity of resulting proteins and thus self-regulate its functions. Thus, regulation of gene expression is a critical step in determining what kind of proteins and how much of each protein is expressed in a cell.

Follwing are the two main stages of gene expression.

Transcription

The production of messenger RNA (mRNA) by the enzyme RNA polymerase, and the processing of the resulting mRNA molecule.

Translation: the use of mRNA to direct protein synthesis, and the subsequent post-translational processing of the protein molecule.

Some genes are responsible for the production of other forms of RNA that play a role in translation, including transfer RNA (tRNA) and ribosomal RNA (rRNA).

A structural gene involves a number of different components:

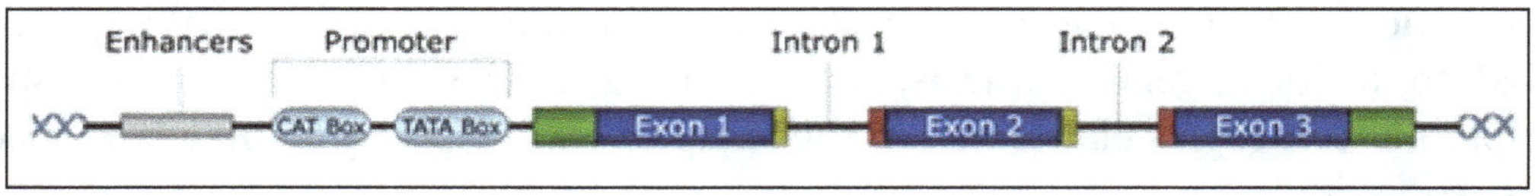

- Exons: Exons code for amino acids and collectively determine the amino acid sequence of the protein product. It is these portions of the gene that are represented in final mature mRNA molecule.
- Introns: Introns are portions of the gene that do not code for amino acids, and are removed (spliced) from the mRNA molecule before translation.

Control Regions

- Start site: A start site for transcription.
- A promoter: A region a few hundred nucleotides 'upstream' of the gene (toward the 5' end). It is not transcribed into mRNA, but plays a role in controlling the transcription of the gene. Transcription factors bind to specific nucleotide sequences in the promoter region and assist in the binding of RNA polymerases.
- Enhancers. Some transcription factors (called activators) bind to regions called 'enhancers' that increase the rate of transcription. These sites may be thousands of nucleotides from the coding sequences or within an intron. Some enhancers are conditional and only work in the presence of other factors as well as transcription factors.
- Silencers. Some transcription factors (called repressors) bind to regions called 'silencers' that depress the rate of transcription.

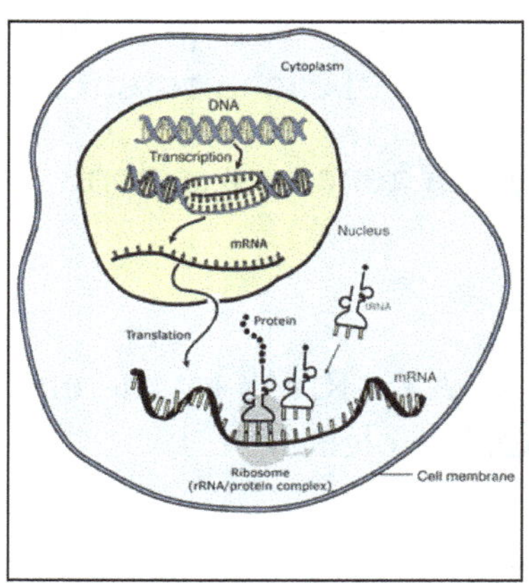

Transcription

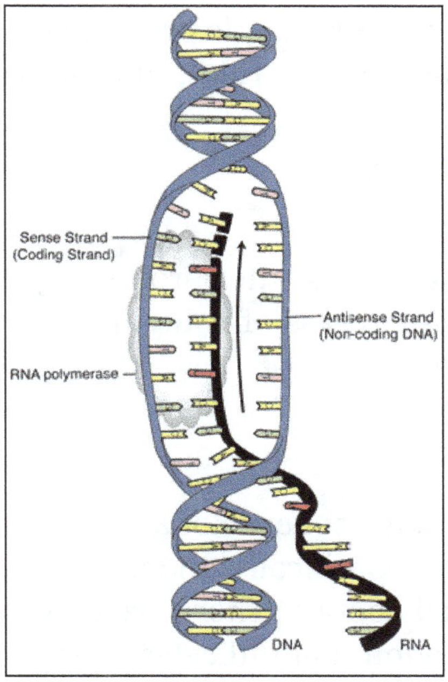

Transcription is the process of RNA synthesis, controlled by the interaction of promoters and enhancers. Several different types of RNA are produced, including messenger RNA (mRNA), which specifies the sequence of amino acids in the protein product, plus transfer RNA (tRNA) and ribosomal RNA (rRNA), which play a role in the translation process.

Transcription involves four steps:

- Initiation: The DNA molecule unwinds and separates to form a small open complex. RNA polymerase binds to the promoter of the template strand (also known as the 'sense strand' or 'coding strand'). The synthesis of RNA proceeds in a 5' to 3' direction, so the template strand must be 3' to 5'.
- Elongation: RNA polymerase moves along the template strand, synthesising an mRNA molecule. In prokaryotes RNA polymerase is a holoenzyme consisting of a number of subunits, including a sigma factor (transcription factor) that recognises the promoter. In eukaryotes there are three RNA polymerases: I, II and III. The process includes a proofreading mechanism.
- Termination: In prokaryotes there are two ways in which transcription is terminated. In ?-dependent termination, a protein? Is responsible for disrupting the complex involving the template strand, RNA polymerase and RNA molecule. Independent termination, a loop forms at the end of the RNA molecule, causing it to detach itself. Termination in eukaryotes is more complicated, involving the addition of additional adenine nucleotides at the 3' of the RNA transcript (a process referred to as polyadenylation).

- Processing: After transcription the RNA molecule is processed in a number of ways: introns are removed and the exons are spliced together to form a mature mRNA molecule consisting of a single protein-coding sequence. RNA synthesis involves the normal base pairing rules, but the base thymine is replaced with the base uracil.

Translation

After transcription (and, in eukaryotes, after processing), an mRNA molecule is ready to direct protein synthesis. The process of using information in an mRNA to build a polypeptide is called translation.

Genetic Code

During translation, the nucleotide sequence of an mRNA is translated into the amino acid sequence of a polypeptide. Specifically, the nucleotides of the mRNA are read in triplets (groups of three) called codons. There are 616161 codons that specify amino acids. One codon is a "start" codon that indicates where to start translation. The start codon specifies the amino acid methionine, so most polypeptides begin with this amino acid. Three other "stop" codons signal the end of a polypeptide. These relationships between codons and amino acids are called the genetic code.

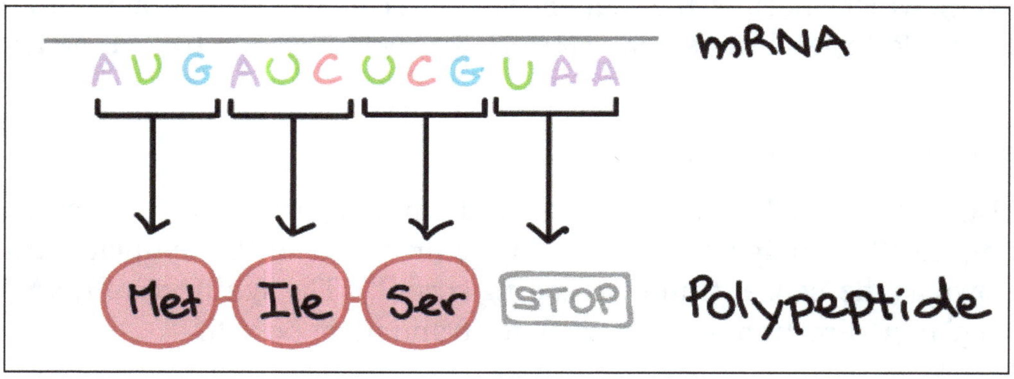

Steps of Translation

Translation takes place inside of structures known as ribosomes. Ribosomes are molecular machines whose job is to build polypeptides. Once a ribosome latches on to an mRNA and finds the "start" codon, it will travel rapidly down the mRNA, one codon at a time. As it goes, it will gradually build a chain of amino acids that exactly mirrors the sequence of codons in the mRNA.

How does the ribosome "know" which amino acid to add for each codon? As it turns out, this matching is not done by the ribosome itself. Instead, it depends on a group of specialized RNA molecules called transfer RNAS (tRNAs). Each tRNA has a three nucleotides sticking out at one end, which can recognize (base-pair with) just one or a few particular codons. At the other end, the tRNA carries an amino acid – specifically, the amino acid that matches those codons.

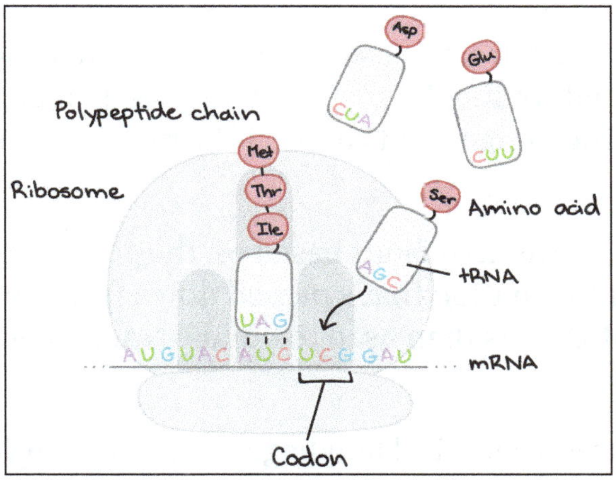

There are many tRNAs floating around in a cell, but only a tRNA that matches (base-pairs with) the codon that's currently being read can bind and deliver its amino acid cargo. Once a tRNA is snugly bound to its matching codon in the ribosome, its amino acid will be added the end of the polypeptide chain.

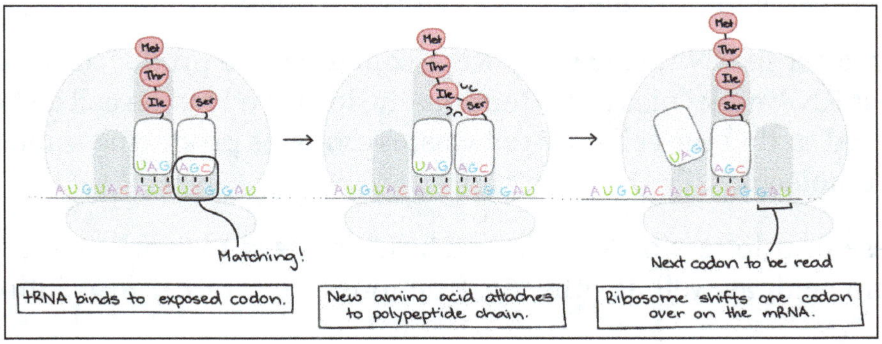

This process repeats many times, with the ribosome moving down the mRNA one codon at a time. A chain of amino acids is built up one by one, with an amino acid sequence that matches the sequence of codons found in the mRNA. Translation ends when the ribosome reaches a stop codon and releases the polypeptide.

GeneCalling

In the field of genomics, an important mRNA transcriptional profiling technique is GeneCalling. It measures the levels of cDNA, which represents the levels of gene expression relative to specific transcripts. The differences in gene expression between healthy tissues and tissues that responds to disease or drug use are examined and comparison of the two made using this technology. This technique can be applied for the study of both plant and animal tissues.

Genetic Code

The genetic code is the code our body uses to convert the instructions contained in our DNA the essential materials of life. It is typically discussed using the "codons" found in mRNA, as mRNA is the messenger that carries information from the DNA to the site of protein synthesis.

Everything in our cells is ultimately built based on the genetic code. Our hereditary information – that is, the information that's passed down from parent to child – is stored in the form of DNA. That DNA is then used to build RNA, proteins, and ultimately cells, tissues, and organs.

Like binary code, DNA uses a chemical language with just a few letters to store information in a very efficient manner. While binary uses only ones and zeroes, DNA has four letters – the four nucleotides Adenine, Cytosine, Guanine, and Thymine/Uracil.

Thymine and Uracil are very similar to each other, except that "Thymine" is slightly more stable and is used in DNA. Uracil is used in RNA, and has all the same properties of Thymine except that it is slightly more prone to mutate.

This doesn't matter in RNA, since new RNA copies can be produced from DNA at any time, and most RNA molecules are intentionally destroyed by the cell a short time after they're produced so that the cell does not waste resources producing unneeded proteins from old RNA molecules.

Together, these four letters of A, C, G, and T/U are used to "spell" coded instructions for each amino acid, as well as other instructions like "start transcription" and "stop transcription."

Instructions for "start," "stop," or for a given amino acid are "read" by the cell in three-letter blocks called "codons." When we talk about "codons," we usually mean codons in mRNA – the "messenger RNA" that is made by copying the information in DNA.

For that reason, we talk about codons made of RNA, which uses Uracil, instead of the original DNA code which uses Thymine.

Each amino acid is represented in our genetic instructions by one or more codons, as seen below:

One of the most remarkable evidences for the common descent of all life on Earth from a single ancestor is the fact that all organisms use the same genetic code to translate DNA into amino acids.

There are a few slight exceptions to be found, but the genetic code is similar enough across organisms that when a gene from a plant or jellyfish is injected into a mammal cell, for example, the mammal cell will read the gene in the same way and build the same product as the original plant or jellyfish.

First letter	Second letter: U		C		A		G		Third letter
U	UUU	Phenylalanine (Phe)	UCU	Serine (Ser)	UAU	Tyrosine (Tyr)	UGU	Cysteine (Cys)	U
	UUC		UCC		UAC		UGC		C
	UUA	Leucine (Leu)	UCA		UAA	Stop	UGA	Stop	A
	UUG		UCG		UAG	Stop	UGG	Tryptophan (Trp)	G
C	CUU	Leucine (Leu)	CCU	Proline (Pro)	CAU	Histidine (His)	CGU	Arginine (Arg)	U
	CUC		CCC		CAC		CGC		C
	CUA		CCA		CAA	Glutamine (Gln)	CGA		A
	CUG		CCG		CAG		CGG		G
A	AUU	Isoleucine (Ile)	ACU	Threonine (Thr)	AAU	Asparagine (Asn)	AGU	Serine (Ser)	U
	AUC		ACC		AAC		AGC		C
	AUA		ACA		AAA	Lysine (Lys)	AGA	Arginine (Arg)	A
	AUG	Methionine (Met)	ACG		AAG		AGG		G
G	GUU	Valine (Val)	GCU	Alanine (Ala)	GAU	Aspartic acid (Asp)	GGU	Glycine (Gly)	U
	GUC		GCC		GAC		GGC		C
	GUA		GCA		GAA	Glutamic acid (Glu)	GGA		A
	GUG		GCG		GAG		GGG		G

Function of Genetic Code

The genetic code allows cells to contain a mind-boggling amount of information.

Consider this: A microscopic fertilized egg cell, following the instructions contained in its genetic code, can produce a human or elephant which even has similar personality and behaviors to those of its parents. There is a lot of information in there.

The development of the genetic code was vital because it allowed living things to reliably produce products necessary for their survival – and pass instructions for how to do the same onto the next generation.

When a cell seeks to reproduce, one of the first things it does is make a copy of its DNA. This is the "S" phase of the cell cycle, which stands for the "Synthesis" of a new copy of the cell's DNA.

The information encoded in DNA is preserved by the specific pairing of DNA bases with each other. Adenine will only bond with Thymine, Cytosine with Guanine, etc.

That means that when a cell wants to copy its DNA, all it has to do is part the two strands of the double helix and line up the nucleotides that the bases of the existing DNA "want" to pair with.

This specific base pairing ensures that the new partner strand will contain the same sequence of base pairs – the same "code" – as the old partner strand. Each resulting double helix contains one strand of old DNA paired with one strand of new DNA.

These new double helixes will be inherited by two daughter cells. When it's time for those daughter cells to reproduce, each strand of these new double helices, act as templates for a new double helix.

When the time comes for a cell to "read" the instructions contained within its DNA, it uses the same principle of specific pair bonding. RNA is very similar to DNA, and each RNA base bonds specifically to one DNA base. Uracil binds to Adenine, Cytosine to Guanine, etc.

This means that, just like DNA replication, the information in DNA is accurately transferred to RNA as long as the resulting RNA strand is composed of the bases that bind specifically with the bases in the DNA.

Sometimes, the RNA strand itself can be the end product. Structures made of RNA perform important functions in our selves, including assembling proteins, regulating gene expression, and catalyzing the formation of proteins.

In fact, some scientists think that the first life on Earth might have been composed mainly of RNA. This is because RNA can store information in its base pairs just like DNA, but can also perform some enzymatic and regulatory functions.

In most cases, however, the RNA goes on to be transcribed into a protein. Using the amino acid "building blocks of life," our cells can build almost protein machines for almost any purpose, from muscle fibers to neurotransmitters to digestive enzymes.

In protein transcription, the RNA codons that were transcribed from the DNA are "read" by a ribosome. The ribosome finds the appropriate transfer RNA (tRNA) with "anti-codons" that are complimentary to the codons in the messenger RNA (mRNA) that has been transcribed from the DNA.

Ribosomes catalyze the formation of peptide bonds between the amino acids as they "read" each codon in the mRNA. At the end of the process, you have a string of amino acids as specified by the DNA – that is, a protein.

Other building blocks of life, such as sugars and lipids, are in turn created by proteins. In this way the information contained in the DNA is transformed into all of the materials of life, using the genetic code.

Types of Genetic Mutations

Because the genetic code contains the information to make the stuff of life, errors in an organism's DNA can have catastrophic consequences. Errors can happen during DNA replication if the wrong base pair is added to a DNA strand, if a base is skipped, or if an extra base is added.

Rarely, these errors may actually be helpful – the "mistaken" version of the DNA may work better than the original, or have an entirely new function. In that case, the new version may become more successful, and its carrier may out compete carriers of the old version in the population. This spread of new traits throughout a population is how evolution progresses.

Silent Mutations and Redundant Coding

In some cases, genetic mutations may not have any effect at all on the end product of a protein. This is because most amino acids are connected to more than one codon.

Glycine, for example, is coded for by the codons GGA, GGC, GGG, and GGU. A mutation resulting in the wrong nucleotide being used for the last letter of the glycine codon, then, would make no difference. A codon starting in "GG" would still code for glycine, no matter what letter was used last.

The use of multiple codons for the same amino acid is thought to be a mechanism evolved over time to minimize the chance of a small mutation causing problems for an organism.

Missense Mutation

In a missense mutation, the substitution of one base pair for an incorrect base pair during DNA replication results in the wrong amino acid being used in a protein.

This may have a small effect on an organism, or a large one – depending on how important the amino acid is to the function of its protein, and what protein is effected.

This can be thought of like furniture construction. How bad would it be if you used the wrong piece to bolt a chair leg in place? If you used a screw instead of a nail, the two are probably similar enough that the chair leg would stay on – but if you tried to use, say, a seat cushion to connect the leg to the chair, your chair wouldn't work very well.

Nonsense Mutation

A nonsense mutation occurs when the incorrect base pair is used during DNA replication – but where the resulting codon does not code for an incorrect amino acid.

Instead, this error creates a stop codon or another piece of information that is indecipherable to the cell. As a result, the ribosome stops working on that protein and all subsequent codons are not transcribed.

Nonsense mutations lead to incomplete proteins, which may function very poorly or not at all. Imagine if you stopped assembling a chair halfway through.

Deletion

In a deletion mutation, one or more DNA bases are not copied during DNA replication. Deletion mutations come in a huge range of sizes– a single base pair may be missing, or a large piece of a chromosome may be missing.

Smaller mutations are not always less harmful. The loss of just one or two bases can result in a frame shift mutation that impairs a crucial gene, as discussed under "frame shift mutations" below.

By contrast, larger deletion mutations may be fatal– or may only result in disability, as in Di George Syndrome and other conditions that result from the deletion of part of a chromosome.

The reason for this is that DNA is very much like computer source code – one piece of code might be crucial for the system to turn on at all, while other pieces of code might just ensure that a website looks pretty or loads quickly.

Depending on the function of the piece of code that is deleted or otherwise mutated, a small change can have catastrophic consequences – or a seemingly large corruption of code one can result in a system that is just a bit glitchy.

Insertion

An insertion mutation occurs when one or more nucleotides are erroneously added to a growing DNA strand during DNA replication. On rare occasions, long stretches of DNA may be incorrectly added in the middle of a gene.

Like a missense mutation, the impact of this can vary. The addition of an unnecessary amino acid in a protein may make the protein only slightly less efficient; or it may cripple it.

Consider what would happen to your chair if you added a random piece of wood to it that the instructions did not call for. The results could vary a lot depending on the size, shape, and placement of the extra piece.

Duplication

A duplication mutation occurs when a segment of DNA is accidentally replicated two or more times. Like the other mutations listed above, these may have mild effects – or they may be catastrophic.

To imagine if your chair had two backs, two seats, or eight legs. A small duplication and the chair may still be useable, if a little odd-looking or uncomfortable. But if the chair had, for example, six seats attached to each other, it may rapidly become useless for its intended purpose.

Frame-shift Mutation

A frame-shift mutation is a subtype of insertion, deletion, and duplication mutations. In a frame-shift mutation, one or two amino acids are deleted or inserted – resulting in a shifting of the "frame" which the ribosome uses to tell where one codon stops and the next begins.

This type of error can be especially dangerous because it causes all codons that occur after the error to be misread. Typically, every amino acid added to the protein after the frame-shift mutation is wrong.

Imagine if you were reading a book – but at some point during the writing, a programming error happened such that every subsequent letter shifted one letter later in the alphabet.

This is approximately what happens in a frame-shift mutation.

Antitermination

Antitermination is a process in which Rna Polymerase does not recognize a normal Transcription termination signal at the end of a Gene or Operon during transcription and so transcribes the DNA beyond the terminator. Antitermination provides a means of regulating the Expression of genes.

Processive Antitermination

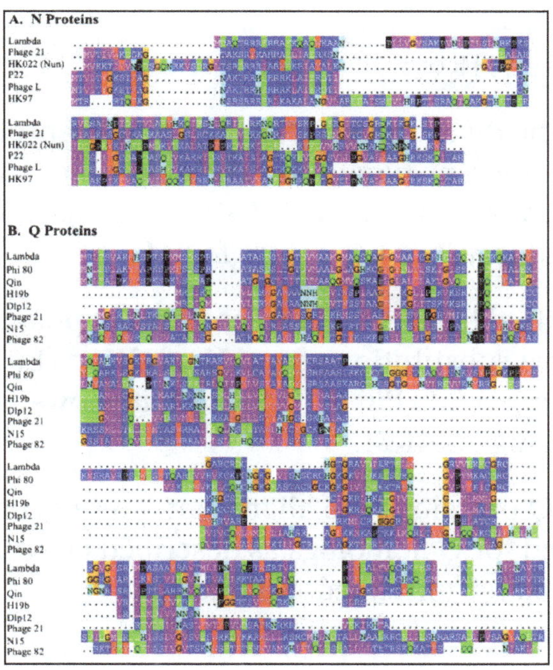

Antitermination in λ is induced by two quite distinct mechanisms. The first is the result of interaction between λ N protein and its targets in the early phage transcripts, and the second is the result of an interaction between the λ Q protein and its target in the late phage promoter. We describe the N mechanism first. Lambda N, A small basic protein of the arginine-rich motif (ARM) family of RNA binding proteins, binds to a 15-nucleotide stem-loop called BOXB. We will capitalize the names of sites in RNA and italicize the names of the corresponding DNA sequences; e.g. BOXB. *Box B* is found twice in the λ chromosome, once in each of the two early operons. It is close to the start point of the P_L operon transcript and just downstream of the first translated gene of the P_R operon. Neither the distance between the transcription start site and *box B*, nor the nature of the promoter (at least in the case of sigma-70-dependent promoters), nor the nature of the terminator is relevant to N action. Although the *box B* sequence is not well conserved in other bacteriophages of the λ family, most of these phages encode proteins that are analogous to λ N and have sequences capable of forming BOXB-like structures in their P_L and P_R operons. In some cases, it has been shown that these structures are recognized by the cognate N analogs. It is believed that this accounts for the phage specificity of N-mediated antitermination.

(A) Alignment of phage N proteins and the HK022 Nun protein. The color groupings reflect the frequency of amino acid substitutions in evolutionarily related protein domains: an amino acid is more likely to be replaced by one in the same color group than by one in a different color group in related proteins. The amino-proximal ARM regions were aligned by eye and according to the structures of the P22 and λ ARMs complex to their cognate *nut* sites and the remainder of the proteins were aligned by Crustal W. The dots indicate gaps introduced to improve the alignment. Aside from the ARM regions, the proteins fall into three very distantly related (or unrelated) families:

(i) λ and phage 21;

(ii) P22, phage L, and HK97;

(iii) HK022 Nun. The divergence of Nun from the N proteins is unsurprising because of their different functions. The sequence database was searched for additional N homologs with the PSI-BLAST program, using each of the listed sequences as a query, but none were found.

Two N proteins were omitted from the alignment: that of phage H19b, because it differs by only three conservative substitutions from N of HK97 (E60D, K80E, and R100K), and that of lambdoid phage φ80 (Phi 80), because it shows no resemblance to any of the other N proteins, lacking an ARM.

(B) Alignment of phage Q proteins. The alignments were generated by Crustal W and the database was searched for Q homologs as described above. These proteins fall into three very distantly related (or unrelated) families:

(i) λ and Qin;

(ii) H19b, Dlp12, and phage 21.

(iii) N15 and phage 82. Qin and Dlp12 are defective lambdoid pro phages of *E. coli*, but it is likely that their Q proteins are active. The Q proteins of phages HK022 and P22 were omitted from the alignment because of their close similarity to that of λ.

A putative and possibly defective Q, encoded by a sequence located upstream of Shiga-like toxin I genes in an E. coli isolate and found by a BLAST search of the translated nucleotide sequence database, was omitted from the alignment because of its close similarity to the Q of phage H19B.

The structures of complexes between the ARMs of λ N (residues 1 to 22) and P22 N and their cognate BOXBs have recently been determined by nuclear magnetic resonance. The two complexes, although similar, show differences that account for the specificity of N-BOXB recognition. Upon binding, λ and P22 ARMs adopt a bent α-helix conformation that packs against the BOXB hairpin through hydrophobic and ionic interactions. Residues in the amino-proximal segments make multiple base, ribose, and phosphate contacts in the 5′-ascending stem of BOXB without disrupting its regular A-form. The two stem-proximal residues of both RNA loops form a sheared G · A base pair which is contacted by R7 of λ N and R19 of P22 N In the λ complex, the fourth residue (G) of the GAAGA loop is extruded and the remaining residues form a GNRA fold similar to the base-stacked GAAA tetra loop reported in a number of important RNA structures. P22 N also creates a GNRA fold, but this is accomplished by extrusion of the third residue (C) of the GACAA loop. In contrast to the extruded G residue in BOXB-λ, which is not close to λ N, the extruded C residue in BOXB-P22 makes contacts with residues in the carboxyl-proximal segment of P22 N. The structure of the λ N complex is stabilized by an important stacking interaction between W18 and the second residue (A) in the BOXB- λ loop. This interaction is not found in the P22 complex. Formation of the GNRA fold is essential for λ N binding, and mutations in loop residues 1, 3, and 5 that prevent tetra loop formation block N binding. The structure of the complex between the BOXB of phage 21 and its cognate N protein must be substantially different than the complexes described above because the BOXB-21 loop (TCTAACCG) cannot be folded into a GNRA tetra loop. However, HK022 Nun protein which recognizes λ BOXB. Probably does so in a way that resembles that of λ. If so, A3, R7, and W18 of λ N, all of which make base contacts in BOXB, probably correspond to S27, R31, and Y42 of Nun.

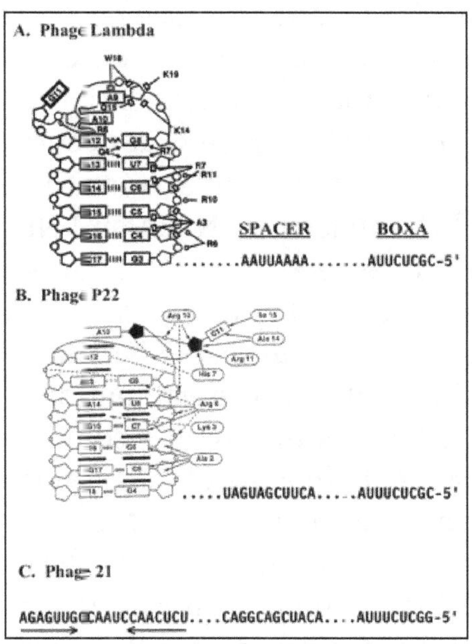

BOXA and BOXB RNAs and their interaction with the ARM of their cognate N proteins

The amino acid-nucleotide interactions are shown to the left except for BOXB of phage 21, for which the structure of the complex is unknown. The sequences of BOXA and BOXA-BOXB spacer are shown to the right. The dots to the left and right of the spacer sequences are for alignment.

(A) λ N-ARM-BOXB complex (adapted from reference with permission of the publisher). Open circles, pentagons, and rectangles represent phosphates, riboses, and bases, respectively. Watson-Crick base pairs (||||) are indicated. The zigzag line denotes a sheared G · A base pair. Open circles, open rectangles, and arrowheads depict ionic, hydrophobic, and hydrogen-bonding interactions, respectively. Guanine-11, indicated by a bold rectangle, is extruded from the BOXB loop.

(B) P22 N-ARM-BOXB complex (adapted from reference with permission of the publisher). Open circles, pentagons, rectangles, and ovals represent phosphates, riboses, bases, and amino acids, respectively. The solid pentagons indicate riboses with a C2′-*endo* pucker. Base stacking, intermolecular hydrogen bonding or electrostatic interactions (<-----), intermolecular hydrophobic or van der Waals interactions (←), intra molecular hydrogen bonds (––––) and Watson-Crick base pairs (|||||) are indicated. Cytosine-11 is extruded from the loop. Note that the amino-terminal amino acid residue in the complex corresponds to Asn-14 in the complete protein, and the displayed amino acids are numbered accordingly.

(C) NUTL site of phage 21. The arrows indicate the inverted sequence repeats of BOXB.

Promoter proximal to each of the *box B* sequences is *boxA*, which is also important for antitermination by *λ* N. *Box A*-like sequences are also found in corresponding positions in several other lambdoid phages. Together, the two boxes constitute a *nut* (for N utilization) site, which contains all of the *cis*-acting elements required for N-dependent antitermination. In the presence of N, polymerase becomes termination resistant after transcription of *nut*. Antitermination can still be detected after polymerase has read through thousands of base pairs and many sequential terminators. This implies that the N-dependent modification to polymerase that occurs after transcription of *nut* is stable. BOXA is the loading site for the *E. coli* S10 (or NusE) and NusB proteins. In association with two additional *E. coli* factors, NusA and NusG, an antitermination complex containing N and *λ* NUT is formed with RNAP. It is clear that N is the active factor in the complex, since at sufficiently high concentration, N suppresses transcription termination in vitro in the absence of *nut* or the Nus factors. However, the additional components increase the stability of the antitermination complex and reduce the level of N that is needed. The RNAP α subunit might also have a specific role in N-mediated antitermination. Mutations that alter the carboxyl-terminal domain of the RNAP α subunit have been reported to enhance or inhibit N antitermination in vivo, according to the nature and location of the mutation. However, deletion of the carboxyl-terminal domain of α does not affect N-dependent antitermination in vitro, perhaps indicating that some regulatory component is missing from the reaction.

The formation of the complete antitermination complex can occur in discrete steps. Initially, NusA binds to an N-BOXB complex. This binding requires extrusion of the loop residue of BOXB. Thus, a BOXB tetra loop mutant (GAAGA → GAAA) binds N but does not form an N-BOXB-NusA complex, as demonstrated by super shift experiments. A core complex of BOXB, N, NusA, and RNAP can read through terminators located close to the NUT site in vitro. In the absence of N and NUT, NusA binds to the elongation complex near the 3′-OH terminus of the nascent RNA chain and enhances pausing and termination.

Processive antitermination requires the complete antitermination complex. The assembly of NusB, S10, and NusG onto the core complex involves nt 2 to 7 of λ BOXA (CGCUCUUACACA), as well as the carboxyl-terminal region of N, which interacts with RNAP. The role of NusG in the N antitermination reaction is not clear. NusG binds to termination factor Rho and to RNAP. It stimulates the rate of transcription elongation and is required for the activity of certain Rho-dependent terminators. NusG is a component of the complete antitermination complex and enhances N antitermination in vitro. However, alteration of λ BOXA to a variant called BOXA consensus (CGCUCUUUAACA) allows NusB and S10 to assemble in the absence of NusG. Furthermore, depletion of NusG has no effect on λ N antitermination in vivo, and unlike *nusA*,*nusB*, and *nusE*, no point mutations in *nusG* that block N activity have been isolated. A NusG homolog, RfaH, enhances elongation of several transcripts in *E. coli* and *S. typhimurium*. The possibility that RfaH and NusG are redundant for N antitermination has not yet been tested, although for several other functions, the two proteins are not interchangeable.

The function of BOXA in λ N-mediated antitermination is likewise not entirely clear. Point mutations in *boxA* that decrease or increase antitermination efficiency have been isolated. On the other hand, deletion of the *boxA* region does not inhibit antitermination in vivo. Instead, antitermination no longer requires NusB. To account for this, it has been proposed that BOXA is not directly required for antitermination but instead is the site of interplay between inhibitory and anti-inhibitory factors. According to this model, *boxA* point mutations that reduce antitermination eliminate binding of the anti-inhibitor but not the inhibitor. *boxA* deletions eliminate binding of the inhibitor, and therefore, the anti-inhibitor, presumably NusB, is no longer needed. This notion is supported by an experiment in which high-level transcription of an antitermination-defective*boxA* point mutant activated growth in *trans* of a *λ* phage carrying the same mutation in a *nusB* mutant host, presumably by titrating the inhibitor. In a similar experiment, high-level transcription of a consensus BOXA inhibited growth of a phage carrying a wild-type *boxA*, probably by titrating NusB. However, the role of NusB is likely to go beyond that of an anti-inhibitor, and that of BOXA is likely to go beyond that of a site for the interplay of inhibitory and anti-inhibitory factors. In vitro studies with purified proteins show that point mutations in *boxA* impede the assembly of the antitermination complex even in the absence of a known inhibitor and NusB stimulates processive antitermination in such a system. In addition, the role of BOXA in antitermination of Rho-dependent terminators in bacterial rRNA operons appears to be more central than it is in λ, raising the possibility that λ BOXA contributes to antitermination in a way that is at least partially independent of BOXB.

Surprisingly, the λ *nut* sites are also components in a transcription termination pathway. In this pathway, N is replaced by Nun, a protein encoded by a relative of λ, phage HK022. Nun converts antitermination into termination. Other components of the two pathways, notably NusA, NusB, S10, NusG, BOXA, and BOXB, are shared. The sequence similarity of Nun to proteins of the N family, although weak, includes the amino-proximal ARM region. This is unsurprising because Nun, like λ N, binds specifically to BOXB and requires the same BOXB nucleotides for biological function. In vivo, Nun terminates transcription just distal to the *nut* sites. In vitro, Nun arrests RNAP translocation at several positions downstream of *nut*. The arrested elongation complexes contain the 3′ ends of the nascent transcripts in the polymerase active center, and this site remains enzymatically active: the 3′ nucleotide can be removed by pyrophosphorolysis and restored by addition of the appropriate nucleoside triphosphate. However, forward and backward translocation of RNAP is blocked. The Nus factors increase the efficiency of transcription arrest but are not essential if the concentration of Nun is elevated. Nun-dependent release of arrested RNAP from the template and transcript has not been observed in a purified transcription system, presumably because a factor(s) is missing. The differences between N and Nun that lead to their opposed biological activities are unknown. However, the amino-proximal regions, which contain the ARMs, can be interchanged between the two proteins without altering their functions. Therefore, the functional differences are in the carboxyl-terminal 50 to 75% of the proteins. Of particular note is the presence of three C-terminal His residues, specific to Nun. These residues form part of a Zn^{2+} binding motif that is required for Nun activity. The carboxyl-terminal regions of N and Nun may bind different RNAP subunits; certain *rpoC*(β′) mutations block Nun but not N activity.

A second phage-encoded factor, λ Q protein, induces antitermination in the λ late operon. Lambda Q, like λ N, has functional analogs in other phages (Fig. 1). These late antiterminators probably act by a similar mechanism, although some are only distantly related or unrelated to λ Q. Initially, Q binds to a region within the λ P_R' promoter. Interaction with RNAP can be detected when the transcription complex pauses at +16, downstream of a site similar to the extended −10 sequence of some sigma-70 promoters. The presence of the sigma-70 subunit of RNAP holoenzyme is essential for pausing and for Q-mediated antitermination: RNAP core enzyme that has been artificially paused at +16 by omission of the appropriate nucleoside triphosphate cannot be modified by Q. In addition, sigma-70 mutants that are unable to support Q-mediated antitermination have been isolated. However, once Q has interacted with RNAP holoenzyme, sigma-70 is no longer needed for stable association of Q with the elongation complex. Although antitermination by Q is enhanced by NusA in vitro, it is not clear if the Q reaction has additional requirements in vivo. How Q modifies RNAP function is likewise unknown.

Processive antitermination can be mediated by RNA as well as proteins. Coliphage HK022, alone among the known lambdoid phages, does not encode an analog to λ N. Instead, it promotes antitermination of early phage transcription through the direct action of transcribed sequences called *put* (for polymerase utilization) sites. There are two closely related *put* sites, one located in the P_L operon and the other located in the P_R operon, roughly corresponding to the positions of the *nut* sequences in λ and in other λ relatives.*put* sites act in *cis* to promote read through of downstream terminators in the absence of all HK022 proteins. The *put* transcripts are predicted to form two stem-loops separated by a single unpaired nucleotide. This prediction is supported by mutational studies and the pattern of sensitivity of the two RNAs to cleavage with single- and double-strand-specific endoribonucleases. RNA structure is critical to antitermination because mutations that prevent the formation of base pairs in the stems reduce function, and these mutations can be suppressed by additional mutations that restore base pairing. Like λ N and Q, the PUT sequences suppress polymerase pausing and promote processive antitermination in a purified in vitro transcription system. In contrast to λ N, no phage or auxiliary bacterial factors are required. The only mutations known to block PUT-mediated antitermination change highly conserved amino acids located in a cysteine-rich amino-proximal domain of the RNAP subunit. Strains carrying these mutations are unable to support lytic growth of HK022 but are normal in all other respects tested, including lytic growth of λ and other λ relatives. The phage-restricted phenol types conferred by these mutations suggests that they alter a domain of RNAP-β' that interacts specifically with nascent PUT RNA in the transcription elongation complex, but this idea has not been directly tested. The stability of the putative PUT-RNAP interaction and the nature of the PUT-induced modification to the elongation complex are unknown.

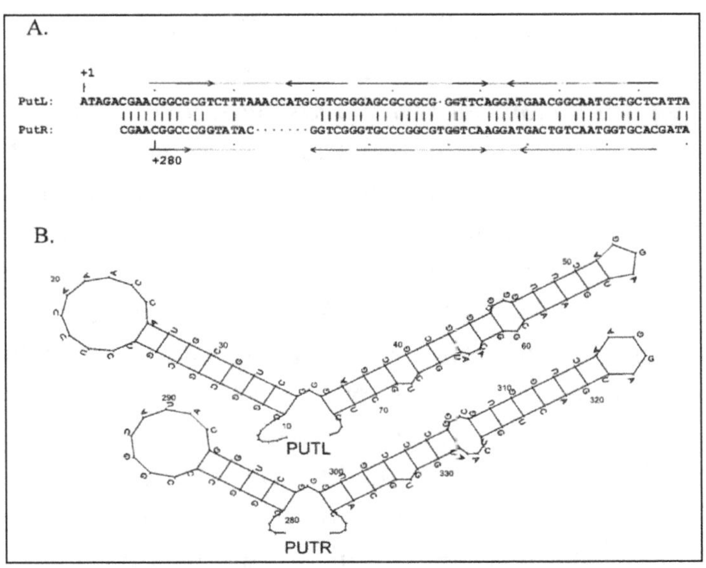

HK022 *put* sites and folded PUT RNAs.

(A) Alignment of *putL* and *putR*. The numbers give distances from the start sites of the P_L and P_R promoters, respectively, and the pairs of arrows indicate inverted sequence repeats.

(B) Folded PUTL and PUTR RNAs. The structures, which were generated by energy minimization as described have been partially confirmed by genetic and biochemical studies.

So far, factor-independent antitermination is unique to phage HK022. Both factor-independent and -dependent modes of antitermination are efficient, processive, and well-suited to their tasks. Indeed, the isolation and characterization of a hybrid phage that contains the early promoters and *put* sites of phage HK022 fused to the P_L and P_R operons of λ show that the HK022 antitermination pathway suppresses the λ terminators to the extent necessary for normal lytic growth and lysogen formation by the hybrid. In addition, late gene expression in phage HK022 is activated by a Q-dependent antitermination pathway that is very closely related to that of λ. The relative advantages of the PUT and N-NUT antitermination mechanisms and the evolutionary reasons for their adoption by different closely related phages are obscure. Recent work suggests that the activity of λ N-dependent antitermination is auto regulated so as to couple the lysogeny-lysis decision of infected cells to their growth phase. It is not clear if the *put*-dependent antitermination activity of HK022 is regulated. However, HK022 has devoted the resources saved by the dispensability of N to the production of Nun, a protein that prevents the growth of a potential competitor by coopting a component of its antitermination system.

Processive antitermination was first discovered in a bacteriophage, but examples have since been found in bacterial operons. The *E. coli rrn* operons are regulated by an antitermination mechanism that is dependent on sites that are closely related to λ *boxA* and located promoter proximal to the 16S and 23S structural genes in each operon. The sequences of the *rrn* BOXA sites are more similar to the bacteriophage consensus than is that of λ, and they bind NusB-S10 more efficiently. Although stem-loop structures analogous to BOXB are found promoter proximal to the BOXA sites, they are not essential for antitermination. An*rrn* BOXA sequence confers full antitermination activity against Rho-dependent but not against intrinsic terminators. BOXA also increases the rate of transcription elongation by RNAP. Point mutations in BOXA induce premature transcription termination. *rrn* antitermination requires NusB in vivo, as shown by a NusB depletion experiment. NusA stimulates the elongation rate of *rrn* RNA chains carrying BOXA. A role for NusA is further suggested by the observation that the *nusA10*(Cs) mutation inhibits both antitermination and the rate of transcription elongation in an*rrn* operon. The role of other Nus factors in*rrn* regulation in vivo is not clear. In vitro, an antitermination complex that includes NusA, NusB, S10, and NusG forms at the BOXA sequence of *rrnG*, but these components are not sufficient for antitermination by themselves. An additional factor or factors that can be supplied by a cellular extract are required, but their identities are unknown.

A second bacterial elongation control pathway depends on the RfaH protein, a NusG homolog. RfaH and a *cis*-acting promoter-proximal sequence element, *ops* (for operon polarity suppressor, also called JUMP start), increase the expression of several bacterial operons. The products of these operons affect the production and transport of components located on the outside of the inner membrane, such as lipopolysaccharide core, exopolysaccharide, F pili, and hemolysin. These operons are relatively long, with several genes and intergenic terminators. RfaH and *ops* appear to function together, since mutational inactivation of both elements does not have an additive effect on gene expression. The two elements are thought to act by suppressing termination. First, transcription is increased distal to the promoter but not proximal to a terminator. Second, stimulation of gene expression is not promoter specific. Third, an *rfaH* mutation can be suppressed by a mutation that reduces the activity of transcription termination factor Rho. Purified RfaH stimulates transcription promoter distal to an intrinsic terminator when added to a crude bacterial extract programmed with an*ops*-containing template. However, it has not been shown that the increase results from elongation of transcripts that would otherwise have been terminated. Nevertheless, the requirement for a*cis*-acting site, the ability of the site to act at a distance from terminators, and the homology of RfaH and NusG suggest common elements with the mechanisms of N-dependent antitermination.

Phage P4 has an entirely different mechanism of controlling elongation. It encodes a protein, Psu (for polarity suppressor), that reduces termination by *E. coli* Rho factor. Unlike λ N and Q, Psu does not require *cis*-acting sites to antiterminate and is specific for Rho-dependent terminators. Extracts of cells that contain Psu are deficient in termination at Rho-dependent terminators, and termination can be restored by adding Rho to the extracts. Psu does not act by reducing the level of Rho protein, but it interferes, directly or indirectly, with Rho action. The importance of Rho inactivation in the life cycle of P4 is unclear. Psu stimulates lytic growth of P4, but this is likely to be the result of incorporation of Psu into the P4 capsid rather than (or in addition to) activation of transcription of essential genes that lie downstream of Rho-dependent terminators. The only known protein that is similar to Psu is encoded by a P4 relative, retronphage φR73.

How do terminators and antiterminators act? Do the antitermination pathways described here have common steps? We cannot yet answer these questions, but a brief discussion of what we think we know about the structure and stability of the elongation complex should limit the possibilities and provide a basis for speculation. The active bacterial elongation complex consists of core RNAP, template, and RNA product. The 3′ end of the RNA is engaged in the active site of the enzyme, the following ~8 nt are hybridized to the template strand of the DNA, and the next ~9 nt remain closely associated with RNAP. About 17 nt of the non-template DNA strand are separated from the template strand in the transcription bubble. Elongation complexes can also contain NusA and NusG. These proteins, which increase the stability of the N-mediated antitermination complex, have different effects on elongation. NusA decreases and NusG increases the elongation rate, and both proteins alter termination efficiency in a terminator-specific manner.

An elongation complex, unless located at a terminator, is extraordinarily stable, even when translocation is prevented by removal of substrates. Recent observations suggest that this stability depends mainly on interactions between RNAP and the RNA-DNA hybrid as well as between polymerase and the downstream duplex DNA template. Nascent RNA emerging from the hybrid region and upstream duplex DNA do not appear to be required. The strength of the RNA-DNA hybrid is believed to assure the lateral stability of the complex. Reducing the strength of the RNA-DNA bonds, for example by incorporation of nucleotide analogs, favors backsliding of RNAP on the template, with consequent disengagement of the 3′ RNA end from the active site, and concerted retreat of the RNA-DNA hybrid region from the 3′ end. Such a disengaged complex retains its resistance to dissociation and is capable of resuming elongation if the original or a newly created 3′ end reengages with the active site.

Intrinsic terminators consist of a guanine- and cytosine-rich RNA hairpin stem immediately followed by a short uracil-rich segment within which termination can occur. If termination does not occur at this point, polymerase continues to elongate the transcript with normal processivity until it reaches the next terminator. Neither the stem nor the uracil-rich segment is sufficient for termination, although either can transiently slow elongation. The weakness of base pairing between rU and dA destabilizes the RNA-DNA hybrid in the uracil-rich segment, and this probably contributes to termination. Formation of the hairpin stem as nascent terminator RNA emerges from polymerase destabilizes the RNA-DNA hybrid and interrupts contacts between the emerging nascent RNA and RNAP. It might also interfere with the stabilizing interactions between RNAP and the hybrid or those between RNAP and the downstream region of the template. Cross-linking of nucleic acid to RNAP suggests that both the downstream DNA and the nascent RNA that emerges from the hybrid region, and within which the terminator hairpin might form, are located close to the same regions of the enzyme. Conversely, modifications that render RNAP termination resistant could prevent the terminator stem from destabilizing one or more of these targets, at least while the 3′ end of the RNA is within the uracil-rich segment of the terminator.

The λ N and Q proteins and HK022 PUT RNA also suppress Rho-dependent terminators which, in contrast to intrinsic terminators, lack a precisely determined termination point. Rho is an RNA-dependent ATPase that binds to cytosine-rich, unstructured regions in nascent RNA and acts preferentially to terminate elongation complexes that are paused at nearby downstream sites. Rho possesses RNA-DNA helicase activity, and this activity is directional, unwinding DNA paired to the 3′ end of the RNA molecule. This corresponds to the location of the hybrid and of RNAP in an active ternary elongation complex. The ability of antiterminators to suppress Rho-dependent and -independent terminators suggests that they prevent a step that is common to both classes. Given the helicase activity of Rho, a likely candidate for this step is disruption of the RNA-DNA hybrid. However, other candidates, such as destabilization of RNAP-template or RNAP-hybrid interactions, are also plausible. Alternatively, the ability of N,

Q, and PUT to suppress RNAP pausing, suggests that they prevent Rho-dependent termination by accelerating polymerase away from Rho bound at upstream RNA sites. This explanation raises the problem of why NusG, which also accelerates polymerase, enhances rather than suppresses Rho-dependent termination.

Antitermination Control of Gene Expression

Bacteria have evolved many different complex mechanisms to control both transcription and translation of genes in response to environmental changes. In many cases, transcription is controlled at the level of initiation by DNA-binding proteins that either inhibit (repressors) or stimulate (activators) initiation. In addition, transcription can be regulated at the level of elongation. In some cases, transcription of a gene or operon will terminate prematurely in the absence of the action of a positive regulatory molecule. In these cases, antitermination factors allow transcription to read through termination signals and to generate full-length transcripts.

Two fundamentally different mechanisms for antitermination has been described. In one case, RNA polymerase is modified so as to allow it to read through transcription terminators. This type of mechanism controls phage development and expression of rRNA operons. The second mechanism involves trans-acting factors that interact with RNA and prevent formation of the terminator structure. This mechanism is very similar to attenuation, but antitermination can be distinguished from attenuation in that the action of the regulatory molecule results in transcription read through, with the default pathway being premature termination. In attenuation, the action of the regulatory molecule induces transcription termination, and the default pathway is read through.

Three distinct mechanisms that regulate gene expression by antitermination will be discussed here. These mechanisms differ primarily in the type of biomolecule used as the regulator. The first mechanism uses antiterminator proteins that are activated to bind RNA targets in response to environmental stimuli. In the second mechanism, transfer RNA is used as the regulator. In this case, the degree of aminoacylation (or charging) of the tRNA is used to sense the availability of the cognate amino acid within the cell, to induce expression of genes involved in metabolism of this amino acid. Finally, in the case of the Escherichia coli tryptophanase operon, it appears that ribosomes are used as the regulatory molecule. Thus, bacteria have evolved a large number of mechanisms to use different biomolecules to all perform the same task—that is, to alter the conformation of the nascent mRNA to signal RNA polymerase whether it should terminate prematurely or continue transcription of the particular structural gene.

RNA-Binding Protein-Mediated Antitermination: The Sac/Bgl Family of Antiterminator Proteins

Expression of several catabolic operons in bacteria is regulated by antitermination involving RNA-binding proteins. These proteins prevent formation of Rho-independent transcription terminators in the nascent mRNA upstream of the regulated gene(s). One such system in E. coli and several in Bacillus subtilis, appear to be highly related based on similarities of their antiterminator proteins as well as their RNA targets. In addition, several other systems function similarly but appear to have arisen independently. A general model for this mechanism is shown in figure below.

A general model for antitermination control by the Sac/Bgl family of antiterminator proteins. Under non inducing conditions, transcription starts at the promoter (designated by the arrow) and terminates prematurely, often in a leader region prior to the structural genes. In the presence of inducer, the antiterminator protein is activated to bind to the RAT (ribonucleic antiterminator) RNA. This binding stabilizes an RNA secondary structure involving the RAT, which prevents formation of the overlapping terminator, and transcription continues into the structural genes.

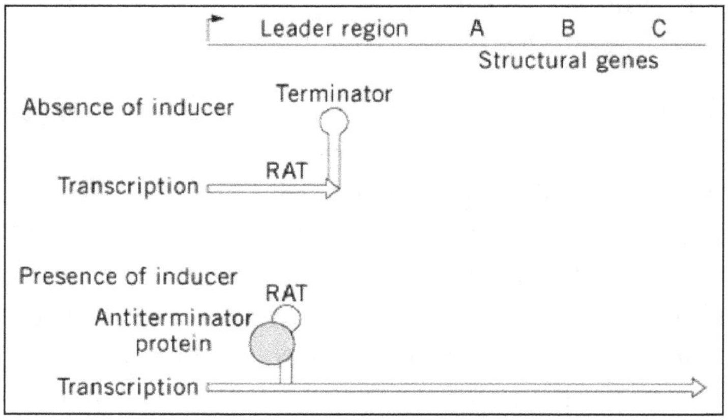

The E. Coli Bgl Operon

The E. coli bglGFB operon: Encodes all the functions necessary for the regulated uptake and utilization of aromatic b-glucosides. The operon is cryptic in wild-type strains but can become functional through spontaneous mutations. When functional, expression of this operon is regulated by antitermination mediated by the BglG protein in response to the levels of b-glucosides. In the absence of inducer, BglG does not bind RNA, and most transcripts terminate at one of two Rho-independent transcription terminators present in the leader region upstream of bglG and between bglG and bglF. When b-glucoside levels are high, BglG binds to an RNA target, named RAT for ribonucleic antiterminator, just upstream of the terminators. This binding stabilizes an alternative antiterminator RNA structure, which prevents formation of the terminator, thus allowing transcription to continue and the operon to be expressed.

The RNA-binding activity of BglG is: Regulated by phosphorylation mediated by BglF. In the absence of b-glucosides, BglF phosphorylates BglG, which prevents it from dimerizing and binding to the RAT. In the presence of b-glucosides, BglF dephosphorylates BglG, which now dimerizes and binds to the RAT. Phosphorylation of both b-glucosides and BglG is accomplished by transfer of the phosphate group from the same phosphorylated residue, Cys24, in BglF. These results suggest that, under conditions in which b-glucoside levels are high, the phosphate group can be transferred from BglG back to Cys24 in BglF. A model has been proposed in which unliganded BglF phosphorylates BglG, and b-glucoside binding induces BglF to undergo a conformational change that activates it to dephosphorylate BglG.

A similar system for b-glucoside utilization: Exists in the related Gram-negative enteric bacterium Erwinia chrystanthemi, although in this case the arb operon is not cryptic. ArbG shows high sequence similarity to BglG, suggesting that it functions analogously in antitermination control of the E. chrystanthemi arb genes. Antitermination also appears to control b-glucoside operons in several Gram-Positive Bacteria as well. A putative b-glucoside (bgl) operon has also been identified in B. subtilis and may be regulated by a similar antitermination mechanism. In addition, a protein, BglR, with homology to BglG also controls b-glucoside usage in Lactococcus lactis.

The B. Subtilis Sac Genes

Expression of two sucrose utilization operons in B. subtilis, sacPA and sacB, is induced by sucrose via transcription antitermination mediated by the RNA-binding proteins SacT and SacY, respectively. SacT and SacY show extensive sequence similarity to each other, as well as to BglG from E. coli. The antitermination mechanisms that control these genes also appear to be quite similar to that described above for the E. coli bgl operon. Rho-independent transcription terminators exist in leader regions upstream of both sacPA and sacB and prevent transcription of the structural genes in the absence of the inducer, which is sucrose. In the presence of sucrose, SacT and SacY are activated to bind RAT sequences in the sacPA and sacB leader transcripts, respectively, and allow transcription to read through into the structural genes. Like BglG in E. coli, both of these antiterminator proteins are phosphorylated. In the case of SacY, phosphorylation negatively regulates RNA-binding activity and appears to be mediated by SacX. SacT is phosphorylated by HPr, a component of the phosphoenolpyruvate phosphotransferase system, but the role of this phosphorylation in sucrose-mediated antitermination is less clear.

Recently, the protein structure of the RNA-binding domain of SacY has been determined by both NMR and X-ray crystallography. The domain exists as a dimer, with each monomer consisting of a four-stranded antiparallel beta-sheet. Several amino acid residues have been identified through genetic, biochemical, and preliminary NMR studies as being important for RNA binding. These residues are clustered on the surface of one side of the protein structure.

Other Examples of Bgl/Sac Type Antiterminators

In addition to the bgl and sac systems described above, several other operons are regulated by RNA-binding antiterminator proteins with homology to BglG, SacY, and SacT. LicT regulates the licS gene, which is involved in b-glucan utilization in B. subtilis. There is also a RAT sequence overlapping a potential Rho-independent terminator upstream of licS.

In Lactobacillus casei, the lactose (lac) operon is regulated in response to lactose levels by LacT, which shows sequence homology to the other members of the Bgl/Sac family of antiterminators. The 5′ -leader region of the lac mRNA contains a region with sequence similarity to the RAT sequence, as well as a potential stem-loop structure resembling a Rho-independent terminator.

Antiterminators with No Similarity to the Bgl/Sac Family

Several other systems are regulated by RNA-binding antiterminator proteins that are unrelated to those of the Bgl/Sac family; furthermore, these proteins do not appear to be related to each other. These regulatory systems thus appear to have arisen independently.

In B. subtilis, both the glp regulon, which is involved in usage of glycerol-3-phosphate, and a histidine-utilization (hut) operon are regulated by RNA-binding antiterminator proteins; GlpP and HutP, respectively. The amino acid sequences of these antiterminator proteins are not similar to any other antiterminator proteins. Further, the mechanisms by which these antiterminator proteins function appear to be different from those described above, because there are no clear antiterminator RNA secondary structures near the terminators in these operons.

The amidase (ami) operon of Pseudomonas aeruginosa is regulated by antitermination in response to short-chain aliphatic amides, such as acetamide. The amiR gene encodes an antiterminator protein (AmiR), which is negatively regulated by AmiC, apparently through formation of an AmiC-AmiR complex. Acetamide destabilizes the AmiC-AmiR complex, leading to antitermination and expression of the operon. AmiR interacts with an RNA target in the 5-leader region of the ami mRNA that contains a Rho-independent terminator. However, no clear antiterminator RNA secondary structure is predicted. AmiR binding has been suggested to function in antitermination by interfering directly with formation of the terminator stem-loop structure.

In addition to all the catabolic operons described above, one anabolic operon has been shown to be regulated by antitermination. Expression of the nas operon of Klebsiella pneumoniae, which encodes enzymes required for nitrate assimilation in this bacterium, is induced by nitrate or nitrite. The NasR protein mediates transcription antitermination through a terminator in the leader region of the operon. This protein shows weak homology with AmiR in the carboxyl-terminal region.

Transfer RNA-Mediated Antitermination

An interesting variation on the antitermination mechanism involves the use of tRNA as the regulatory molecule. This mechanism regulates a large number of aminoacyl-tRNA synthetase genes in Gram-Positive Bacteria and several amino acid biosynthetic operons, including the ilv-leu, in B. subtilis, and the his trp operons in Lactococcus lactis. Expression of these genes is induced specifically by starvation for the corresponding amino acid. In the case of the amino acid operons, insufficient levels of the amino acid leads to increased expression of the corresponding biosynthetic operon. For the amino-acyl-tRNA synthetase genes, increasing the level of the synthetase is thought to allow more efficient charging of the cognate tRNA when the corresponding amino acid pool is low.

A long (approx. 300-nucleotide) untranslated leader region exists upstream of the structural gene(s) of these operons that contains several conserved features, including three stem-loop structures preceding a Rho-independent transcription terminator. Hence, in the absence of the inducing signal, transcription terminates prematurely in the leader region prior to the coding sequences. In addition to the conserved secondary structures, there is an important conserved 14-nucleotide sequence known as the T-box present in each leader region; hence these genes are known as the T-box family. An alternate arrangement of the leader region involving base-pairing between a portion of the T-box and a conserved sequence in the 5′ side of the terminator stem has been proposed to form an antiterminator structure that allows transcription to read through into the structural genes.

Model for ant termination control by tRNA. Under conditions with adequate levels of the cognate amino acid (aa), the charged tRNA does not interact with the leader region, and the terminator forms. Under conditions of starvation for the appropriate amino acid, the uncharged tRNA interacts with the leader region via base-pairing between the anticodon and the specifier sequence, and by base-pairing between the CCA sequence at the acceptor end of the tRNA with the side bulge of the antiterminator in the leader. These interactions stabilize formation of the antiterminator conformation of the leader transcript, resulting in induction of expression of the gene. The tRNA is shown as the shaded cloverleaf structure, and a boxed "A.A." attached to the tRNA indicates it is aminoacylated.

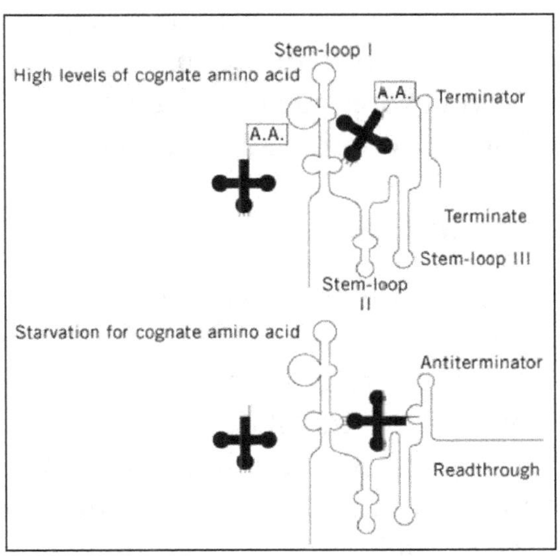

Another important conserved feature: The leader region of these genes is the presence of a triplet sequence corresponding to a codon for the appropriate amino acid for each operon. For example, in tyrS, which encodes tyrosyl-tRNA synthetase, the leader contains a UAC tyrosine codon, while the ilv-leu operon leader contains a CUC leucine codon. This triplet is always present in a bulged sequence in Stem-loop and has been shown to be critical for induction in several systems. It was the presence of these triplets that led to the hypothesis that tRNAs play a role in this regulatory mechanism. This triplet was designated the "specifier sequence" because, in the case of the B. subtilis tyrS gene, altering the sequence to correspond to a codon for another amino acid switched induction to respond to starvation for the new amino acid. Other experiments demonstrated that translation of this codon was not involved in induction and that uncharged tRNA was the inducer. In addition, a second interaction between the CCA sequence at the 3′ end of the uncharged tRNA and the complementary UGG sequence in the T-box have been shown to be important.

A model for tRNA-regulated antitermination: Under starvation conditions for the corresponding amino acid, the cognate uncharged tRNA interacts with two sites in the leader region, to induce formation of the antiterminator structure and allow transcription to read through into the coding region. Aminoacylation of this tRNA is predicted to interfere with the interaction at the CCA end and prevent the charged tRNA from binding; the leader transcript then folds into the conformation with the terminator, halting transcription. It is not known if factors in addition to tRNA are required for antitermination. To date, however, it has not been possible to reconstitute tRNA-mediated antitermination in vitro, and several other lines of evidence also suggest that other factors may be involved in this mechanism.

In addition to the antitermination mechanism: Described above, processing of the leader RNA has been shown to play a role in regulating expression of the B. subtilis thrS gene. Cleavage occurs in the loop of the antiterminator near the T-box sequence and is more efficient under threonine starvation conditions, suggesting that bound tRNA induces both antitermination and RNA processing. This processing increases the stability of the mRNA, which would allow for increased translation and production of the threonyl-tRNA synthetase. Thus induction of expression of this gene in response to threonine starvation occurs at both the level of transcription antitermination and mRNA stability.

The E. coli Tryptophanase Operon

E. coli and several other microorganisms have the capacity to degrade tryptophan as a source of carbon, nitrogen, and/or energy. The degradative tryptophanase operon (tnaCAB) of E. coli is regulated by catabolite repression and by an antitermination mechanism. Antitermination involves translation of a cis-acting 24-residue leader peptide (tnaC) containing a critical Trp codon, one or more RNA polymerase pause sites between tnaC and tnaA, and Rho termination facto. While the precise antitermination mechanism responsible for controlling the tna operon is not firmly established, all of the data are consistent with the following model. During growth in a medium lacking both tryptophan and a catabolite-repressing carbon source, transcription initiation is efficient. As transcription proceeds, translation of the leader peptide occurs as soon as the coding sequence becomes available. Once the translating ribosome reaches the UGA stop codon, ribosome release exposes a rut (Rho utilization) site that immediately follows the stop codon. Rho then binds to the rut site and begins to translocate in the 3′-direction, until it encounters paused RNA polymerase, ultimately leading to transcription termination upstream of tnaA. When cells are growing with inducing levels of tryptophan, TnaC, or a complex of TnaC with an unidentified protein, prevents ribosome release at the tnaC stop codon, thereby masking the rut site and, hence, blocking Rho interaction with the transcript. Eventually RNA polymerase would overcome the pause signal and transcribe the structural genes encoding tryptophanase (tnaA) and a tryptophan permease (tnaB). This model assumes that there is a fundamental difference between the TnaC peptide, the TnaC peptide-protein complex, in cells growing with or without tryptophan. It was proposed that such a complex under inducing conditions would prevent ribosome release, reminiscent of characterized translation attenuation mechanisms. The tryptophanase operon of Proteus vulgaris is thought to be regulated by a mechanism essentially identical to that of E. coli.

Artificial Gene Synthesis

Artificial gene synthesis is the chemical synthesis of a DNA sequence that represents one or more genes. While site-directed mutagenesis is regularly used to probe biological hypotheses by incorporating single base mutations, artificial gene synthesis provides a method to efficiently produce long stretches of natural and non-natural nucleic acid sequences, broadening the scope of biological experiments. Sequences that are hard to isolate from natural sources can be routinely generated in the lab, even entirely non-natural gene sequences can be synthesised. Not only does gene synthesis have implications for synthetic biology, it also introduces the possibility of developing genes containing modified nucleotides or even novel base pairs that could allow for the expansion of the genetic code.

An initial challenge of gene synthesis was the production of long stretches of nucleotide sequence: genes can range from several hundred to several thousand base pairs in length. Current solid-phase oligonucleotide synthesis technology is limited to the generation of oligos 200 nucleotides in length. Two major methods exist to circumvent this issue: polymerase cycling assembly and synthesis by ligation.

Polymerase Cycling Assembly (PCA)

Polymerase cycling assembly (PCA) exploits single stranded templates rather than relying on total synthesis of both DNA strands. Short oligos are synthesised that alternate between both strands of the gene, overlapping by regions of 20-30 base pairs. Then, to give a superficial explanation, a polymerase uses dNTPs to "fill in the gaps" on one gene strand using the complementary bases on the other. A subsequent, entirely separate polymerase step then amplifies the constructed gene by traditional polymerase chain reaction (PCR).

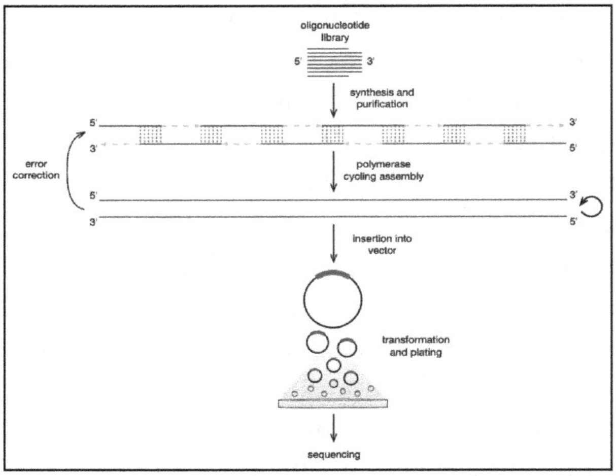

Gene synthesis by polymerase cycling assembly (PCA).

With PCA, the commonly used explanation of the polymerase "filling in the gaps" is misleading: the gene is actually assembled step by step. Overlapping oligos anneal at their termini to form discrete overhanging dimers. The overhangs act as templates to elongate the complementary oligo by polymerase action generating fully double-stranded DNA that makes up a small section of the gene. These extended dsDNA sections are then melted and annealed to another overlapping oligo further along the gene, forming more single stranded overhangs. Further polymerase action turns these dimers into longer dsDNA fragments and this process continues until the entire gene is assembled.

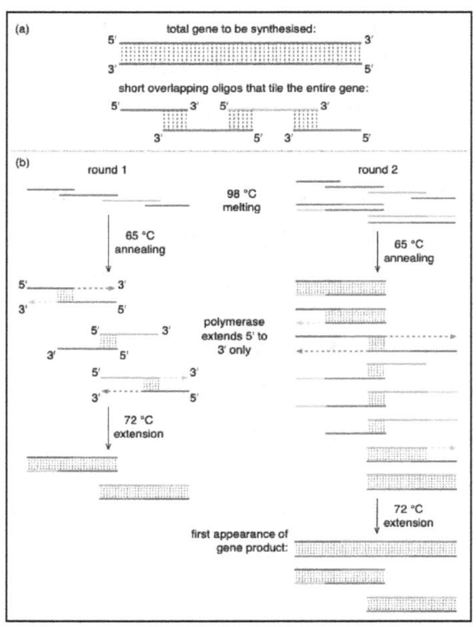

Gene synthesis by PCA – 4 oligos.

As shown in Figure above, many combinations of dimers can anneal and not all result in successful extension. For examples of any more than 4 oligos, the various iterations of annealing and elongation get too complicated to represent diagrammatically. One consistent point however is that polymerases extend 5' to 3' only, so although many oligo dimers will overlap and anneal, only those with the potential to extend 5' to 3' will be elongated. This limitation implies that the terminal oligos that cover the ends of the gene must be on the gene strand with the 5' terminus. Otherwise, if a terminal oligo were on the strand with the 3' terminus, the complementary oligo would have to be extended 3' to 5' to complete the gene, which is impractical for standard polymerase methodologies.

PCA eliminates the need for the whole gene to be synthesised solid phase. Once assembled, the gene product can be amplified by traditional PCR to generate a suitable concentration and purified by filtration to remove any remaining primers.

Error Correction

Unfortunately, the process is not without errors. Insertion and deletion mutations (frameshift mutations) can be introduced at the solid phase synthesis stage and chain extension by polymerase risks further substitutions, insertions and deletions. Fortunately, natural DNA repair enzymes exist to correct these errors. However, since synthetic DNA has no indication of which strand has the correct sequence nucleases that rely on positive identification of the correct template strand cannot be used. Instead, it is necessary to use endonucleases that completely excise stretches of DNA that contain a mutation.

Since PCR cannot discriminate between correct sequences and those with errors, any mutations that have accumulated in the solid phase synthesis or polymerase extension of one gene strand will be matched in the reverse complement during amplification: the mutation is "invisible". Therefore, the mixture of correct and incorrect gene products is melted and reannealed. It is statistically unlikely that mutations will occur in the same place in the same way on two random strands. As such, random reannealing to a strand which likely doesn't have the complementary mutation generates a mismatch as shown in figure below. This mismatch perturbs the secondary structure of the gene.

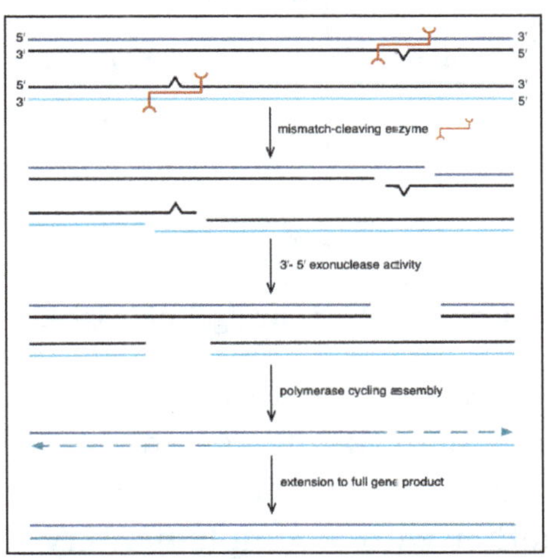

Error correctionRandom errors present in dsDNA sequence will present as mismatches.

Certain mismatch repair enzymes such as T7 Endonuclease I recognise these structural perturbations and cleave the first, second or third phosphodiester bond 3' to the mismatch (dependent on the specific enzyme) on both strands. A single strand exonuclease then chews up the overhanging single stranded DNA 3' to 5'. This process produces dsDNA of varying sizes that all overlap and between them contain the correct sequence spanning the entire gene. Subsequent PCA and PCR amplification generates the corrected gene product.

Gene Synthesis by Ligation

While PCA facilitates the synthesis of large genes and genomes, even after error correction the product often has many errors present in the sequence. Gene synthesis by ligationhas a lower error rate than PCA, but is limited to the synthesis of smaller genes since the fragment oligos must span the entire sequence of both the sense and antisense strands: this is impractical for longer genes.

However, if the goal is to synthesise a smaller gene (< 2 kbp), gene synthesis by ligation will likely generate a product with fewer errors because:

- It removes the need for a polymerase that can often introduce mutations;
- Strands are subject to hybridisation selection – since all oligos overlap entirely, errors in their syntheses will produce mismatches which mean they are less likely to anneal and form a ligated product.

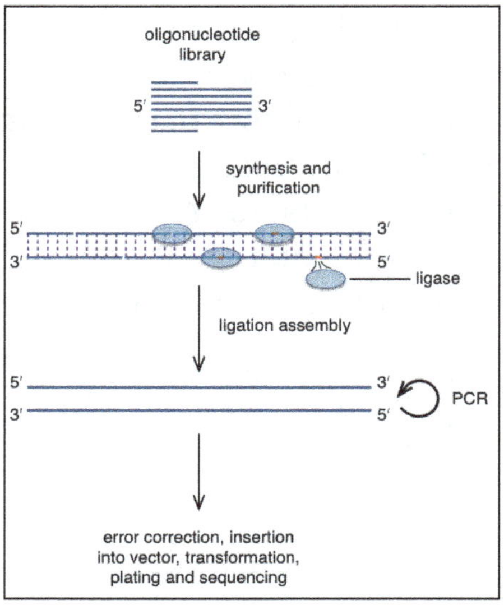

Gene synthesis by ligationOligos.

Solid phase synthesis generates the constituent fragments which are then purified ready for ligation. These oligos span the entire gene on both strands, missing only the phosphodiester bonds that link neighbouring oligos. Fragments are designed so two adjacent oligos are held together with a complementary template (the splint) then ligatedenzymatically.

Three major methods of ligation-based synthesis exist:

- Shotgun ligation – the gene is synthesised en masse by simply mixing the constituent oligos together and adding a ligase. The reaction can occur at room temperature though it usually requires extended reaction times. Use of thermo stable ligases and high purity oligos have now made this one-pot reaction more efficient.
- Ligase chain reaction (LCR) – LCR originally superseded shotgun ligation as a more efficient method with shorter reaction times and the ability to ligate longer genes (still limited to 2 kbp). This method follows a similar set of melting and annealing steps to PCR though differs by using a ligase instead of a polymerase to ligate templated oligos until the entire gene is assembled.
- Solid-support based ligation-mediated oligonucleotide assembly – The initial dsDNA oligo is immobilised on a solid support and iterative rounds of annealing and ligation steadily assemble the entire gene, which is cleaved from the support at the end of the synthesis.

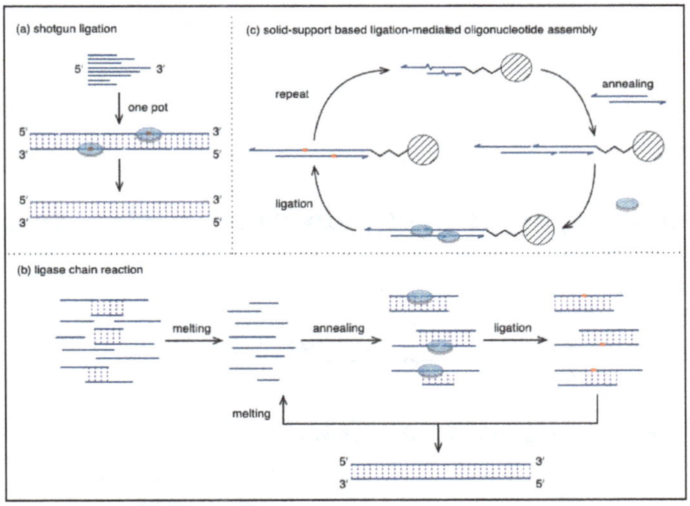

Three methods of ligation-based gene synthesis.

While the error rate of ligation based synthesis is lower, it may still be necessary to perform error correction to yield a legitimate number of viable genes. Furthermore, PCA is the dominant methodology used for artificial gene synthesis. Ligation-based techniques require polyacrylamide gel electrophoresis (PAGE) purification and 5' phosphorylation, making the whole method costly and labor intensive. PCA and associated error correction is preferred as these steps are eliminated.

Further methods of gene synthesis exist including assembly directly on to the plasmid relying on bacterial homologous recombination (Sequence- and Ligation- Independent Cloning – SLIC) or in vivo assembly of transformed oligonucleotides by the yeast Saccharomyces cerevisiae. PCA and ligation based methods remain by far the most common methods of gene synthesis.

Gene Silencing

The term gene silencing is commonly used to describe the "switching off" of a gene by a mechanism without genetic modification. The term gene silencing refers to an epigenetic phenomenon, the heritable inactivation of gene expression that does not involve any changes to the deoxyribonucleic acid (DNA) sequence. While this phenomenon has initially been studied in transgenic plants, its relevance in the regulation of endogenous plant genes has become increasingly apparent.

Below some cellular components are mentioned where gene silencing occurred:

- Chromatin and heterochromatin.
- Dicer.
- dsRNA.
- Histones.
- MicroRNA.
- siRNA.
- Transposons.

Gene silencing has following two major subdivisions by which genes are regulated:

1. Transcriptional gene silencing (TGS);
2. Posttranscriptional gene silencing (PTGS).

Transcriptional Gene Silencing (TGS)

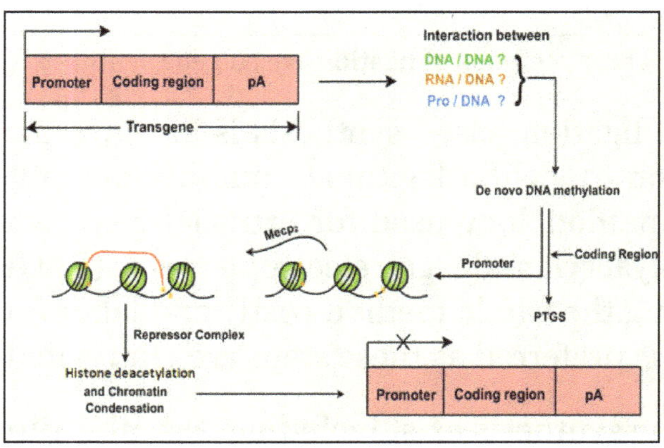

Schematic representation of DNA methylation – mediated transcriptional gene silencing (TGS).

Transcriptional gene silencing is the product of chromosomal histone modifications, creating an environment of heterochromatin, which is surrounded to a gene that makes it inaccessible to transcriptional machinery (RNA polymerase, transcription factors, etc.). TGS blocks primary transcription from nuclear DNA and is in most cases associated with DNA methylation and chromatin condensation in nearly all organisms that possess a DNA methylation system.

Post-transcriptional Gene Silencing (PTGS)

Post-transcriptional gene silencing is the product of transcribed mRNA of a specific gene being silenced. When mRNA was destructed, then translation to form an active gene product (in most cases, a protein) will be prevented. A general process of post-transcriptional gene silencing is by RNAi. PTGS involves a cytoplasmic, target sequence-specific RNA degradation process that is possibly activated by double-stranded RNA (dsRNA). This dsRNA is independent of ongoing translation.

TGS can be transmitted generation to generation by meiosis whereas PTGS is usually lost during meiosis. In PTGS, double stranded RNA is interred into a cell and gets chopped up by the enzyme known as dicer to form siRNA. siRNA then binds to the RNA-induced silencing complex (RISC) and is unwound. The antisense. RNA complex with RISC protein and binds to its corresponding mRNA, which is then cleaved by the enzyme slicer rendering it inactive.

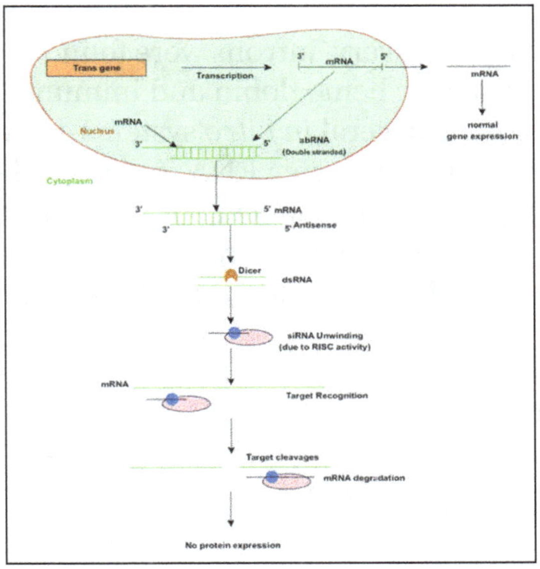

Schematic representation of post translational gene silencing (PTGS).

RNA Splicing

For most eukaryotic genes (and some prokaryotic ones), the initial RNA that is transcribed from a gene's DNA template must be processed before it becomes a mature messenger RNA (mRNA) that can direct the synthesis of protein. One of the steps in this processing, called RNA splicing, involves the removal or "splicing out" of certain sequences referred to as intervening sequences, or introns. The final mRNA thus consists of the remaining sequences, called exons, which are connected to one another through the splicing process. RNA splicing was initially discovered in the 1970s, overturning years of thought in the field of gene expression.

Early Studies in Bacteria

Gene regulation was first studied most thoroughly in relatively simple bacterial systems. Most bacterial RNA transcripts do not undergo splicing; these transcripts are said to be co-linear, with DNA directly encoding them. In other words, there is a one-to-one correspondence of bases between the gene and the mRNA transcribed from the gene (excepting 5′ and 3′ noncoding regions). However, in 1977, several groups of researchers who were working with adenoviruses that infect and replicate in mammalian cells obtained some surprising results. These scientists identified a series of RNA molecules that they termed "mosaics," each of which contained sequences from noncontiguous sites in the viral genome. These mosaics were found late in viral infection. Studies of early infection revealed long primary RNA transcripts that contained all of the sequences from the late RNAs, as well as what came to be called the intervening sequences (introns).

Subsequent to the adenoviral discovery, introns were found in many other viral and eukaryotic genes, including those for hemoglobin and immunoglobulin. Splicing of RNA transcripts was then observed in several *in vitro* systems derived from eukaryotic cells, including removal of introns from transfer RNA in yeast cell-free extracts. These observations solidified the hypothesis that splicing of large initial transcripts did, in fact, yield the mature mRNA. Other hypotheses proposed that the DNA template in some way looped or assumed a secondary structure that allowed transcription from noncontiguous regions.

Occurance of Splicing

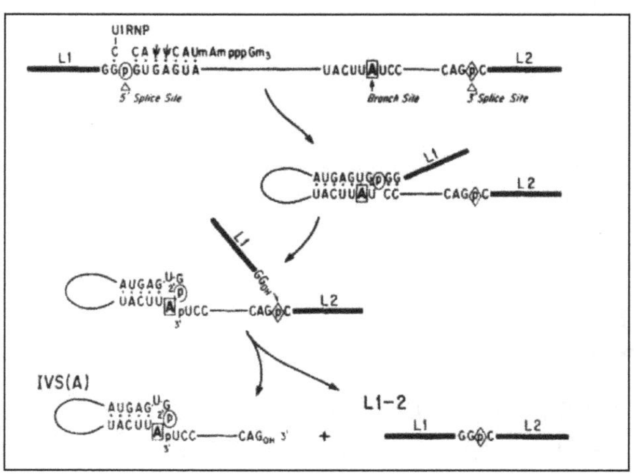

Pre-mRNA splicing.

The biochemical mechanism by which splicing occurs has been studied in a number of systems and is now fairly well characterized. Introns are removed from primary transcripts by cleavage at conserved sequences called splice sites. These sites are found at the 5′ and 3′ ends of introns. Most commonly, the RNA sequence that is removed begins with the dinucleotide GU at its 5′ end, and ends with AG at its 3′ end. These consensus sequences are known to be critical, because changing one of the conserved nucleotides results in inhibition of splicing. Another important sequence occurs at what is called the branch point, located anywhere from 18 to 40 nucleotides upstream from the 3′ end of an intron. The branch point always contains an adenine, but it is otherwise loosely conserved. A typical sequence is YNYYRAY, where Y indicates a pyrimidine, N denotes any nucleotide, R denotes any purine, and A denotes adenine. Rarely, alternate splice site sequences are found that begin with the dinucleotide AU and end with AC; these are spliced through a similar mechanism.

Splicing occurs in several steps and is catalyzed by small nuclear ribonucleoproteins (snRNPs, commonly pronounced "snurps"). First, the pre-mRNA is cleaved at the 5′ end of the intron following the attachment of a snRNP called U1 to its complementary sequence within the intron. The cut end then attaches to the conserved branch point region downstream through pairing of guanine and adenine nucleotides from the 5′ end and the branch point, respectively, to form a looped structure known as a lariat. The bonding of the guanine and adenine bases takes place via a chemical reaction known as transesterification, in which a hydroxyl (OH) group on a carbon atom of the adenine "attacks" the bond of the guanine nucleotide at the splice site. The guanine residue is thus cleaved from the RNA strand and forms a new bond with the adenine.

Next, the snRNPs U2 and U4/U6 appear to contribute to positioning of the 5′ end and the branch point in proximity. With the participation of U5, the 3′ end of the intron is brought into proximity, cut, and joined to the 5′ end. This step occurs by transesterification; in this case, an OH group at the 3′ end of the exon attacks the phosphodiester bond at the 3′ splice site. The adjoining exons are covalently bound, and the resulting lariat is released with U2, U5, and U6 bound to it.

In addition to consensus sequences at their splice sites, eukaryotic genes with long introns also contain exonic splicing enhancers (ESEs). These sequences, which help position the splicing apparatus, are found in the exons of genes and bind proteins that help recruit splicing machinery to the correct site. Most splicing occurs between exons on a single RNA transcript, but occasionally trans-splicing occurs, in which exons on different pre-mRNAs are ligated together.

The splicing process occurs in cellular machines called spliceosomes, in which the snRNPs are found along with additional proteins. The primary variety of spliceosome is one of the most plentiful structures in the cell, and recently, a secondary type of spliceosome has been identified that processes a minor category of introns. These introns are referred to as U12-type introns because they depend upon the action of a snRNP called U12 (the common introns described above are called U2-type introns). The role of U12-type introns is not yet defined, but their persistence throughout evolution and conservation between homologous genes of widely divergent species suggests an important functional basis.

Self-Splicing and Alternative Splicing

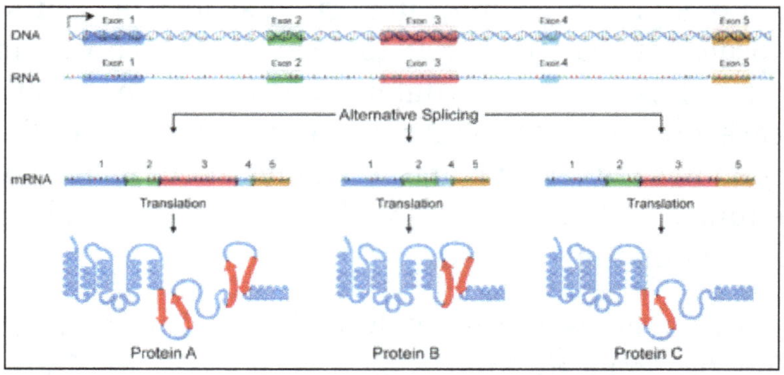

A schematic representation of alternative splicing.

Some RNA molecules have the capacity to splice themselves; the initial discovery of this self-splicing ability in the protozoan *Tetrahymena thermophila* was recognized with the Nobel Prize in 1989. The self-splicing introns found in *T. thermophila* are now referred to as Group I introns; this class also includes other protozoan ribosomal RNA genes, some fungal mitochondrial genes, and some phage genes. Group I introns all fold into a complex secondary structure with nine loops and employ transesterification reactions as described above. On the other hand, Group II self-splicing introns are found in mitochondrial genes and are excised by a mechanism that bears similarities to pre-mRNA splicing, including the production of lariats. For this reason, it has been proposed that perhaps pre-mRNA introns and splicing mechanisms evolved from the Group II introns.

Early in the course of splicing research, yet another surprising discovery was made; specifically, researchers noticed that not only was pre-mRNA punctuated by introns that needed to be excised, but also that alternative patterns of splicing within a single pre-mRNA molecule could yield different functional mRNAs.

The first example of alternative splicing was defined in the adenovirus in 1977 and demonstrated that one pre-mRNA molecule could be spliced at different junctions to result in a variety of mature mRNA molecules, each containing different combinations of exons.

Shortly afterward, alternative splicing was found to occur in cellular genes as well, with the first example identified in the *IgM* gene, a member of the immunoglobulin superfamily. Another example of a gene with an impressive number of alternative splicing patterns is the *Dscam* gene from *Drosophila*, which is involved in guiding embryonic nerves to their targets during formation of the fly's nervous system. Examination of the *Dscam* sequence reveals such a large number of introns that differential splicing could, in theory, create a staggering 38,000 different mRNAs. This ability to create so many mRNAs may provide the diversity necessary for forming a complex structure such as the nervous system. In fact, the existence of multiple mRNA transcripts within single genes may account for the complexity of some organisms, such as humans, that have relatively few genes (approximately 20,000). For example, work from Wang *et al.* (2008) suggests that more than 90% of human genes are alternatively spliced.

Alternative Splicing

Alternative splicing of precursor mRNA is an essential mechanism to increase the complexity of gene expression, and it plays an important role in cellular differentiation and organism development. Regulation of alternative splicing is a complicated process in which numerous interacting components are at work, including cis-acting elements and trans-acting factors, and is further guided by the functional coupling between transcription splicing. Additional molecular features, such as chromatin structure, RNA structure and alternative transcription initiation or alternative transcription termination, collaborate with these basic components to generate the protein diversity due to alternative splicing. All these factors contributing to this one fundamental biological process add up to a mechanism that is critical to the proper functioning of cells. Any corruption of the process may lead to disruption of normal cellular function and the eventuality of disease. Cancer is one of those diseases, where alternative splicing may be the basis for the identification of novel diagnostic and prognostic biomarkers, as well as new strategies for therapy. Thus, an in-depth understanding of alternative splicing regulation has the potential not only to elucidate fundamental biological principles, but to provide solutions for various diseases.

Molecular Mechanisms of Alternative Spicing

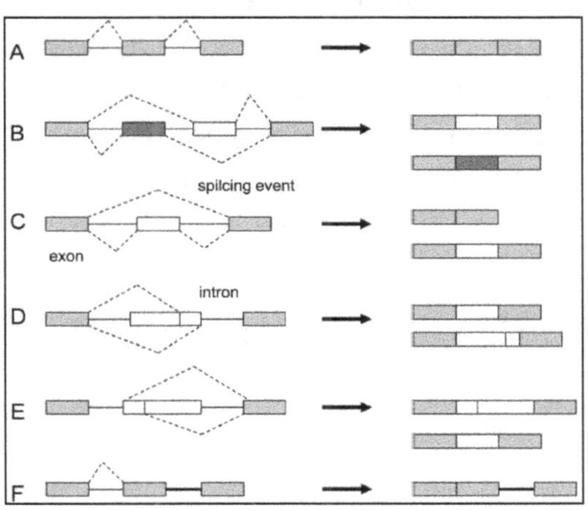

Five main types of alternative splicing events are depicted. (A) Constitutive splicing; (B) mutually exclusive exons; (C) cassette alternative exon; (D) alternative 3′ splice site; (E) alternative 5′ splice site; and (F) intron retention.

Systematic analyses of ESTs and microarray data have so far revealed seven main types of alternative splicing. The most prevalent pattern (~30%) is the cassette-type alternative exon (exon skipping) in vertebrates and invertebrates, while in lower metazoans, it is intron retention. Intron retention in human transcripts is positioned primarily in the un-translated regions (UTRs) and has been associated with weaker splice sites, short intron length and the regulation of cis-regulatory elements.

Alternative selection of 5′ or 3′ splice sites within exon sequences (~25%) may lead to subtle changes in the coding sequence, and an additional layer of complexity arises with mutually exclusive alternative exons. One example of a transcript that undergoes alternative splicing, which generates variation in the protein. Differences in the splicing machinery in different cell types and unique cis-acting elements in the *FGF-R2* pre-mRNA lead to altered tissue specific choices that create either *FGF-R2IIIb* or *FGF-R2IIIc* mature transcripts.

The protein expression is further regulated by alternative polyadenylation of mRNA, which influences the coding potential or the 3′UTR length by modifying the binding availability of microRNA or RNA. Of note, it has been demonstrated that each type of alternative splicing can operate in a stochastic manner, and different splice-site identification and processing mechanisms do not necessarily occur at the same frequencies among all biological kingdoms.

The mechanisms outlined above are just one indication of the complexity, as numerous molecules are involved in alternative splicing in a coordinated manner. Even the basic nucleotide components and the essential molecules that recognize them can introduce diversity in the synthesis of mature transcripts.

Two major steps constitute the basic process of splicing: Assembly of the spliceosome followed by the actual splicing of pre-mRNA. The spliceosome is mainly composed of U1, U2 small nuclear ribonucleic proteins (snRNPs) and the U4/U6.U5 tri-snRNP, and configure in identify a core set of splicing signals: The 5′ splice site, the branch point sequence and the 3′ splice site. Specific spliceosomal complexes (E, A, B and others) and eight evolutionarily conserved DExD/H-type RNA- dependent ATPases/helicases assemble in a proposed stepwise manner and execute multiple splicing steps that result in exon ligation and intron excision. Numerous steps in the pathway are reversible.

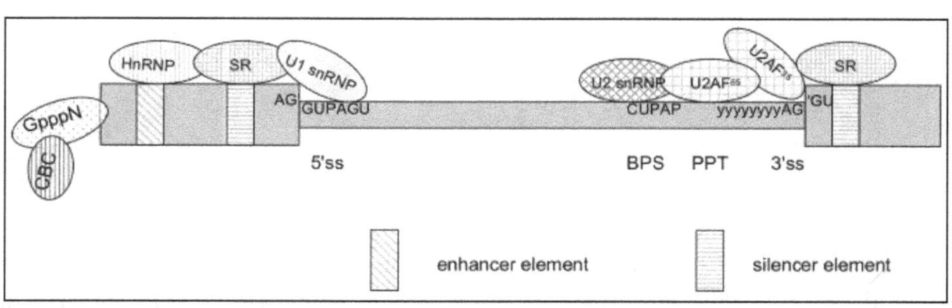

Schematic representation of the sequence elements and proteins at 5′ and 3′ exon-intron boundaries in an RNA transcript.

The exons that end up in the mature mRNA during the process of alternative splicing is entirely defined by the interaction between cis-acting elements and trans-acting factors. Cis-acting elements include exonic splicing enhancers (ESEs) and intronic splicing enhancers (ISE) that are bound by positive trans-acting factors, such as SR proteins (serine/arginine-rich family of nuclear phosphoproteins), whereas exonic splicing silencers (ESSs) and intronic splicing silencers are bound by negative acting factors, such as heterogeneous nuclear ribonucleoproteins (hnRNPs). The collaboration between these elements results in the promotion or inhibition of splicesome assembly of the weak splice sites, respectively. In general, the cis-acting elements function additively. The enhancing elements tend to play dominant roles in constitutive splicing, while the silencers are relatively more important in the control of alternative splicing. Enhancer activity has been shown to be abolished by a stable stem-loop structure as short as 7 base pairs in an RNA transcript owing to the mechanisms of physical competition, long-range RNA pairing, a structure splice code and co-transcription splicing. Furthermore, the specificity of cis-acting enhancer elements for introns or exons has been investigated. In these experiments, an ESE was found to act as an ISE depending on its location in an exon or intron.

HnRNPs are highly conserved from nematodes to mammals and have several critical roles in pre-mRNA maturation. Their function is to bind to the ESS to the exclusion of SR proteins. A looping out pre-mRNA leads to exonic sequestration from the rest of pre-mRNA transcript. HnRNPs A/B are a family of RNA-binding proteins, its diversification roles in the modulation of alternative splicing have evolved based on differing affinities for their cognate nucleic acids. HnRNP H and F serve to alter the proteolipid protein (PLP/DM20) ratio via the variation in the recruitment of U1 snRNP. Similarly, the antagonistic role of hnRNP M to the splicing factor Nova-1 generates alternatively spliced dopamine receptor pre-mRNAs, which create isoforms associated with diverse key physical functions, such as control, reward, learning and memory. In addition, hnRNP L and phosphorylation of ser513 have been recently shown to be involved in the regulation of alternative splicing through dynamic membrane depolarization and Ca^{2+}/calmodulin-dependent protein kinase IV activation.

In addition to the coupling of SR proteins to enhancer elements, SR proteins interact with U1 snRNP and the 35 kDa subunit of the heterodimeric factor, U2AF. The second subunit of U2AF, $U2AF^{65}$, binds SF1 and the pyrimidine tract simultaneously, on the basis of the arginine/serine (RS)-rich domain, which results in recognition and stability of the branch point, as well as polypyrimidine tract sequences. Approximately 10–12 serines in the N-terminal region of the RS domain are rapidly phosphorylated by the binding of SR-specific protein kinase to serine/arginine-rich splicing factor 1 with an unusually high affinity. This continuous phosphorylation/dephosphorylation cycle of SR proteins facilitates the shuttling of SR proteins between the nucleus and the cytoplasm, and is critically required for the regulation of alternative splicing by growth signals transduced to the nucleus. SR proteins have also been proposed to participate in post-splicing activities, such as mRNA nuclear export, nonsense-mediated decay (NMD) and mRNA translation.

In general, positive or negative splice-site recognition is regulated through various mechanisms, such as the local concentration or activity of splicing regulatory factors, under diverse physiological or pathological conditions.

Coupling of Alternative Splicing to Transcription

A large number of components associated with the physical interaction between splicing and transcription have been purified, with particular attention on the carboxyl terminal domain (CTD) of the large subunit of RNAPII. The CTD consists of 52 tandem repeats of the heptapeptide YSPTSPS in mammals (26 tandem repeats in yeast), which act as a special platform to recruit different factors to the nascent transcripts via dynamic phosphorylation of serine residues. Kinases that phosphorylate specific CTD serine residues have been identified and are components of the protein apparatus driving the specific function. For example, ser5 phosphorylation is associated with transcription initiation through cyclin-dependent kinase 7 (CDK7) of the general transcription factor IIH (TFIIH), whereas ser2 phosphorylation is preferentially linked with CTD activity at the 3′-end of genes through CDK9 of the positive transcription elongation factor. In addition, phosphorylation of ser7 has been found to facilitate elongation and splicing. Thus, phosphorylation is a mechanism that clearly demonstrates that functional coupling exists between transcription and alternative splicing.

CTD participates in gene expression-related functions ranging from 5′ capping, splicing, poly-adenylation and chromatin remodeling, Of note, mutation and deletion analysis of CTD has revealed multiple defects in mRNA processing therefore, CTD and additional components of the two machineries have emerged as a central element in governing the interactions between transcription and splicing. Taken together, functional coupling appears to maintain an important role in alternative splicing in driving determinative physiological changes, and fine-tune gene expression in mathematical modeling approaches.

Two models have been suggested to explain the co-transcription process of how transcription coupled repair influences alternative splicing. The mechanism of the recruitment model may mainly depend on specific features of CTD (as mentioned above), whereas the kinetic model is based on the different elongation rates of Pol II, which in turn determine the timing of the presentation of splices sites.

Fundamentally, the aforementioned mechanism influences patterns of alternative splicing via the variations in Pol II elongation and recruitment of splicing factors by specific histone marks. Thus, alternative splicing is highly influenced not only by transcription, but also by the chromatin structure, which underscores chromatin as another layer in the regulation of alternative splicing. The resultant mature mRNA is thus a reflection of numerous DNA modifications, such as patterns of histone methylation at exons, modulation of histone modifications and increased DNA methylation at exons, Conversely, a previous study indicated that splicing may mediate chromatin remodeling via deposition of histone marks on DNA or numerous associations between splicing factors and elongation proteins.

Adding additional complexity to the regulation network is alternative transcription initiation (ATI) and alternative transcription termination (ATT) sites. ATI and ATT significantly contribute to the diversity of the human and mouse transcriptomes to a degree that may exceed alternative splicing, when considering the number of possibilities available through alternative nucleotides, isoforms and introns. In contrast to the prevalence of alternative splicing that occurs within coding sequences (CDSs), the dominant class of alternative events, which includes ATI and ATT, occur in UTRs. This discovery reflects the preferential regulation of large distinct groups of genes with different mechanisms, such as strong coupling with alternative splicing in 5′ and 3′UTRs.

Despite the strong correlation between alternative splicing and transcription, alternative transcription mainly results in variations of the transcript number or the 5′/3′ terminal protein variants due to differential transcriptional start or terminal sites. By contrast, alternative splicing associated alterations mostly lie within the protein sequence, potentially affecting almost all areas of protein function.

Trans-splicing

Trans-splicing is a common phenomenon in trypanosomes, nematodes, *Drosophila* and even humans, and refers to the novel and unusual splicing of exons from independent pre-mRNAs. The phenomenon has been explored as a therapeutic option for a variety of genetic diseases, particularly in the treatment of cancer. The carcinoembryonic antigen (CEA), for example, is associated with a variety of neoplastic processes and was exploited as a target for trans-splicing. A CEA RNA-targeting trans-splicing ribozyme was designed to perform RNA replacement through a trans-splicing reaction specifically in CEA expressing cells. The activity of the ribozyme simultaneously reduced CEA expression and introduced the thymidine kinase gene, which rendered the cells sensitive to ganciclovir treatment. RNA trans-splicing has also been utilized for the potential treatment of neurodegenerative diseases through a novel technology, splic osome mediated trans-splicing (SMaRT). SMaRT was successfully used *in vivo* to re-engineer tau mRNA transcripts to include E10, and therefore, offers the opportunity potential to correct tau missplicing and treat the underlying disease.

Epitranscriptome

An epitranscriptome is a set of functionally appropriate RNA modifications. There are nearly one hundred known modifications of RNA; the most common modification in internal mRNA is N6-methyladenosine (m^6A). Transfer RNAs (tRNAs), mRNAs, and ribosomal RNAs (rRNAs) are the three types of RNAs that are post-transcriptionally modified.

DNA and RNA helix.

In all living organisms, translation is a fundamental process. The arrangement and operation of translation process is very expensive and it consumes 40% of cellular energy, which has led to the need for strict regulation of protein production in many aspects.

The regulation of translation is generally related to the need of regulated and non-regulated proteins. Nucleotide modifications are also important for the translation process. RNA base modifications are detected by appropriate sequencing methods.

At a genome level, the study of m^6A is possible with the help of recent advances in sequencing technology and the design of m^6A antibodies. The sequencing technologies are listed below.

m^6A-sequencing

The m6A is the most available base in eukaryotic mRNA. The m6A protocol can be carried out on mRNA. The first step is to acquire a better quality of RNA sample. Polyadenylated RNA is enhanced by oligo-dT selection while the m6A-seq protocol is executed through RNA. Polyadenylated RNA is divided into small fragments with a length of 100 nt each.

Polyadenylated RNA fragmentation is performed chemically to achieve efficient fragmentation. These small fragments are used to reduce the resolution to 200 nt. Remapping of smaller fragments into the genome would become tedious because of further fragmentation.

Post-fragmentation, the sample is put through immunoprecipitation. This small amount of fragmented mRNA is retained aside to perform input control process. In the following data analysis, small fragments are retained at basic level status. The rest of the sample is immunoprecipitated by incubating it with a particular m6A antibody that pulls fragments containing multiple m6A residues.

Recombinant protein agarose bead, which can bind with the antibodies, is used to incubate the RNA-antibody mixture in the subsequent process. Finally, the unbound fragments are eradicated, bound fragments are removed by addition of m6A to the mixture, and the bead-antibody-RNA complex deposited in solid form.

During immune precipitation, the RNA fragments and beads are combined together in the absence of antibodies. Bioanalyzer analysis and concentration measurements are used to estimate the quality of the material throughout the remaining protocol. Good result can be obtained when better quality starting material is used.

MeRIP-sequencing

There exist some differences between m6A sequencing and MeRIP-sequencing strategies. The main difference is the category of bead: Magnetic Dynabeads are used in MeRIP sequencing strategy, which is first paired with antibodies and then combined with ribo-minus treated RNA. Here, immune precipitation process is performed twice.

Buffer composition involved in fragmentation also differs. Solvent extraction elution is used instead of competition elution. However, the fundamental concept involved in these two methods is similar to each other.

PA-m6A-sequencing

The PA-m6A sequencing name was coined from m6A sequencing strategy in combination with a PA sequencing strategy. In this method, PA-m6A has the ability of undergoing high-resolution mapping with a mammalian transcriptome m6A. It can be observed that the m6A sequence modification sites are more accurately defined when compared to conventional sequencing methods. The resolution has been reduced from 200 to 23 nt.

PA-m6A sequence is versatile when compared to the single nucleotide approaches. Photoactivatable ribonucleoside-enhanced crosslinking and immunoprecipitation (PARCLIP) strategy is the base for PA-m6A strategy. In PARCLIP strategy, photoactivatable ribonucleosides are integrated into an mRNA and then they are subjected to immuneoprecipitation process. It is similar to m6A protocol, but immunoprecipitation is performed with full length RNA and then this RNA is fragmented post-crosslinking through UV irradiation.

Single Molecule Sequencing

Single molecule sequencing has the ability to provide the base level resolution of m6A sites without the conclusion drawn on the basis of motif. The Single Molecule Real-Time (SMRT) technology is the most common platform for this method of sequencing.

This method was developed by Pacific Biosciences and Oxford Nano pore Technologies. Continuous monitoring of the process, in which DNA polymerase is integrated with fluorescent nucleotides, helps in detecting both genetic and epigenetic information at the same time.

Single molecule sequencing offers some advantages as it can be used to detect RNA modification that includes m6A sites. Single nucleotide resolution is to detect the m6A sites in RNA, which is achieved by using DNA polymerase rather than reverse transcriptase.

It can be used as an enzyme within a ZMW and in allowing direct complementary DNA (cDNA) synthesis in real time. Reverse transcription appears at standard speeds. At the instant of base incorporation, designed m6A sites exhibit important growth of inter pulse duration (IPD).

Single Nucleotide Methods

The above techniques are not useful in the case of determining nucleotide level modification, due to a lower determination capability. Site-specific cleavage and radioactive-labeling followed by ligation-assisted extraction and thin-layer chromatography method can be used to overcome this disadvantage. Therefore, the m6A modification is non-stoichiometric.

References

- Gene-synthesis, Artificial-gene-synthesis-63: atdbio.com, Retrieved 26, June 2020
- Genetic-code: biologydictionary.net, Retrieved 28, June 2020
- Epitranscriptome-Sequencing-Technologies, life-sciences: news-medical.net, Retrieved 18, July 2020
- Gene-Expression-An-Overview: news-medical net Retrieved 18 April 2020
- Antitermination-control-of-gene-expression-molecular-biology: what-when-how.com, Retrieved 19, May 2020

7

Sub-disciplines of Molecular Biology

Molecular Biology is an interdisciplinary field of biology which makes it essential to understand its related fields. Some of the topics related to the study of molecular biology such as molecular modeling, molecular engineering and molecular genetics are explained in a detailed manner. The significant sub-disciplines of molecular biology have been thoroughly discussed in this chapter.

Molecular Modelling

Molecular modelling encompasses all theoretical methods and computational techniques used to model or mimic the behaviour of molecules. The techniques are used in the fields of computational chemistry, drug design, computational biology and materials science for studying molecular systems ranging from small chemical systems to large biological molecules and material assemblies. The simplest calculations can be performed by hand, but inevitably computers are required to perform molecular modelling of any reasonably sized system. The common feature of molecular modelling techniques is the atomistic level description of the molecular systems. This may include treating atoms as the smallest individual unit (the Molecular mechanics approach), or explicitly modeling electrons of each atom (the quantum chemistry approach).

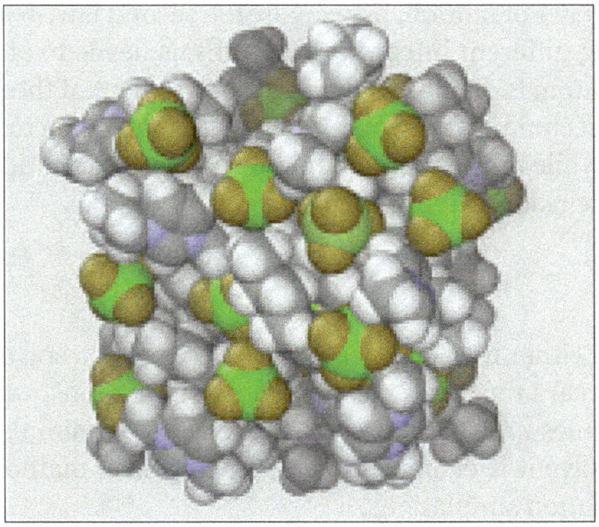

Modeling of ionic liquid.

Molecular Mechanics

Molecular mechanics is one aspect of molecular modelling, as it refers to the use of classical mechanics/Newtonian mechanics to describe the physical basis behind the models. Molecular models typically describe atoms (nucleus and electrons collectively) as point charges with an associated mass. The interactions between neighbouring atoms are described by spring-like interactions (representing chemical bonds) and van der Waals forces. The Lennard-Jones potential is commonly used to describe van der Waals forces. The electrostatic interactions are computed based on Coulomb's law. Atoms are assigned coordinates in Cartesian space or in internal coordinates, and can also be assigned velocities in dynamical simulations. The atomic velocities are related to the temperature of the system, a macroscopic quantity. The collective mathematical expression is known as a potential function and is related to the system internal energy (U), a thermodynamic quantity equal to the sum of potential and kinetic energies. Methods which minimize the potential energy are known as energy minimization techniques (e.g., steepest descent and conjugate gradient), while methods that model the behaviour of the system with propagation of time are known as molecular dynamics.

$$E = E_{\text{bonds}} + E_{\text{angle}} + E_{\text{dihedral}} + E_{\text{non-bonded}}$$

$$E_{\text{non-bonded}} = E_{\text{electrostatic}} + E_{\text{van der Waals}}$$

This function, referred to as a potential function, computes the molecular potential energy as a sum of energy terms that describe the deviation of bond lengths, bond angles and torsion angles away from equilibrium values, plus terms for non-bonded pairs of atoms describing van der Waals and electrostatic interactions. The set of parameters consisting of equilibrium bond lengths, bond angles, partial charge values, force constants and van der Waals parameters are collectively known as a force field. Different implementations of molecular mechanics use different mathematical expressions and different parameters for the potential function. The common force fields in use today have been developed by using high level quantum calculations and/or fitting to experimental data. The technique known as energy minimization is used to find positions of zero gradient for all atoms, in other words, a local energy minimum. Lower energy states are more stable and are commonly investigated because of their role in chemical and biological processes. A molecular dynamics simulation, on the other hand, computes the behaviour of a system as a function of time. It involves solving Newton's laws of motion, principally the second law, $\mathbf{F} = m\mathbf{a}$. Integration of Newton's laws of motion, using different integration algorithms, leads to atomic trajectories in space and time. The force on an atom is defined as the negative gradient of the potential energy function. The energy minimization technique is useful for obtaining a static picture for comparing between states of similar systems, while molecular dynamics provides information about the dynamic processes with the intrinsic inclusion of temperature effects.

Variables

Molecules can be modeled either in vacuum or in the presence of a solvent such as water. Simulations of systems in vacuum are referred to as *gas-phase* simulations, while those that include the presence of solvent molecules are referred to as *explicit solvent* simulations. In another type of simulation, the effect of solvent is estimated using an empirical mathematical expression; these are known as *implicit solvation* simulations.

Coordinate Representations

Most force fields are distance-dependent, making the most convenient expression for these Cartesian coordinates. Yet the comparatively rigid nature of bonds which occur between specific atoms, and in essence, defines what we mean by the molecule itself, make an internal coordinate system the most logical representation. In some fields the IC representation (bond length, angle between bonds, and twist angle of the bond as shown in the figure) is known as the Z-matrix or torsion angle representation. Unfortunately, continuous motions in Cartesian space often require discontinuous angular branches in internal coordinates making it relatively hard to work with force fields in the internal coordinate representation and conversely a simple displacement of an atom in Cartesian space may not be a straight line trajectory due to the prohibitions of the interconnected bonds. Thus, it is very common for computational optimization programs to flip back and forth between representations during their iterations; this can dominate the calculation time of the potential itself and in long chain molecules introduce cumulative numerical inaccuracy. While all conversion algorithms produce mathematically identical results, they differ in speed and numerical accuracy. Currently, the fastest and most accurate torsion to Cartesian conversion is the Natural Extension Reference Frame (NERF) method.

Applications

Molecular modelling methods are now routinely used to investigate the structure, dynamics, surface properties and thermodynamics of inorganic, biological and polymeric systems. The types of biological activity that have been investigated using molecular modelling include protein folding, enzyme catalysis, protein stability, conformational changes associated with biomolecular function, and molecular recognition of proteins, DNA, and membrane complexes.

Molecular Engineering

Molecular engineering is an emerging field of study concerned with the design and testing of molecular properties, behavior and interactions in order to assemble better materials, systems, and processes for specific functions. This approach, in which observable properties of a macroscopic system are influenced by direct alteration of a molecular structure, falls into the broader category of "bottom-up" design.

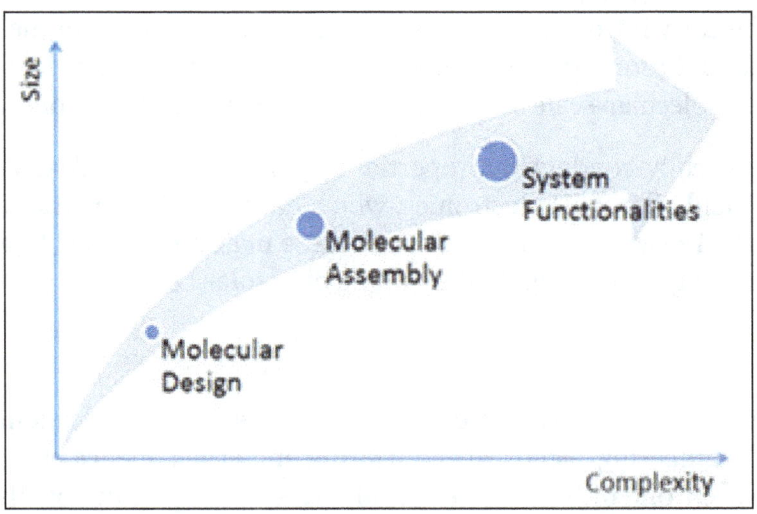

Molecular engineering deals with material development efforts in emerging technologies that require rigorous rational molecular design approaches towards systems of high complexity.

Molecular engineering is highly interdisciplinary by nature, encompassing aspects of chemical engineering, materials science, bioengineering, electrical engineering, physics, and chemistry. There is also considerable overlap with nanotechnology, in that both are concerned with the behavior of materials on the scale of nanometers or smaller. Given the highly fundamental nature of molecular interactions, there are a plethora of potential application areas, limited perhaps only by one's imagination and the laws of physics. However, some of the early successes of molecular engineering have come in the fields of immunotherapy, synthetic biology, and printable electronics (molecular engineering applications).

Molecular engineering is a dynamic and evolving field with complex target problems; breakthroughs require sophisticated and creative engineers who are conversant across disciplines. A rational engineering methodology that is based on molecular principles is in contrast to the widespread trial-and-error approaches common throughout engineering disciplines. Rather than relying on well-described but poorly-understood empirical correlations between the makeup of a system and its properties, a molecular design approach seeks to manipulate system properties directly using an understanding of their chemical and physical origins. This often gives rise to fundamentally new materials and systems, which are required to address outstanding needs in numerous fields, from energy to healthcare to electronics. Additionally, with the increased sophistication of technology, trial-and-error approaches are often costly and difficult, as it may be difficult to account for all relevant dependencies among variables in a complex system. Molecular engineering efforts may include computational tools, experimental methods, or a combination of both.

History

Molecular engineering was first mentioned in the research literature in 1956 by Arthur R. von Hippel, who defined it as "... a new mode of thinking about engineering problems. Instead of taking prefabricated materials and trying to devise engineering applications consistent with their macroscopic properties, one builds materials from their atoms and molecules for the purpose at hand." This concept was echoed in Richard Feynman's seminal 1959 lecture *There's Plenty of Room at the Bottom*, which is widely regarded as giving birth to some of the fundamental ideas of the field of nanotechnology. In spite of the early introduction of these concepts, it was not until the mid-1980s with the publication of *Engines of Creation: The Coming Era of Nanotechnology* by Drexler that the modern concepts of nano and molecular-scale science began to grow in the public consciousness.

The discovery of electrically-conductive properties in polyacetylene by Alan J. Heeger in 1977 effectively opened the field of organic electronics, which has proved foundational for many molecular engineering efforts. Design and optimization of these materials has led to a number of innovations including organic light-emitting diodes and flexible solar cells.

Applications

Molecular design has been an important element of many disciplines in academia, including bioengineering, chemical engineering, electrical engineering, materials science, mechanical engineering and chemistry. However, one of the on-going challenges is in bringing together the critical mass of manpower amongst disciplines to span the realm from design theory to materials production, and from device design to product development. Thus, while the concept of rational engineering of technology from the bottom-up is not new, it is still far from being widely translated into R&D efforts.

Molecular engineering is used in many industries. Some applications of technologies where molecular engineering plays a critical role:

Consumer Products

- Antibiotic surfaces.
- Cosmetics (eg. rheological modification with small molecules in shampoo).
- Cleaning products (eg. nanosilver in laundry detergent).
- Consumer electronics (organic light-emitting diode displays (OLED).
- Electrochromic windows (eg. windows in Dreamliner 787).
- Zero emission vehicles (eg. advanced fuel cells/batteries).
- Self-cleaning surfaces (eg. super hydrophobic surface coatings).

Energy Harvesting and Storage

- Flow batteries - Synthesizing molecules for high-energy density electrolytes and highly-selective membranes in grid-scale energy storage systems.
- Lithium-ion batteries - Creating new molecules for use as electrode binders, electrolytes, electrolyte additives, or even for energy storage directly in order to improve energy density (using materials such as graphene, silicon nanorods, and lithium metal), power density, cycle life, and safety.
- Solar cells - Developing new materials for more efficient and cost-effective solar cells including organic, quantum dot or perovskite-based photovoltaics.
- Photocatalytic water splitting - Enhancing the production of hydrogen fuel using solar energy and advanced catalytic materials such as semiconductor nanoparticles.

Environmental Engineering

- Water desalination (eg. new membranes for highly-efficient low-cost ion removal).
- Soil remediation (eg. catalytic nanoparticles that accelerate the degradation of long-lived soil contaminants such as chlorinated organic compounds).
- Carbon sequestration (eg. new materials for CO_2 adsorption).
- Wastewater treatment.

Immunotherapy

Synthetic Biology

- CRISPR - Faster and more efficient gene editing technique.

- Gene delivery/gene therapy - Designing molecules to deliver modified or new genes into cells of live organisms to cure genetic disorders.
- Metabolic engineering - Modifying metabolism of organisms to optimize production of chemicals (eg. synthetic genomics).
- Protein engineering - Altering structure of existing proteins to enable specific new functions, or the creation of fully artificial proteins.

Techniques and Instruments Used

Molecular engineers utilize sophisticated tools and instruments to make and analyze the interactions of molecules and the surfaces of materials at the molecular and nano-scale. The complexity of molecules being introduced at the surface is increasing, and the techniques used to analyze surface characteristics at the molecular level are ever-changing and improving. Meantime, advancements in high performance computing have greatly expanded the use of computer simulation in the study of molecular scale systems.

Computational and Theoretical Approaches

- Computational chemistry.
- High performance computing.
- Molecular dynamics.
- Molecular modeling.
- Statistical mechanics.
- Theoretical chemistry.

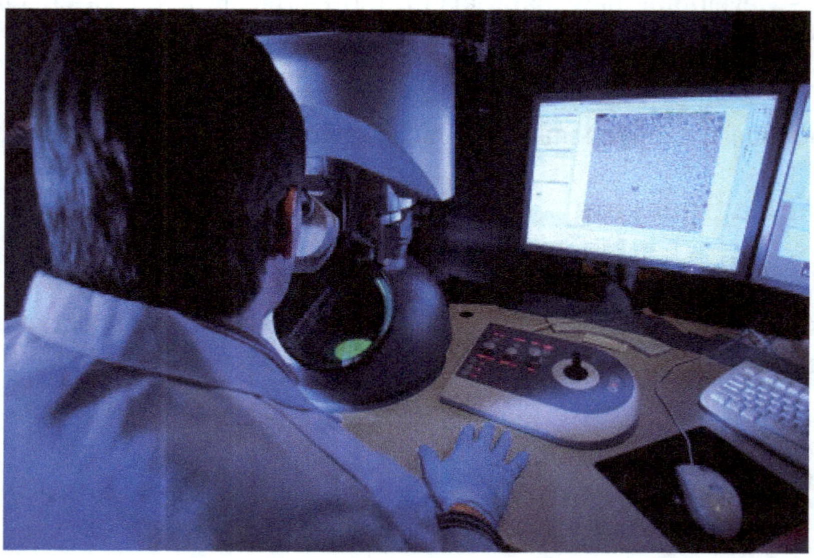

An EMSL scientist using the environmental transmission electron microscope at Pacific Northwest National Laboratory. The ETEM provides in situ capabilities that enable atomic-resolution imaging and spectroscopic studies of materials under dynamic operating conditions. In contrast to traditional operation of TEM under high vacuum, EMSL's ETEM uniquely allows imaging within high-temperature and gas environments.

Microscopy

- Atomic Force Microscopy (AFM).
- Scanning Electron Microscopy (SEM).
- Transmission Electron Microscopy (TEM).

Molecular Characterization

- Dynamic light scattering (DLS).
- Matrix-assisted laser desorption/ionization (MALDI) spectrocosopy.
- Nuclear magnetic resonance (NMR) spectroscopy.
- Size exclusion chromatography (SEC).

Spectroscopy

- Ellipsometry.
- 2D X-Ray Diffraction (XRD).
- Raman Spectroscopy/Microscopy.

Surface Science

- Glow Discharge Optical Emission Spectrometry.
- Time of Flight-Secondary Ion Mass Spectrometry (ToF-SIMS).
- X-Ray Photoelectron Spectroscopy (XPS).

Synthetic Methods

- DNA synthesis.
- Nanoparticle synthesis.
- Organic synthesis.
- Peptide synthesis.
- Polymer synthesis.

Other Tools

- Focused Ion Beam (FIB).
- Profilometer.
- UV Photoelectron Spectroscopy (UPS).
- Vibrational Sum Frequency Generation.

Research / Education

At least three universities offer graduate degrees dedicated to molecular engineering: the University of Chicago, the University of Washington, and Kyoto University. These programs are interdisciplinary institutes with faculty from several research areas.

The academic journal Molecular Systems Design & Engineering publishes research from a wide variety of subject areas that demonstrates "a molecular design or optimisation strategy targeting specific systems functionality and performance."

Molecular Genetics

Molecular genetics is the field of biology and genetics that studies the structure and function of genes at a molecular level. The study of chromosomes and gene expression of an organism can give insight into heredity, genetic variation, and mutations. This is useful in the study of developmental biology and in understanding and treating genetic diseases.

Techniques in Molecular Genetics

Amplification

Gene amplification is a procedure in which a certain gene or DNA sequence is replicated many times in a process called DNA replication.

Polymerase Chain Reaction

> The main genetic components of the polymerase chain reaction (PCR) are DNA nucleotides, template DNA, primers and Taq polymerase. DNA nucleotides make up the DNA template strand for the specific sequence being amplified, primers are short strands of complementary nucleotides where DNA replication starts, and Taq polymerase is a heat stable enzyme that jump-starts the production of new DNA at the high temperatures needed for reaction.

Cloning DNA in Bacteria

> Cloning is the process of creating many identical copies of a sequence of DNA. The target DNA sequence is inserted into a cloning vector. Because this vector originates from a self-replicating virus, plasmid, or higher organism cell when the appropriate size DNA is inserted the "target and vector DNA fragments are then ligated" and create a recombinant DNA molecule.

The recombinant DNA molecules are then put into a bacterial strain (usually *E. coli*) which produces several identical copies by transformation. Transformation is the DNA uptake mechanism possessed by bacteria. However, only one recombinant DNA molecule can be cloned within a single bacteria cell, so each clone is of just one DNA insert.

Separation and Detection

In separation and detection DNA and mRNA are isolated from cells and then detected simply by the isolation. Cell cultures are also grown to provide a constant supply of cells ready for isolation.

Cell cultures:

> A cell culture for molecular genetics is a culture that is grown in artificial conditions. Some cell types grow well in cultures such as skin cells, but other cells are not as productive in cultures. There are different techniques for each type of cell, some only recently being found to foster growth in stem and nerve cells. Cultures for molecular genetics are frozen in order to preserve all copies of the gene specimen and thawed only when needed. This allows for a steady supply of cells.

DNA isolation:

> DNA isolation extracts DNA from a cell in a pure form. First, the DNA is separated from cellular components such as proteins, RNA, and lipids. This is done by placing the chosen cells in a tube with a solution that mechanically, chemically, breaks the cells open. This solution contains enzymes, chemicals, and salts that breaks down the cells except for the DNA. It contains enzymes to dissolve proteins, chemicals to destroy all RNA present, and salts to help pull DNA out of the solution. Next, the DNA is separated from the solution by being spun in a centrifuge, which allows the DNA to collect in the bottom of the tube. After this cycle in the centrifuge the solution is poured off and the DNA is resuspended in a second solution that makes the DNA easy to work with in the future. This results in a concentrated DNA sample that contains thousands of copies of each gene. For large scale projects such as sequencing the human genome, all this work is done by robots.

mRNA isolation:

> Expressed DNA that codes for the synthesis of a protein is the final goal for scientists and this expressed DNA is obtained by isolating mRNA (Messenger RNA).

First, laboratories use a normal cellular modification of mRNA that adds up to 200 adenine nucleotides to the end of the molecule (poly(A) tail). Once this has been added, the cell is ruptured and its cell contents are exposed to synthetic beads that are coated with thymine string nucleotides. Because Adenine and Thymine pair together in DNA, the poly(A) tail and synthetic beads are attracted to one another, and once they bind in this process the cell components can be washed away without removing the mRNA. Once the mRNA has been isolated, reverse transcriptase is employed to convert it to single-stranded DNA, from which a stable double-stranded DNA is produced using DNA polymerase. Complementary DNA (cDNA) is much more stable than mRNA and so, once the double-stranded DNA has been produced it represents the expressed DNA sequence scientists look for.

Genetic Screens

This technique is used to identify which genes or genetic mutations produce a certain phenotype. A mutagen is very often used to accelerate this process. Once mutants have been isolated, the mutated genes can be molecularly identified.

Forward saturation genetics is a method for treating organisms with a mutagen, then screens the organism's offspring for particular phenotypes. This type of genetic screening is used to find and identify all the genes involved in a trait.

Reverse genetics:

> Reverse genetics determines the phenotype that results from a specifically engineered gene. In some organisms, such as yeast and mice, it is possible to induce the deletion of a particular gene, creating what's known as a gene "knockout" - the laboratory origin of so-called "knockout mice" for further study. In other words this process involves the creation of transgenic organisms that do not express a gene of interest. Alternative methods of reverse genetic research include the random induction of DNA deletions and subsequent selection for deletions in a gene of interest, as well as the application of RNA interference.

Gene Therapy

A mutation in a gene can cause encoded proteins and the cells that rely on those proteins to malfunction. Conditions related to gene mutations are called genetic disorders. However, altering a patient's genes can sometimes be used to treat or cure a disease as well. Gene therapy can be used to replace a mutated gene with the correct copy of the gene, to inactivate or knockout the expression of a malfunctioning gene, or to introduce a foreign gene to the body to help fight disease. Major diseases that can be treated with gene therapy include viral infections, cancers, and inherited disorders, including immune system disorders.

Gene therapy delivers a copy of the missing, mutated, or desired gene via a modified virus or vector to the patient's target cells so that a functional form of the protein can then be produced and incorporated into the body. These vectors are often siRNA. Treatment can be either in vivo or ex vivo. The therapy has to be repeated several times for the infected patient to continually be relieved, as repeated cell division and cell death slowly randomizes the body's ratio of functional-to-mutant genes. Gene therapy is an appealing alternative to some drug-based approaches, because gene therapy repairs the underlying genetic defect using the patients own cells with minimal side effects. Gene therapies are still in development and mostly used in research settings. All experiments and products are controlled by the U.S. FDA and the NIH.

Classical gene therapies usually require efficient transfer of cloned genes into the disease cells so that the introduced genes are expressed at sufficiently high levels to change the patient's physiology. There are several different physicochemical and biological methods that can be used to transfer genes into human cells. The size of the DNA fragments that can be transferred is very limited, and often the transferred gene is not a conventional gene. Horizontal gene transfer is the transfer of genetic material from one cell to another that is not its offspring. Artificial horizontal gene transfer is a form of genetic engineering.

The Human Genome Project

The Human Genome Project is a molecular genetics project that began in the 1990s and was projected to take fifteen years to complete. However, because of technological advances the progress of the project was advanced and the project finished in 2003, taking only thirteen years. The project was started by the U.S. Department of Energy and the National Institutes of Health in an effort to reach six set goals. These goals included:

1. identifying 20,000 to 25,000 genes in human DNA (although initial estimates were approximately 100,000 genes),
2. determining sequences of chemical base pairs in human DNA,
3. storing all found information into databases,
4. improving the tools used for data analysis,
5. transferring technologies to private sectors,
6. addressing the ethical, legal, and social issues (ELSI) that may arise from the projects.

The project was worked on by eighteen different countries including the United States, Japan, France, Germany, and the United Kingdom. The collaborative effort resulted in the discovery of the many benefits of molecular genetics. Discoveries such as molecular medicine, new energy sources and environmental applications, DNA forensics, and livestock breeding, are only a few of the benefits that molecular genetics can provide.

Molecular Microbiology

Molecular microbiology is the branch of microbiology devoted to the study of the molecular basis of the physiological processes that occur in microorganisms.

Bacteria

Mainly because of their relative simplicity, ease of manipulation and growth in vitro, and importance in medicine, bacteria were instrumental in the development of molecular biology. The complete genome sequence for a large number of bacterial species is now available. A list of sequenced prokaryotic genomes is available. Molecular microbiology techniques are currently being used in the development of new genetically engineered vaccines, in bioremediation, biotechnology, food microbiology, probiotic research, antibacterial development and environmental microbiology.

Many bacteria have become model organisms for molecular studies.

Molecular techniques have had a direct influence on the clinical practice of medical microbiology. In many cases where traditional phenotypic methods of microbial identification and typing are insufficient or time-consuming, molecular techniques can provide rapid and accurate data, potentially improving clinical outcomes. Specific examples include:

- 16s rRNA sequencing to provide bacterial identifications.
- Pulsed Field Gel Electrophoresis for strain typing of epidemiologically related organisms.
- Direct detection of genes related to resistance mechanisms, such as mecA gene in Staphylococcus aureus.

Signaling

Bacteria possess diverse proteins and RNA that can sense changes to their intracellular and extracellular environment. The signals received by these macromolecules are transmitted to key genes or proteins, which alter their activities to suit the new conditions.

Viruses

Viruses are important pathogens of humans and animals. Their genomes are relatively small. For these reasons they were among the first organisms to be fully sequenced. The complete DNA sequence of the Epstein-Barr virus was completed in 1984. Bluetongue virus (BTV) has been in the forefront of molecular studies for last three decades and now represents one of the best understood viruses at the molecular and structural levels. Other viruses such as Papillomavirus, Coronavirus, Caliciviruses, Paramyxoviruses and Influenza virus have also been extensively studied at the molecular level.

Bacterial viruses, or bacteriophages, are estimated to be the most widely distributed and diverse entities in the biosphere. Bacteriophages, or "phage", have been fundamental in the development of the science of molecular biology and became "model organisms" for probing the basic chemistry of life. The first DNA-genome project to be completed was the phage Φ-X174 in 1977. Φ29 phage, a phage of Bacillus, is a paradigm for the study of several molecular mechanisms of general biological processes, including DNA replication and regulation of transcription.

Gene Therapy

Some viruses are used as vectors for gene therapy. Virus vectors have been developed that mediate stable genetic modification of treated cells by chromosomal integration of the transferred vector genomes. Gammaretroviral and lentiviral vectors, for example, can be utilized in clinical gene therapy aimed at the long-term correction of genetic defects, e.g., in stem and progenitor cells. Gammaretroviral and lentiviral vectors have so far been used in more than 300 clinical trials, addressing treatment options for various diseases.

Fungi

Yeasts and molds are eukaryotic microorganisms classified in the kingdom Fungi.

Technology

Polymerase chain reaction (PCR) is used in microbiology to amplify (replicate many times) a single DNA sequence. If required, the sequence can also be altered in predetermined ways. Quantitative PCR is used for the rapid detection of microorganisms and is currently employed in diagnostic clinical microbiology laboratories, environmental analysis, food microbiology, and many other fields. The closely related technique of quantitative PCR permits the quantitative measurement of DNA or RNA molecules and is used to estimate the densities of the reference pathogens in food, water and environmental samples. Quantitative PCR provides both specificity and quantification of target microorganisms.

Gel electrophoresis is used routinely in microbiology to separate DNA, RNA, or protein molecules using an electric field by virtue of their size, shape or electric charge.

Southern blotting, northern blotting, western blotting and Eastern blotting are molecular techniques for detecting the presence of microbial DNA sequences (Southern), RNA sequences (northern), protein molecules (western) or protein modifications (Eastern).

DNA microarrays are used in microbiology as the modern alternative to the "blotting" techniques. Microarrays permit the exploration of thousands of sequences at one time. This technique is used in molecular microbiology to detect the presence of pathogens in a sample (air, water, organ tissue, etc.). It is also used to determine the genetic differences between two microbial strains.

DNA sequencing and genomics have been used for many decades in molecular microbiology studies. Due to their relatively small size, viral genomes were the first to be completely analysed by DNA sequencing. A huge range of sequence and genomic data is now available for a number of species and strains of microorganisms.

RNA interference (RNAi) was discovered as a cellular gene regulation mechanism in 1998, but several RNAi-based applications for gene silencing have already made it into clinical trials. RNA interference (RNAi) technology has formed the basis of novel tools for biological research and drug discovery.

References

- Herold, KE; Rasooly, A (editor) (2009). Lab-on-a-Chip Technology: Biomolecular Separation and Analysis. Caister Academic Press ISBN 978-1-904455-47-9
- Mc Grath S and van Sinderen D (editors). (2007). Bacteriophage: Genetics and Molecular Biology. Caister Academic Press. ISBN 978-1-904455-14-1
- Morris, KV (editor) (2008). RNA and the Regulation of Gene Expression: A Hidden Layer of Complexity. Caister Academic Press. ISBN 978-1-904455-25-7
- Kurth, R; Bannert, N (editors) (2010). Retroviruses: Molecular Biology, Genomics and Pathogenesis. Caister Academic Press ISBN 978-1-904455-55 4
- Sen, K; Ashbolt, NK (editor) (2010). Environmental Microbiology: Current Technology and Water Applications. Caister Academic Press. ISBN 978-1-904455-70-7
- Roy P (2008). "Molecular Dissection of Bluetongue Virus". Animal Viruses: Molecular Biology. Caister Academic Press. ISBN 978-1-904455-22-6
- Campo MS (editor). (2006). Papillomavirus Research: From Natural History To Vaccines and Beyond. Caister Academic Press. ISBN 978-1-904455-04-2
- Fratamico PM and Bayles DO (editor). (2005). Foodborne Pathogens: Microbiology and Molecular Biology. Caister Academic Press. ISBN 978-1-904455-00-4
- Miller, AA; Miller, PF (editor) (2011). Emerging Trends in Antibacterial Discovery: Answering the Call to Arms. Caister Academic Press ISBN 978-1-904455-89-9
- Diaz E (editor). (2008). Microbial Biodegradation: Genomics and Molecular Biology. Caister Academic Press. ISBN 978-1-904455-17-2
- Mayo, B; van Sinderen, D (editor) (2010). Bifidobacteria: Genomics and Molecular Aspects. Caister Academic Press. ISBN 978-1-904455-68-4

8

Molecular Biology: Areas of Application

Molecular biology is a vast area of study which has applications in a wide variety of fields. A few of these are molecular medicine and bioinformatics. This chapter has been carefully written to provide an easy understanding of these areas of application of molecular biology as well as the central dogma of molecular biology.

Molecular Medicine

Molecular medicine is a broad field, where physical, chemical, biological and medical techniques are used to describe molecular structures and mechanisms, identify fundamental molecular and genetic errors of disease, and to develop molecular interventions to correct them. The molecular medicine perspective emphasizes cellular and molecular phenomena and interventions rather than the previous conceptual and observational focus on patients and their organs.

In November 1949, with the seminal paper, "Sickle Cell Anemia, a Molecular Disease", in *Science* magazine, Linus Pauling, Harvey Itano and their collaborators laid the groundwork for establishing the field of molecular medicine. In 1956, Roger J. Williams wrote *Biochemical Individuality*, a prescient book about genetics, prevention and treatment of disease on a molecular basis, and nutrition which is now variously referred to as individualized medicine and orthomolecular medicine. Another paper in *Science* by Pauling in 1968, introduced and defined this view of molecular medicine that focuses on natural and nutritional substances used for treatment and prevention.

Published research and progress was slow until the 1970s' "biological revolution" that introduced many new techniques and commercial applications.

Molecular medicine is a new scientific discipline in European universities. Combining contemporary medical studies with the field of biochemistry, it offers a bridge between the two subjects. At present only a handful of universities offer the course to undergraduates. With a degree in this discipline the graduate is able to pursue a career in medical sciences, scientific research, laboratory work and postgraduate medical degrees.

Subjects

Core subjects are similar to biochemistry courses and typically include gene expression, research methods, proteins, cancer research, immunology, biotechnology and many more besides. In some universities molecular medicine is combined with another discipline such as chemistry, functioning as an additional study to enrich the undergraduate program.

Bioinformatics

Bioinformatics is an interdisciplinary field that develops methods and software tools for understanding biological data. As an interdisciplinary field of science, bioinformatics combines computer science, statistics, mathematics, and engineering to analyze and interpret biological data. Bioinformatics has been used for *in silico* analyses of biological queries using mathematical and statistical techniques.

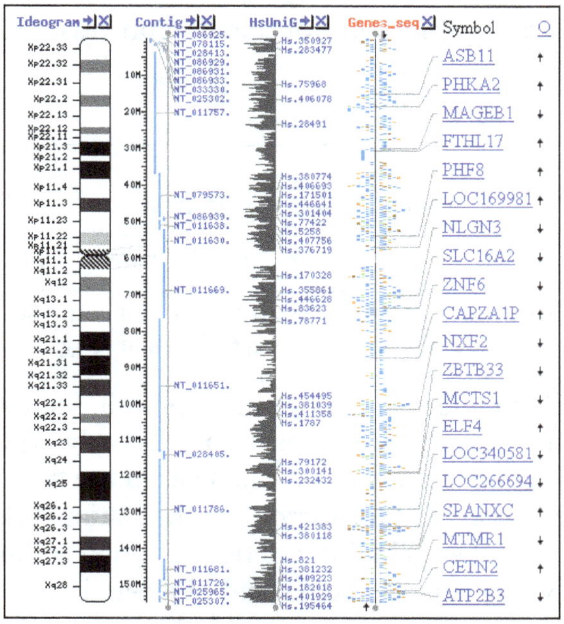

Map of the human X chromosome.

Bioinformatics is both an umbrella term for the body of biological studies that use computer programming as part of their methodology, as well as a reference to specific analysis "pipelines" that are repeatedly used, particularly in the field of genomics. Common uses of bioinformatics include the identification of candidate genes and nucleotides (SNPs). Often, such identification is made with the aim of better understanding the genetic basis of disease, unique adaptations, desirable properties (esp. in agricultural species), or differences between populations. In a less formal way, bioinformatics also tries to understand the organisational principles within nucleic acid and protein sequences.

Introduction

Bioinformatics has become an important part of many areas of biology. In experimental molecular biology, bioinformatics techniques such as image and signal processing allow extraction of useful results from large amounts of raw data. In the field of genetics and genomics, it aids in sequencing and annotating genomes and their observed mutations. It plays a role in the text mining of biological literature and the development of biological and gene ontologies to organize and query biological data. It also plays a role in the analysis of gene and protein expression and regulation. Bioinformatics tools aid in the comparison of genetic and genomic data and more generally in the understanding of evolutionary aspects of molecular biology. At a more integrative level, it helps analyze and catalogue the biological pathways and networks that are an important part of systems biology. In structural biology, it aids in the simulation and modeling of DNA, RNA, proteins as well as biomolecular interactions.

History

Historically, the term *bioinformatics* did not mean what it means today. Paulien Hogeweg and Ben Hesper coined it in 1970 to refer to the study of information processes in biotic systems. This definition placed bioinformatics as a field parallel to biophysics (the study of physical processes in biological systems) or biochemistry (the study of chemical processes in biological systems).

Sequences

```
5'ATGACGTGGGGA3'
3'TACTGCACCCCT5'
```

Sequences of genetic material are frequently used in bioinformatics and are easier to manage using computers than manually.

Computers became essential in molecular biology when protein sequences became available after Frederick Sanger determined the sequence of insulin in the early 1950s. Comparing multiple sequences manually turned out to be impractical. A pioneer in the field was Margaret Oakley Dayhoff, who has been hailed by David Lipman, director of the National Center for Biotechnology Information, as the "mother and father of bioinformatics." Dayhoff compiled one of the first protein sequence databases, initially published as books and pioneered methods of sequence alignment and molecular evolution. Another early contributor to bioinformatics was Elvin A. Kabat, who pioneered biological sequence analysis in 1970 with his comprehensive volumes of antibody sequences released with Tai Te Wu between 1980 and 1991.

Goals

To study how normal cellular activities are altered in different disease states, the biological data must be combined to form a comprehensive picture of these activities. Therefore, the field of bioinformatics has evolved such that the most pressing task now involves the analysis and interpretation of various types of data. This includes nucleotide and amino acid sequences, protein domains, and protein structures. The actual process of analyzing and interpreting data is referred to as computational biology. Important sub-disciplines within bioinformatics and computational biology include:

- Development and implementation of computer programs that enable efficient access to, use and management of, various types of information.
- Development of new algorithms (mathematical formulas) and statistical measures that assess relationships among members of large data sets. For example, there are methods to locate a gene within a sequence, to predict protein structure and/or function, and to cluster protein sequences into families of related sequences.

The primary goal of bioinformatics is to increase the understanding of biological processes. What sets it apart from other approaches, however, is its focus on developing and applying computationally intensive techniques to achieve this goal. Examples include: pattern recognition, data mining, machine learning algorithms, and visualization. Major research efforts in the field include sequence alignment, gene finding, genome assembly, drug design, drug discovery, protein structure alignment, protein structure prediction, prediction of gene expression and protein–protein interactions, genome-wide association studies, the modeling of evolution and cell division/mitosis.

Bioinformatics now entails the creation and advancement of databases, algorithms, computational and statistical techniques, and theory to solve formal and practical problems arising from the management and analysis of biological data.

Over the past few decades, rapid developments in genomic and other molecular research technologies and developments in information technologies have combined to produce a tremendous amount of information related to molecular biology. Bioinformatics is the name given to these mathematical and computing approaches used to glean understanding of biological processes.

Common activities in bioinformatics include mapping and analyzing DNA and protein sequences, aligning DNA and protein sequences to compare them, and creating and viewing 3-D models of protein structures.

Relation to Other Fields

Bioinformatics is a science field that is similar to but distinct from biological computation and computational biology. Biological computation uses bioengineering and biology to build biological computers, whereas bioinformatics uses computation to better understand biology. Bioinformatics and computational biology have similar aims and approaches, but they differ in scale: bioinformatics organizes and analyzes basic biological data, whereas computational biology builds theoretical models of biological systems, just as mathematical biology does with mathematical models.

Analyzing biological data to produce meaningful information involves writing and running software programs that use algorithms from graph theory, artificial intelligence, soft computing, data mining, image processing, and computer simulation. The algorithms in turn depend on theoretical foundations such as discrete mathematics, control theory, system theory, information theory, and statistics.

Sequence Analysis

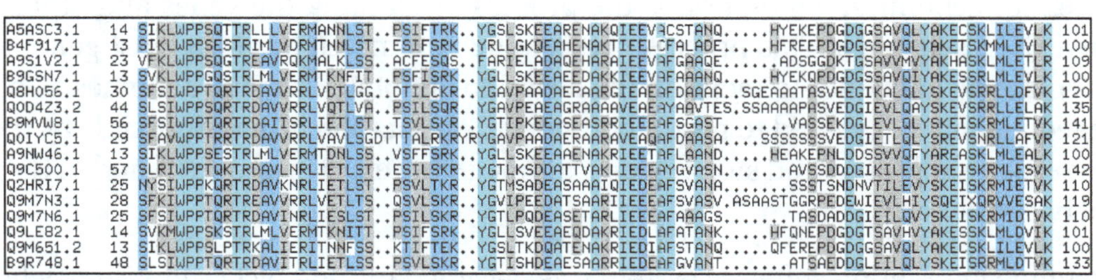

The sequences of different genes or proteins may be aligned side-by-side to measure their similarity. This alignment compares protein sequences containing WPP domains.

Since the Phage Φ-X174 was sequenced in 1977, the DNA sequences of thousands of organisms have been decoded and stored in databases. This sequence information is analyzed to determine genes that encode proteins, RNA genes, regulatory sequences, structural motifs, and repetitive sequences. A comparison of genes within a species or between different species can show similarities between protein functions, or relations between species (the use of molecular systematics to construct phylogenetic trees). With the growing amount of data, it long ago became impractical to analyze DNA sequences manually. Today, computer programs such as BLAST are used daily to search sequences from more than 260 000 organisms, containing over 190 billion nucleotides. These programs can compensate for mutations (exchanged, deleted or inserted bases) in the DNA sequence, to identify sequences that are related, but not identical. A variant of this sequence alignment is used in the sequencing process itself. The so-called shotgun sequencing technique (which was used, for example, by The Institute for Genomic Research (TIGR) to sequence the first bacterial genome, *Haemophilus influenzae*) does not produce entire chromosomes. Instead it generates the sequences of many thousands of small DNA fragments (ranging from 35 to 900 nucleotides long, depending on the sequencing technology). The ends of these fragments overlap and, when aligned properly by a genome assembly program, can be used to reconstruct the complete genome. Shotgun sequencing yields sequence data quickly, but the task of assembling the fragments can be quite complicated for larger genomes. For a genome as large as the human genome, it may take many days of CPU time on large-memory, multiprocessor computers to assemble the fragments, and the resulting assembly usually contains numerous gaps that must be filled in later. Shotgun sequencing is the method of choice for virtually all genomes sequenced today, and genome assembly algorithms are a critical area of bioinformatics research.

Following the goals that the Human Genome Project left to achieve after its closure in 2003, a new project developed by the National Human Genome Research Institute in the U.S appeared. The so-called ENCODE project is a collaborative data collection of the functional elements of the human genome that uses next-generation DNA-sequencing technologies and genomic tiling arrays, technologies able to generate automatically large amounts of data with lower research costs but with the same quality and viability.

Another aspect of bioinformatics in sequence analysis is annotation. This involves computational gene finding to search for protein-coding genes, RNA genes, and other functional sequences within a genome. Not all of the nucleotides within a genome are part of genes. Within the genomes of higher organisms, large parts of the DNA do not serve any obvious purpose.

Genome Annotation

In the context of genomics, annotation is the process of marking the genes and other biological features in a DNA sequence. This process needs to be automated because most genomes are too large to annotate by hand, not to mention the desire to annotate as many genomes as possible, as the rate of sequencing has ceased to pose a bottleneck. Annotation is made possible by the fact that genes have recognisable start and stop regions, although the exact sequence found in these regions can vary between genes.

The first genome annotation software system was designed in 1995 by Owen White, who was part of the team at The Institute for Genomic Research that sequenced and analyzed the first genome of a free-living organism to be decoded, the bacterium *Haemophilus influenzae*. White built a software system to find the genes (fragments of genomic sequence that encode proteins), the transfer RNAs, and to make initial assignments of function to those genes. Most current genome annotation systems work similarly, but the programs available for analysis of genomic DNA, such as the

GeneMark program trained and used to find protein-coding genes in *Haemophilus influenzae*, are constantly changing and improving.

Computational Evolutionary Biology

Evolutionary biology is the study of the origin and descent of species, as well as their change over time. Informatics has assisted evolutionary biologists by enabling researchers to:

- trace the evolution of a large number of organisms by measuring changes in their DNA, rather than through physical taxonomy or physiological observations alone,
- more recently, compare entire genomes, which permits the study of more complex evolutionary events, such as gene duplication, horizontal gene transfer, and the prediction of factors important in bacterial speciation,
- build complex computational models of populations to predict the outcome of the system over time,
- track and share information on an increasingly large number of species and organisms.

Future work endeavours to reconstruct the now more complex tree of life.

The area of research within computer science that uses genetic algorithms is sometimes confused with computational evolutionary biology, but the two areas are not necessarily related.

Comparative Genomics

The core of comparative genome analysis is the establishment of the correspondence between genes (orthology analysis) or other genomic features in different organisms. It is these intergenomic maps that make it possible to trace the evolutionary processes responsible for the divergence of two genomes. A multitude of evolutionary events acting at various organizational levels shape genome evolution. At the lowest level, point mutations affect individual nucleotides. At a higher level, large chromosomal segments undergo duplication, lateral transfer, inversion, transposition, deletion and insertion. Ultimately, whole genomes are involved in processes of hybridization, polyploidization and endosymbiosis, often leading to rapid speciation. The complexity of genome evolution poses many exciting challenges to developers of mathematical models and algorithms, who have recourse to a spectra of algorithmic, statistical and mathematical techniques, ranging from exact, heuristics, fixed parameter and approximation algorithms for problems based on parsimony models to Markov Chain Monte Carlo algorithms for Bayesian analysis of problems based on probabilistic models.

Many of these studies are based on the homology detection and protein families computation.

Pan Genomics

Pan genomics is a concept introduced in 2005 by Tettelin and Medini which eventually took root in bioinformatics. Pan genome is the complete gene repertoire of a particular taxonomic group: although initially applied to closely related strains of a species, it can be applied to a larger context like genus, phylum etc. It is divided in two parts- The Core genome: Set of genes common to all the genomes under study (These are often housekeeping genes vital for survival) and The Dispensable/Flexible Genome: Set of genes not present in all but one or some genomes under study. a bioinformatics tool BPGA can be used to characterize the Pan Genome of bacterial species.

Genetics of Disease

With the advent of next-generation sequencing we are obtaining enough sequence data to map the genes of complex diseases such as diabetes, infertility, breast cancer or Alzheimer's Disease. Genome-wide association studies are a useful approach to pinpoint the mutations responsible for such complex diseases. Through these studies, thousands of DNA variants have been identified that are associated with similar diseases and traits. Furthermore, the possibility for genes to be used at prognosis, diagnosis or treatment is one of the most essential applications. Many studies are discussing both the promising ways to choose the genes to be used and the problems and pitfalls of using genes to predict disease presence or prognosis.

Analysis of Mutations in Cancer

In cancer, the genomes of affected cells are rearranged in complex or even unpredictable ways. Massive sequencing efforts are used to identify previously unknown point mutations in a variety of genes in cancer. Bioinformaticians continue to produce specialized automated systems to manage the sheer volume of sequence data produced, and they create new algorithms and software to compare the sequencing results to the growing collection of human genome sequences and germline polymorphisms. New physical detection technologies are employed, such as oligonucleotide microarrays to identify chromosomal gains and losses (called comparative genomic hybridization), and single-nucleotide polymorphism arrays to detect known *point mutations*. These detection methods simultaneously measure several hundred thousand sites throughout the genome, and when used in high-throughput to measure thousands of samples, generate terabytes of data per experiment. Again the massive amounts and new types of data generate new opportunities for bioinformaticians. The data is often found to contain considerable variability, or noise, and thus Hidden Markov model and change-point analysis methods are being developed to infer real copy number changes.

With the breakthroughs that this next-generation sequencing technology is providing to the field of Bioinformatics, cancer genomics could drastically change. These new methods and software allow bioinformaticians to sequence many cancer genomes quickly and affordably. This could create a more flexible process for classifying types of cancer by analysis of cancer driven mutations in the genome. Furthermore, tracking of patients while the disease progresses may be possible in the future with the sequence of cancer samples.

Another type of data that requires novel informatics development is the analysis of lesions found to be recurrent among many tumors.

Gene and Protein Expression

Analysis of Gene Expression

The expression of many genes can be determined by measuring mRNA levels with multiple techniques including microarrays, expressed cDNA sequence tag (EST) sequencing, serial analysis of gene expression (SAGE) tag sequencing, massively parallel signature sequencing (MPSS), RNA-Seq, also known as "Whole Transcriptome Shotgun Sequencing" (WTSS), or various applications of multiplexed in-situ hybridization. All of these techniques are extremely noise-prone and/or subject to bias in the biological measurement, and a major research area in computational biology involves developing statistical tools to separate signal from noise in high-throughput gene expression studies. Such studies are often used to determine the genes implicated in a disorder: one might compare microarray data from cancerous epithelial cells to data from non-cancerous cells to determine the transcripts that are up-regulated and down-regulated in a particular population of cancer cells.

Analysis of Protein Expression

Protein microarrays and high throughput (HT) mass spectrometry (MS) can provide a snapshot of the proteins present in a biological sample. Bioinformatics is very much involved in making sense of protein microarray and HT MS data; the former approach faces similar problems as with microarrays targeted at mRNA, the latter involves the problem of matching large amounts of mass data against predicted masses from protein sequence databases, and the complicated statistical analysis of samples where multiple, but incomplete peptides from each protein are detected.

Analysis of Regulation

Regulation is the complex orchestration of events by which a signal, potentially an extracellular signal such as a hormone, eventually leads to an increase or decrease in the activity of one or more proteins. Bioinformatics techniques have been applied to explore various steps in this process.

For example, promoter analysis involves the identification and study of sequence motifs in the DNA surrounding the coding region of a gene. These motifs influence the extent to which that region is transcribed into mRNA.

Expression data can be used to infer gene regulation: one might compare microarray data from a wide variety of states of an organism to form hypotheses about the genes involved in each state. In a single-cell organism, one might compare stages of the cell cycle, along with various stress conditions (heat shock, starvation, etc.). One can then apply clustering algorithms to that expression data to determine which genes are co-expressed. For example, the upstream regions (promoters) of co-expressed genes can be searched for over-represented regulatory elements. Examples of clustering algorithms applied in gene clustering are k-means clustering, self-organizing maps (SOMs), hierarchical clustering, and consensus clustering methods.

Structural Bioinformatics

Protein structure prediction is another important application of bioinformatics. The amino acid sequence of a protein, the so-called primary structure, can be easily determined from the sequence on the gene that codes for it. In the vast majority of cases, this primary structure uniquely determines a structure in its native environment. (Of course, there are exceptions, such as the bovine spongiform encephalopathy – a.k.a. Mad Cow Disease – prion.) Knowledge of this structure is vital in understanding the function of the protein. Structural information is usually classified as one of *secondary*, *tertiary* and *quaternary* structure. A viable general solution to such predictions remains an open problem. Most efforts have so far been directed towards heuristics that work most of the time.

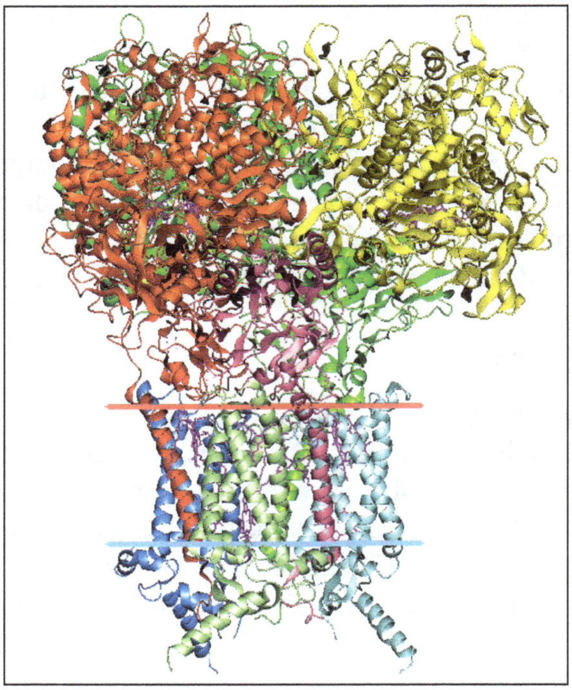

3-dimensional protein structures such as this one are common subjects in bioinformatic analyses.

One of the key ideas in bioinformatics is the notion of homology. In the genomic branch of bioinformatics, homology is used to predict the function of a gene: if the sequence of gene *A*, whose function is known, is homologous to the sequence of gene *B,* whose function is unknown, one could infer that B may share A's function. In the structural branch of bioinformatics, homology is used to determine which parts of a protein are important in structure formation and interaction with other proteins. In a technique called homology modeling, this information is used to predict the structure of a protein once the structure of a homologous protein is known. This currently remains the only way to predict protein structures reliably.

One example of this is the similar protein homology between hemoglobin in humans and the hemoglobin in legumes (leghemoglobin). Both serve the same purpose of transporting oxygen in the organism. Though both of these proteins have completely different amino acid sequences, their protein structures are virtually identical, which reflects their near identical purposes.

Other techniques for predicting protein structure include protein threading and *de novo* (from scratch) physics-based modeling.

Network and Systems Biology

Network analysis seeks to understand the relationships within biological networks such as metabolic or protein–protein interaction networks. Although biological networks can be constructed from a single type of molecule or entity (such as genes), network biology often attempts to integrate many different data types, such as proteins, small molecules, gene expression data, and others, which are all connected physically, functionally, or both.

Systems biology involves the use of computer simulations of cellular subsystems (such as the networks of metabolites and enzymes that comprise metabolism, signal transduction pathways and gene regulatory networks) to both analyze and visualize the complex connections of these cellular processes. Artificial life or virtual evolution attempts to understand evolutionary processes via the computer simulation of simple (artificial) life forms.

Molecular Interaction Networks

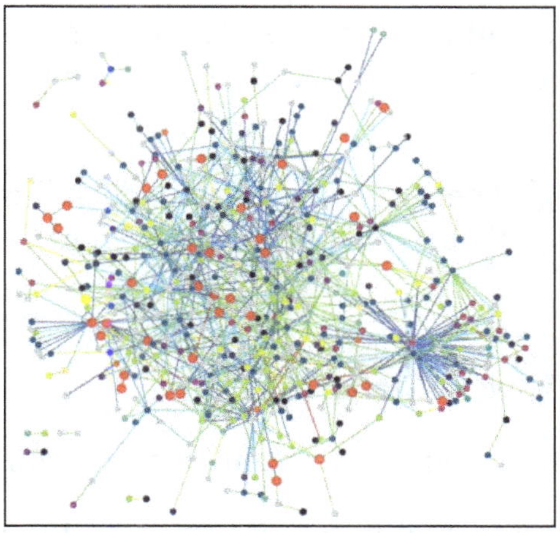

Interactions between proteins are frequently visualized and analyzed using networks. This network is made up of protein–protein interactions from *Treponema pallidum*, the causative agent of syphilis and other diseases.

Tens of thousands of three-dimensional protein structures have been determined by X-ray crystallography and protein nuclear magnetic resonance spectroscopy (protein NMR) and a central question in structural bioinformatics is whether it is practical to predict possible protein–protein interactions only based on these 3D shapes, without performing protein–protein interaction experiments. A variety of methods have been developed to tackle the protein–protein docking problem, though it seems that there is still much work to be done in this field.

Other interactions encountered in the field include Protein–ligand (including drug) and protein–peptide. Molecular dynamic simulation of movement of atoms about rotatable bonds is the fundamental principle behind computational algorithms, termed docking algorithms, for studying molecular interactions.

Others

Literature Analysis

The growth in the number of published literature makes it virtually impossible to read every paper, resulting in disjointed sub-fields of research. Literature analysis aims to employ computational and statistical linguistics to mine this growing library of text resources. For example:

- Abbreviation recognition – identify the long-form and abbreviation of biological terms.
- Named entity recognition – recognizing biological terms such as gene names.
- Protein–protein interaction – identify which proteins interact with which proteins from text.

The area of research draws from statistics and computational linguistics.

High-throughput Image Analysis

Computational technologies are used to accelerate or fully automate the processing, quantification and analysis of large amounts of high-information-content biomedical imagery. Modern image analysis systems augment an observer's ability to make measurements from a large or complex set of images, by improving accuracy, objectivity, or speed. A fully developed analysis system may completely replace the observer. Although these systems are not unique to biomedical imagery, biomedical imaging is becoming more important for both diagnostics and research. Some examples are:

- high-throughput and high-fidelity quantification and sub-cellular localization (high-content screening, cytohistopathology, Bioimage informatics).
- morphometrics.
- clinical image analysis and visualization.
- determining the real-time air-flow patterns in breathing lungs of living animals.
- quantifying occlusion size in real-time imagery from the development of and recovery during arterial injury.
- making behavioral observations from extended video recordings of laboratory animals.
- infrared measurements for metabolic activity determination.
- inferring clone overlaps in DNA mapping, e.g. the Sulston score.

High-throughput Single Cell Data Analysis

Computational techniques are used to analyse high-throughput, low-measurement single cell data, such as that obtained from flow cytometry. These methods typically involve finding populations of cells that are relevant to a particular disease state or experimental condition.

Biodiversity Informatics

Biodiversity informatics deals with the collection and analysis of biodiversity data, such as taxonomic databases, or microbiome data. Examples of such analyses include phylogenetics, niche modelling, species richness mapping, or species identification tools.

Databases

Databases are essential for bioinformatics research and applications. There is a huge number of available databases covering almost everything from DNA and protein sequences, molecular structures, to phenotypes and biodiversity. Databases generally fall into one of three types. Some contain data resulting directly from empirical methods such as gene knockouts. Others consist of predicted data, and most contain data from both sources. There are meta-databases that incorporate data compiled from multiple other databases. Some others are specialized, such as those specific to an organism. These databases vary in their format, way of accession and whether they are public or not. Some of the most commonly used databases are listed below. For a more comprehensive list, please check the link at the beginning of the subsection:

- Used in Motif Finding: GenomeNet MOTIF Search.
- Used in Gene Ontology: ToppGene FuncAssociate, Enrichr, GATHER.
- Used in Gene Finding: Hidden Markov Model.
- Used in finding Protein Structures/Family: PFAM.
- Used for Next Generation Sequencing: (Not database but data format), FASTQ Format.
- Used in Gene Expression Analysis: GEO, ArrayExpress.
- Used in Network Analysis: Interaction Analysis Databases(BioGRID, MINT, HPRD, Curated Human Signaling Network), Functional Networks (STRING, KEGG).
- Used in design of synthetic genetic circuits: GenoCAD.

Please keep in mind that this is a quick sampling and generally most computation data is supported by wet lab data as well.

Software and Tools

Software tools for bioinformatics range from simple command-line tools, to more complex graphical programs and standalone web-services available from various bioinformatics companies or public institutions.

Open-source Bioinformatics Software

Many free and open-source software tools have existed and continued to grow since the 1980s. The combination of a continued need for new algorithms for the analysis of emerging types of biological readouts, the potential for innovative *in silico* experiments, and freely available open code bases have helped to create opportunities for all research groups to contribute to both bioinformatics and the range of open-source software available, regardless of their funding arrangements. The open source tools often act as incubators of ideas, or community-supported plug-ins in commercial applications. They may also provide *de facto* standards and shared object models for assisting with the challenge of bioinformation integration.

The range of open-source software packages includes titles such as Bioconductor, BioPerl, Biopython, BioJava, BioJS, BioRuby, Bioclipse, EMBOSS, .NET Bio, Orange with its bioinformatics add-on, Apache Taverna, UGENE and GenoCAD. To maintain this tradition and create further opportunities, the non-profit Open Bioinformatics Foundation have supported the annual Bioinformatics Open Source Conference (BOSC) since 2000.

An alternative method to build public bioinformatics databases is to use the MediaWiki engine with the *WikiOpener* extension. This system allows the database to be accessed and updated by all experts in the field.

Web Services in Bioinformatics

SOAP- and REST-based interfaces have been developed for a wide variety of bioinformatics applications allowing an application running on one computer in one part of the world to use algorithms, data and computing resources on servers in other parts of the world. The main advantages derive from the fact that end users do not have to deal with software and database maintenance overheads.

Basic bioinformatics services are classified by the EBI into three categories: SSS (Sequence Search Services), MSA (Multiple Sequence Alignment), and BSA (Biological Sequence Analysis). The availability of these service-oriented bioinformatics resources demonstrate the applicability of web-based bioinformatics solutions, and range from a collection of standalone tools with a common data format under a single, standalone or web-based interface, to integrative, distributed and extensible bioinformatics workflow management systems.

Bioinformatics Workflow Management Systems

A Bioinformatics workflow management system is a specialized form of a workflow management system designed specifically to compose and execute a series of computational or data manipulation steps, or a workflow, in a Bioinformatics application. Such systems are designed to:

- provide an easy-to-use environment for individual application scientists themselves to create their own workflows.
- provide interactive tools for the scientists enabling them to execute their workflows and view their results in real-time.
- simplify the process of sharing and reusing workflows between the scientists.
- enable scientists to track the provenance of the workflow execution results and the workflow creation steps.

Some of the platforms giving this service: Galaxy, Kepler, Taverna, UGENE, Anduril.

Education Platforms

Software platforms designed to teach bioinformatics concepts and methods include Rosalind and online courses offered through the Swiss Institute of Bioinformatics Training Portal. The Canadian Bioinformatics Workshops provides videos and slides from training workshops on their website under a Creative Commons license. The 4273π project or 4273pi project also offers open source educational materials for free. The course runs on low cost raspberry pi computers and has been used to teach adults and school pupils. 4273π is actively developed by a consortium of academics and research staff who have run research level bioinformatics using raspberry pi computers and the 4273π operating system.

MOOC platforms also provide online certifications in bioinformatics and related disciplines, including Coursera's Bioinformatics Specialization (UC San Diego) and Genomic Data Science Specialization (Johns Hopkins) as well as EdX's Data Analysis for Life Sciences XSeries (Harvard).

Conferences

There are several large conferences that are concerned with bioinformatics. Some of the most notable examples are Intelligent Systems for Molecular Biology (ISMB), European Conference on Computational Biology (ECCB), and Research in Computational Molecular Biology (RECOMB).

Central Dogma of Molecular Biology

The central dogma of molecular biology is an explanation of the flow of genetic information within a biological system. It was first stated by Francis Crick in 1956 and re-stated in a *Nature* paper published in 1970:

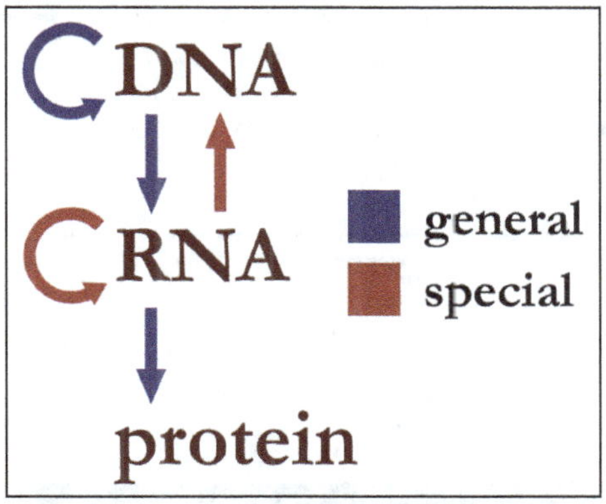

Information flow in biological systems

"The central dogma of molecular biology deals with the detailed residue-by-residue transfer of sequential information. It states that such information cannot be transferred back from protein to either protein or nucleic acid."

— Francis Crick

The central dogma has also been described as "DNA makes RNA and RNA makes protein," a positive statement which was originally termed the sequence hypothesis by Crick. However, this simplification does not make it clear that the central dogma as stated by Crick does not preclude the reverse flow of information from RNA to DNA, only ruling out the flow from protein to RNA or DNA. Crick's use of the word dogma was unconventional, and has been controversial.

The dogma is a framework for understanding the transfer of sequence information between information-carrying biopolymers, in the most common or general case, in living organisms. There are 3 major classes of such biopolymers: DNA and RNA (both nucleic acids), and protein. There are 3×3 = 9 conceivable direct transfers of information that can occur between these. The dogma classes these into 3 groups of 3: 3 general transfers (believed to occur normally in most cells), 3 special transfers (known to occur, but only under specific conditions in case of some viruses or in a laboratory), and 3 unknown transfers (believed never to occur). The general transfers describe the normal flow of biological information: DNA can be copied to DNA (DNA replication), DNA information can be copied into mRNA (transcription), and proteins can be synthesized using the information in mRNA as a template (translation).

Biological Sequence Information

The biopolymers that comprise DNA, FNA, RNA and (poly)peptides are linear polymers (i.e.: each monomer is connected to at most two other monomers). The sequence of their monomers effectively encodes information. The transfers of information described by the central dogma ideally are faithful, deterministic transfers, wherein one biopolymer's sequence is used as a template for the construction of another biopolymer with a sequence that is entirely dependent on the original biopolymer's sequence.

General Transfers of Biological Sequential Information

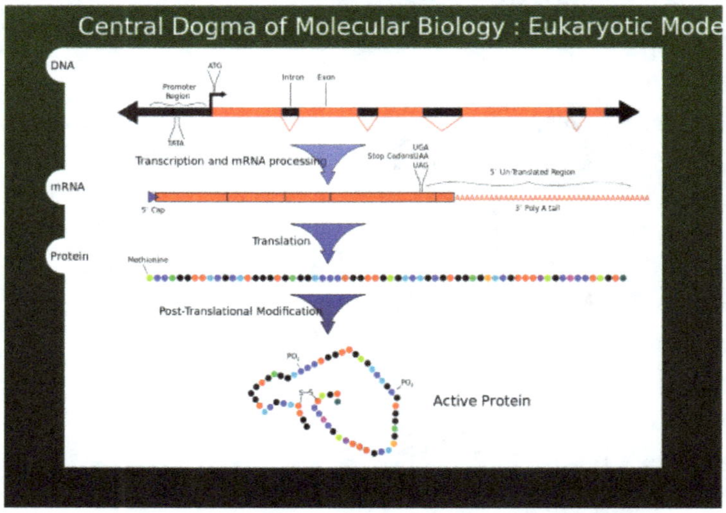

Table of the 3 classes of information transfer suggested by the dogma		
General	**Special**	**Unknown**
DNA → DNA	RNA → DNA	protein → DNA
DNA → RNA	RNA → RNA	protein → RNA
RNA → protein	DNA → protein	protein → protein

DNA Replications

In the sense that DNA replication must occur if genetic material is to be provided for the progeny of any cell, whether somatic or reproductive, the copying from DNA to DNA arguably is the fundamental step in the central dogma. A complex group of proteins called the replisome performs the replication of the information from the parent strand to the complementary daughter strand.

The replisome comprises:

- a helicase that unwinds the superhelix as well as the double-stranded DNA helix to create a replication fork.

- SSB protein that binds open the double-stranded DNA to prevent it from reassociating.
- RNA primase that adds a complementary RNA primer to each template strand as a starting point for replication.
- DNA polymerase III that reads the existing template chain from its 3' end to its 5' end and adds new complementary nucleotides from the 5' end to the 3' end of the daughter chain.
- DNA polymerase I that removes the RNA primers and replaces them with DNA.
- DNA ligase that joins the two Okazaki fragments with phosphodiester bonds to produce a continuous chain.

This process typically takes place during S phase of the cell cycle.

Transcription

Transcription is the process by which the information contained in a section of DNA is replicated in the form of a newly assembled piece of messenger RNA (mRNA). Enzymes facilitating the process include RNA polymerase and transcription factors. In eukaryotic cells the primary transcript is (pre-mRNA). Pre-mRNA must be processed for translation to proceed. Processing includes the addition of a 5' cap and a poly-A tail to the pre-mRNA chain, followed by splicing. Alternative splicing occurs when appropriate, increasing the diversity of the proteins that any single mRNA can produce. The product of the entire transcription process that began with the production of the pre-mRNA chain, is a mature mRNA chain.

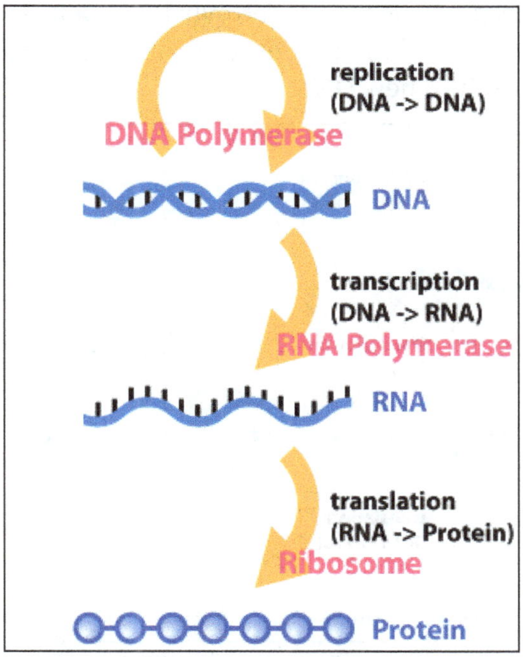

Translation

The mature mRNA finds its way to a ribosome, where it gets translated. In prokaryotic cells, which have no nuclear compartment, the processes of transcription and translation may be linked together without clear separation. In eukaryotic cells, the site of transcription (the cell nucleus) is usually separated from the site of translation (the cytoplasm), so the mRNA must be transported out of the nucleus into the cytoplasm, where it can be bound by ribosomes. The ribosome reads the mRNA triplet codons, usually beginning with an AUG (adenine–uracil–guanine), or initiator methionine codon downstream of the ribosome binding site. Complexes of initiation factors and elongation factors bring aminoacylated transfer RNAs (tRNAs) into the ribosome-mRNA complex, matching the codon in the mRNA to the anti-codon on the tRNA. Each tRNA bears the appropriate amino acid residue to add to the polypeptide chain being synthesised. As the amino acids get linked into the growing peptide chain, the chain begins folding into the correct conformation. Translation ends with a stop codon which may be a UAA, UGA, or UAG triplet.

The mRNA does not contain all the information for specifying the nature of the mature protein. The nascent polypeptide chain released from the ribosome commonly requires additional processing before the final product emerges. For one thing, the correct folding process is complex and vitally important. For most proteins it requires other chaperone proteins to control the form of the product. Some proteins then excise internal segments from their own peptide chains, splicing the free ends that border the gap; in such processes the inside "discarded" sections are called inteins. Other proteins must be split into multiple sections without splicing. Some polypeptide chains need to be cross-linked, and others must be attached to cofactors such as haem (heme) before they become functional.

Special Transfers of Biological Sequential Information

Reverse Transcription

Reverse transcription is the transfer of information from RNA to DNA (the reverse of normal transcription). This is known to occur in the case of retroviruses, such as HIV, as well as in eukaryotes, in the case of retrotransposons and telomere synthesis. It is the process by which genetic information from RNA gets transcribed into new DNA.

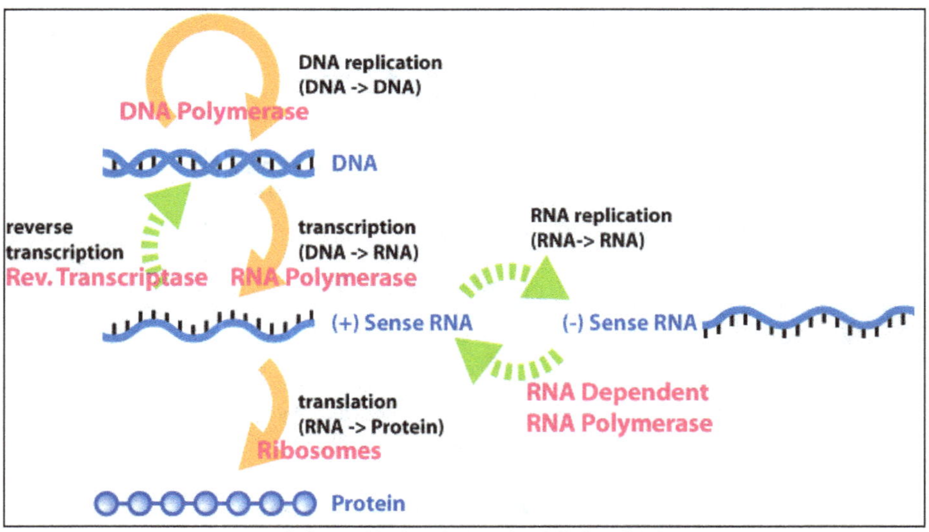

Unusual flow of information highlighted in green

RNA Replication

RNA replication is the copying of one RNA to another. Many viruses replicate this way. The enzymes that copy RNA to new RNA, called RNA-dependent RNA polymerases, are also found in many eukaryotes where they are involved in RNA silencing.

RNA editing, in which an RNA sequence is altered by a complex of proteins and a "guide RNA", could also be seen as an RNA-to-RNA transfer.

Direct Translation from DNA to Protein

Direct translation from DNA to protein has been demonstrated in a cell-free system (i.e. in a test tube), using extracts from *E. coli* that contained ribosomes, but not intact cells. These cell fragments could synthesize proteins from single-stranded DNA templates isolated from other organisms (e,g., mouse or toad), and neomycin was found to enhance this effect. However, it was unclear whether this mechanism of translation corresponded specifically to the genetic code.

Transfers of Information not Explicitly Covered in the Theory

Posttranslational Modification

After protein amino acid sequences have been translated from nucleic acid chains, they can be edited by appropriate enzymes. Although this is a form of protein affecting protein sequence, not explicitly covered by the central dogma, there are not many clear examples where the associated concepts of the two fields have much to do with each other.

Inteins

An intein is a "parasitic" segment of a protein that is able to excise itself from the chain of amino acids as they emerge from the ribosome and rejoin the remaining portions with a peptide bond in such a manner that the main protein "backbone" does not fall apart. This is a case of a protein changing its own primary sequence from the sequence originally encoded by the DNA of a gene. Additionally, most inteins contain a homing endonuclease or HEG domain which is capable of finding a copy of the parent gene that does not include the intein nucleotide sequence. On contact with the intein-free copy, the HEG domain initiates the DNA double-stranded break repair mechanism. This process causes the intein sequence to be copied from the original source gene to the intein-free gene. This is an example of protein directly editing DNA sequence, as well as increasing the sequence's heritable propagation.

Methylation

Variation in methylation states of DNA can alter gene expression levels significantly. Methylation variation usually occurs through the action of DNA methylases. When the change is heritable, it is considered epigenetic. When the change in information status is not heritable, it would be a somatic epitype. The effective information content has been changed by means of the actions of a protein or proteins on DNA, but the primary DNA sequence is not altered.

Prions

Prions are proteins of particular amino acid sequences in particular conformations. They propagate themselves in host cells by making conformational changes in other molecules of protein with the same amino acid sequence, but with a different conformation that is functionally important to the cell. Once the protein has been transconformed to the prion folding it changes function. In turn it can convey information into new cells and reconfigure more functional molecules of that sequence into the alternate prion form. In some types of prion in fungi this change is continuous and direct; the information flow is Protein → Protein.

Some scientists such as Alain E. Bussard and Eugene Koonin have argued that prion-mediated inheritance violates the central dogma of molecular biology. However, Rosalind Ridley in *Molecular Pathology of the Prions* (2001) has written that "The prion hypothesis is not heretical to the central dogma of molecular biology—that the information necessary to manufacture proteins is encoded in the nucleotide sequence of nucleic acid—because it does not claim that proteins replicate. Rather, it claims that there is a source of information within protein molecules that contributes to their biological function, and that this information can be passed on to other molecules."

Natural Genetic Engineering

James A. Shapiro argues that a superset of these examples should be classified as natural genetic engineering and are sufficient to falsify the central dogma. While Shapiro has received a respectful hearing for his view, his critics have not been convinced that his reading of the central dogma is in line with what Crick intended.

Use of the Term "Dogma"

In his autobiography, *What Mad Pursuit*, Crick wrote about his choice of the word *dogma* and some of the problems it caused him:

"I called this idea the central dogma, for two reasons, I suspect. I had already used the obvious word hypothesis in the sequence hypothesis, and in addition I wanted to suggest that this new assumption was more central and more powerful. ... As it turned out, the use of the word dogma caused almost more trouble than it was worth. Many years later Jacques Monod pointed out to me that I did not appear to understand the correct use of the word dogma, which is a belief *that cannot be doubted*. I did apprehend this in a vague sort of way but since I thought that *all* religious beliefs were without foundation, I used the word the way I myself thought about it, not as most of the world does, and simply applied it to a grand hypothesis that, however plausible, had little direct experimental support."

Similarly, Horace Freeland Judson records in *The Eighth Day of Creation*:

"My mind was, that a dogma was an idea for which there was *no reasonable evidence*. You see?!" And Crick gave a roar of delight. "I just didn't *know* what dogma *meant*. And I could just as well have called it the 'Central Hypothesis,' or — you know. Which is what I meant to say. Dogma was just a catch phrase."

References

- Wong, KC (2016). Computational Biology and Bioinformatics: Gene Regulation. CRC Press (Taylor & Francis Group). ISBN 9781498724975

- Nisbet, Robert (14 May 2009). "BIOINFORMATICS". Handbook of Statistical Analysis and Data Mining Applications. John Elder IV, Gary Miner. Academic Press. p. 328. Retrieved 9, May 2020
- Attwood TK, Gisel A, Eriksson NE, Bongcam Rudloff E (2011). Concepts, Historical Milestones and the Cen tral Place of Bioinformatics in Modern Biology: A European Perspective". Bioinformatics – Trends and Methodologies. InTech. Retrieved 8, Jan 2020
- Horace Freeland Judson (1996). "Chapter 6: My mind was, that a dogma was an idea for which there was no reasonable evidence. You see?!". The Eighth Day of Creation: Makers of the Revolution in Biology (25th anniversary edition). Cold Spring Harbor, NY: Cold Spring Harbor Laboratory Press. ISBN 0-87969-477-7
- Brohée, Sylvain; Barriot, Roland; Moreau, Yves. "Biological knowledge bases using Wikis: combining the flexibility of Wikis with the structure of databases". Bioinformatics. Oxford Journals. Retrieved 5, May 2020
- Moody, Glyn (2004). Digital Code of Life: How Bioinformatics is Revolutionizing Science, Medicine, and Business. ISBN 978-0-471-32788-2
- Hye-Jung, E.C.; Jaswinder, K.; Martin, K.; Samuel, A.A; Marco,"A.M (2014). ""Second-Generation Sequencing for Cancer Genome Analysis"". In Dellaire, Graham; Berman, Jason N.; Arceci, Robert J. Cancer Genomics. Boston (US): Academic Press. pp. 13–30. doi:10.1016/B978-0-12-396967-5.00002-5. ISBN 9780123969675

Permissions

All chapters in this book are published with permission under the Creative Commons Attribution Share Alike License or equivalent. Every chapter published in this book has been scrutinized by our experts. Their significance has been extensively debated. The topics covered herein carry significant information for a comprehensive understanding. They may even be implemented as practical applications or may be referred to as a beginning point for further studies.

We would like to thank the editorial team for lending their expertise to make the book truly unique. They have played a crucial role in the development of this book. Without their invaluable contributions this book wouldn't have been possible. They have made vital efforts to compile up to date information on the varied aspects of this subject to make this book a valuable addition to the collection of many professionals and students.

This book was conceptualized with the vision of imparting up-to-date and integrated information in this field. To ensure the same, a matchless editorial board was set up. Every individual on the board went through rigorous rounds of assessment to prove their worth. After which they invested a large part of their time researching and compiling the most relevant data for our readers.

The editorial board has been involved in producing this book since its inception. They have spent rigorous hours researching and exploring the diverse topics which have resulted in the successful publishing of this book. They have passed on their knowledge of decades through this book. To expedite this challenging task, the publisher supported the team at every step. A small team of assistant editors was also appointed to further simplify the editing procedure and attain best results for the readers.

Apart from the editorial board, the designing team has also invested a significant amount of their time in understanding the subject and creating the most relevant covers. They scrutinized every image to scout for the most suitable representation of the subject and create an appropriate cover for the book.

The publishing team has been an ardent support to the editorial, designing and production team. Their endless efforts to recruit the best for this project, has resulted in the accomplishment of this book. They are a veteran in the field of academics and their pool of knowledge is as vast as their experience in printing. Their expertise and guidance has proved useful at every step. Their uncompromising quality standards have made this book an exceptional effort. Their encouragement from time to time has been an inspiration for everyone.

The publisher and the editorial board hope that this book will prove to be a valuable piece of knowledge for students, practitioners and scholars across the globe.

Index